Vertebrate Ecophysiology

An Introduction to its Principles and Applicati

Ecophysiology attempts to clarify the rôle and impo as digestion, excretion and respiration, in the ecologi... habitats. The basic principles and methods that are central to any ecophysiological study are outlined and discussed, including animal capture, blood collection, and the measurement of plasma components and hormone concentrations. Attention is paid to animal welfare and ethical considerations, and the question of stress and how to identify its presence in animals in their natural environment is approached through a series of case studies. Examples are given from a wide range of vertebrates living in deserts, cold climates and oceans, and recent findings on the physiological adaptations of Antarctic birds and mammals are a highlight of the book. This textbook will provide an introduction to the study of ecophysiology for advanced undergraduates and postgraduate students, as well as for researchers in ecology, biodiversity and conservation.

DON BRADSHAW holds the Chair of Zoology, and is Director of the Centre for Native Animal Research, at The University of Western Australia, Perth. He teaches undergraduates in all years, focusing on vertebrate adaptations, comparative endocrinology and ecophysiology. He has previously written two books, *Ecophysiology of Desert Reptiles* (1986) and *Homeostasis of Desert Reptiles* (1997).

To Felicity, for sharing and enriching my life

Vertebrate Ecophysiology

An Introduction to its Principles and Applications

Don Bradshaw

The University of Western Australia

PUBLISHED BY THE PRESS SYNDICATE OF THE UNIVERSITY OF CAMBRIDGE
The Pitt Building, Trumpington Street, Cambridge, United Kingdom

CAMBRIDGE UNIVERSITY PRESS
The Edinburgh Building, Cambridge CB2 2RU, UK
40 West 20th Street, New York, NY 10011-4211, USA
477 Williamstown Road, Port Melbourne, VIC 3207, Australia
Ruiz de Alarcón 13, 28014 Madrid, Spain
Dock House, The Waterfront, Cape Town 8001, South Africa

http://www.cambridge.org

First published 2003

Printed in the United Kingdom at the University Press, Cambridge

Typeface Adobe Garamond 11/13 pt and Frutiger *System* LATEX 2_ε [TB]

A catalogue record for this book is available from the British Library

Library of Congress Cataloguing in Publication data

Bradshaw, S. D. (Sidney Donald)
Vertebrate Ecophysiology : An Introduction to its Principles and Applications / Don Bradshaw.
p. cm.
Includes bibliographical references (p.).
ISBN 0-521-81797-8 – ISBN 0-521-52109-2 (pb.)
1. Vertebrates – Physiology. 2. Vertebrates – Ecology. I. Title.
QL739.2 .B725 2003
571.1′6 – dc21 2002031053

ISBN 0 521 81797 8 hardback
ISBN 0 521 52109 2 paperback

Contents

Introduction

Ecophysiology is a relatively new discipline and seeks to clarify the rôle and importance of physiological processes in the ecological relations of species in their natural habitat. It has its antecedents in the fields now known as 'environmental physiology' and 'physiological ecology', both of which developed strongly in the United States in the 1950s and 1960s. Ecophysiology differs principally from both of these in its emphasis on studying animals unrestrained in their natural environment, rather than in laboratory situations where animals are forcibly constrained and often exposed to stressors that may not be obvious, even to the experimenter.

Ecologists often have little training, or interest, in the study of basic physiological processes (such as digestion, respiration, excretion, etc.) and many would dispute that the primary ecological processes structuring populations – those of birth, growth, food acquisition, recruitment, reproduction and death – require any understanding of the physiological processes and mechanisms occurring in the individual animal. Indeed, Andrewartha and Birch (1954), in their seminal text on population ecology, went even further and argued that evolution had nothing to do with ecology! One only needs to cite Theodosius Dobzhanky's oft quoted aphorism that 'Nothing makes any sense in biology except in the light of evolution' to realise that this was an extreme view no longer shared by either ecologists or evolutionary biologists (Dobzhansky, 1953).

Another major factor distinguishing ecophysiological studies is the emphasis placed on genetics. Animals vary from one to another; the study of individual variation and its evolutionary significance is only now starting to be appreciated (see, for example, Bennett, 1987). The importance of the genetic structuring of the physiological responses of individual animals within a population is very important and this is an area traditionally ignored in physiological studies. Ecophysiology attempts to integrate individual responses (which are rarely uniform) within the context of a single population or species.

Although animals are usually studied singly in the laboratory (which facilitates the elucidation of significant mechanisms) they rarely exist alone in nature and occur in populations that are more or less isolated in space and time. Many populations are polymorphic in nature, and much of this variety and its underlying genetic variability is lost when animals are transferred to the laboratory and exposed to markedly different and uniform conditions of climate and nutrition. This inherent variation found within natural populations is held to be an essential component of the adaptive responses of organisms to environmental pressures and one of the major foci of ecophysiological studies.

Variation is intrinsically difficult to study (our minds function by association through similarity) and this is immediately evident when one considers normal laboratory procedures. Animals are studied in groups that are thought to be large enough to encompass the individual variability found within a given population, and responses are then averaged and compared by using statistical tests, which embody a typological concept (the Gaussian distribution). Typically, animals that behave 'aberrantly' (i.e. fall more than two standard deviations from the 'norm') are excluded from the analysis rather than being made the subject of further analysis to determine the reasons for their discrete behaviour. As Block and Vannier (1994) point out in their interesting paper '. . . the ecophysiological approach must take into account polymorphism in individual responses, which are largely responsible for the adaptive capacity of any given population. In this respect, ecophysiological study yields information which is fundamental for an understanding of the mechanisms underlying adaptive strategies.'

In classical physiology, of course, this process of elimination of variability goes even further by relying on stocks of laboratory species that have been selected over long periods of time and display greatly reduced genetic variability – to the extent that skin transplants are possible between some strains of rats. Clearly, this is a necessary adjunct to the study of mechanism, where the aim is to produce a stable and reproducible preparation, but this begs the question of whether this variation is equally a nuisance to the animal in the pursuit of its normal existence and what rôle, if any, such variation plays at the level of the population.

In this book the basic principles and methods that are central to any ecophysiological study will be outlined and discussed, using concrete examples from the published literature. The very important question of stress, and how to recognise its presence – and then measure its intensity and effect on the vital processes of an animal – will also be addressed. So too will be the question of homeostasis, which is a fundamental organising paradigm in biology and one of central importance in ecophysiology. Are animals in their natural environment maintaining their internal state within limits that permit normal activity, or do they display perturbations, which if prolonged, will lead to a loss of condition and perhaps to pathological states?

Answers to some of these questions will be illustrated by considering, firstly, a wide range of vertebrate animals that inhabit deserts: reptiles, burrowing frogs, Australian kangaroos and wallabies, desert rodents in the Sahara and North America, and desert-inhabiting birds. Although less is known of their ecophysiology, the final chapter will consider marine birds and mammals, such as the albatrosses, penguins, seals and whales, where recent research is revealing fascinating details about the physiological adaptations and compromises that they make in order to survive in challenging environments, such as the Antarctic.

The book is meant as an introduction to the study of ecophysiology that will assist advanced undergraduate and postgraduate students, as well as researchers in ecology, biodiversity and conservation. Primarily, I hope that it will be of interest and assistance to young researchers interested in studying animals in their natural habitat. Writing the book would not have been possible without the assistance of many of my colleagues and former students, foremost amongst these being: Professeur Maurice Fontaine, the late Hubert and Marie-Charlotte Saint Girons, François Lachiver, Claude Grenot, Roland Vernet, Xavier Bonnet, Pierre Jouventin, Henri Weimerskirch, Yvon Le Maho, Patrick Duncan and Barbara Demeneix in France; Ken Nagy, Joe Williams, Dan Costa and Rudy Ortiz in the United States; Phil Withers, Don Edward, Nick Gales, Brian Clay, Terry Miller, Julie McAllister, Stephen Ambrose, Keith Morris, Chris Dickman, Ian Rooke, Ron Wooller, Bob McNeice, Darren Murphy, Juliet King, Dorian Moro, Mitch Ladyman, Stewart Ford, Ernie Stead-Richardson and Jessica Oates in Australia. My sincere thanks go to Danielle Philippe for her invaluable assistance in carrying out the many arduous tasks involved in acquiring and checking references and permissions for the book and in proofreading the index. Thanks also to Mark Gargaklis for the Schnabel spreadsheet in Appendix I. Acknowledgement is also made to WAPET for support received when working on Barrow Island and to the Australian Research Committee for continued funding of my research on Australian animals over many years.

1

Homeostasis: a fundamental organising paradigm in ecophysiology

The concept of 'homoiostasis' is now a central one in many sciences and its widespread use and utility attests to the genius of the American physiologist Walter Cannon, to whom we owe the original insight. Cannon coined the term in 1929 and defined homeostasis as '. . . the coordinated physiological processes which maintain most of the steady states in the organism' (Cannon, 1929). He then went on to employ it with great success in his later books and publications (see Cannon, 1939) and the concept is now a central one in biology as well as in other fields such as engineering, economics and information technology. The idea of a process of self-regulation is based, however, on the earlier studies and speculations of the great French physiologist Claude Bernard, who first suggested that animals regulate and hold constant an internal state or *milieu intérieur* that is quite different from that of the environment around them. As he states in his famous textbook of 'lessons' published in Paris in 1878:

> I believe that I am the first to have proposed this idea that animals in reality possess two environments: an external environment in which the animal is situated and an internal environment in which are found the tissue elements[1] (Bernard, 1878).

The more recent concept of an idea, or theory, functioning as a paradigm comes from the work of the American philosopher Thomas Kuhn, who coined this term to describe the way in which whole scientific communities suddenly change the way in which they interpret and describe phenomena. In studying the ways in which the ideas of the obscure sixteenth century astronomer

[1] My translation.

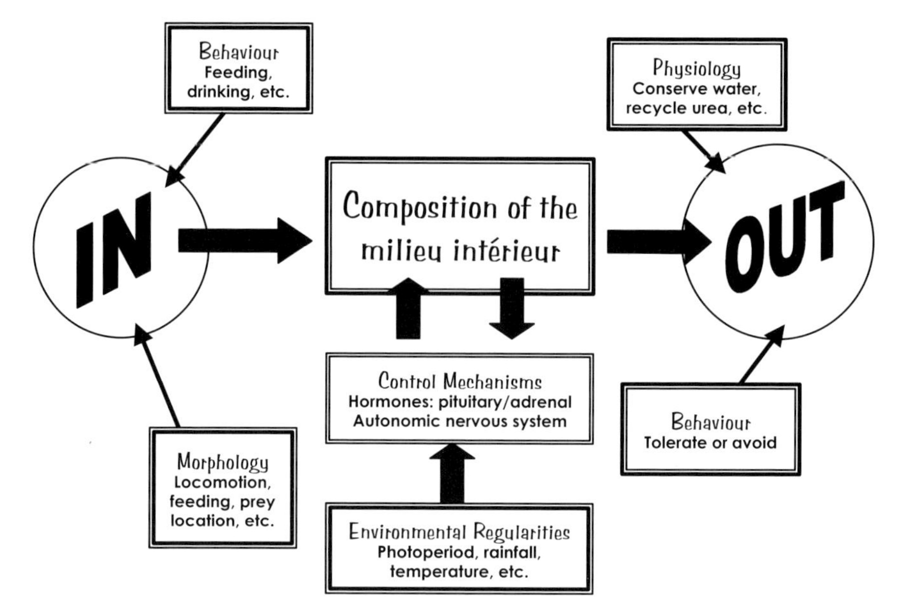

Figure 1.1. Schema illustrating the essential components of the process of homeostasis in a living system and the means by which the *milieu intérieur* is regulated.

Nicolas Copernicus revolutionised our understanding of the universe, and the Earth's position in relation to the sun (Kuhn, 1976). Kuhn developed his general theory of 'scientific revolutions' and coined the term 'paradigm' to describe '. . . a coherent, universally-recognised scientific explanation, or theory, of a hitherto unresolved set of data' (Kuhn, 1962). He describes the process by which scientists are quite content to accept, sometimes for long periods of time, explanations that are often contradicted by published data, and then, quite suddenly, they are supplanted by a new explanation or set of explanations. A good example of such a scientific revolution in recent times is the acceptance of Alfred Wegener's once heretical ideas on drifting continents (Wegener, 1966) and the central rôle now played by the concept of plate tectonics in geology.

Homeostasis is certainly one of the most durable of these paradigms and, as yet, shows no signs of being supplanted. It helps to focus on the myriad dynamic processes that occur within a living organism, and the plethora of interactions that occur constantly with the surrounding environment, and place them in a meaningful context. This is best illustrated diagrammatically, and Figure 1.1 attempts to portray the processes that are involved in the homeostatic maintenance of a constant internal state in a vertebrate animal. The constancy of the *milieu intérieur* is maintained through the interplay of fluxes, both in and out of the body, of essential elements and molecules, such as water, oxygen, carbon dioxide, sodium, glucose, nitrogen, etc. Both behavioural and physiological processes in turn influence these. Animals need to seek their food, and morphological adaptations and behaviours that control food acquisition have a major impact on influxes of water and essential nutrients. The extent to which these resources are ultimately made available to the body, however, depends on many physiological factors, such as the rate of passage of the food through the gut, the efficiency of digestive enzymes, and the efficacy of absorptive processes in the small intestine.

Effluxes, or outfluxes, of temperature, water, CO_2, and molecules such as urea, sodium and potassium are again influenced by both behavioural and physiological processes. Behavioural changes can markedly influence rates of heat gain and loss in animals, especially ectotherms such as reptiles that use the sun to maintain their body temperature constant when active during the day (see Bradshaw, 1986). Physiological processes are also very much involved in regulating heat loss from the body of animals such as mammals, where heat flow from the interior to the exterior of the body is modulated by varying blood flow through the dermis and hence modifying its conductance. Although many lower vertebrates lose much of their body water via evaporation from the skin, birds and mammals are able to produce a hyperosmotic urine that is more concentrated than their body fluids, and the kidneys are thus the major site of water conservation in these animals. The development of impressive concentrating mechanisms with large medullae in the kidneys of desert rodents (see Figure 1.2) has long been interpreted as an adaptation for the conservation of water but, as we shall see in Chapter 6, ecophysiological studies of the animals in their own environment suggest that they are never short of water and this interpretation may be too simplistic.

The composition of the internal environment or *milieu intérieur* is also monitored and regulated constantly by elements of the autonomic nervous system and by hormones, especially those elaborated by the pituitary and adrenal glands. The pituitary gland produces a large number of protein and peptide hormones whose secretion is controlled in turn by 'releasing factors' secreted in the hypothalamus of the brain and transported by a discrete portal blood system to the pituitary. These releasing factors, which are peptide hormones themselves, activate gene expression in the special cells of the anterior pituitary (adenohypophysis), each of which is dedicated to the secretion of a separate hormone (some examples are the two gonadotrophins that stimulate the gonads to secrete sex hormones, follicle stimulating hormone (FSH)

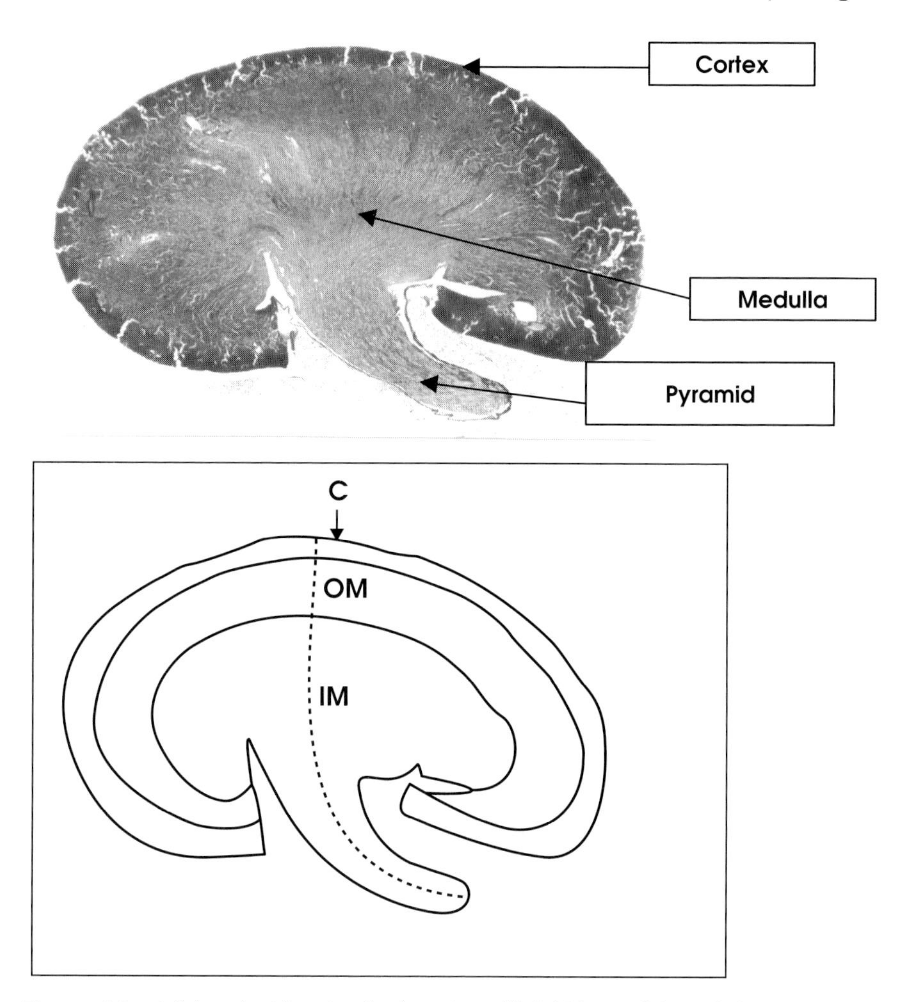

Figure 1.2. Mid-sagittal longitudinal section of left kidney of the Lakeland Downs short-tailed mouse (*Leggadina lakedownensis*) from Thevenard Island, Western Australia, showing zones of cortex (C), outer medulla (OM) and inner medulla (IM) identified by staining. Scale: 1 cm = 1 mm. (Photo courtesy of Dr Dorian Moro.)

and luteinising hormone (LH); thyroid-stimulating hormone (TSH), which stimulates the thyroid gland to secrete the hormones thyroxine and tri-iodothyronine; adrenocorticotrophic hormone (ACTH), which controls both the size and the secretory activity of the adrenal glands; growth hormone (GH); and prolactin). The hierarchical arrangement of brain, pituitary and effector endocrine glands is shown diagrammatically in Figure 1.3 for the main hormones regulating reproduction.

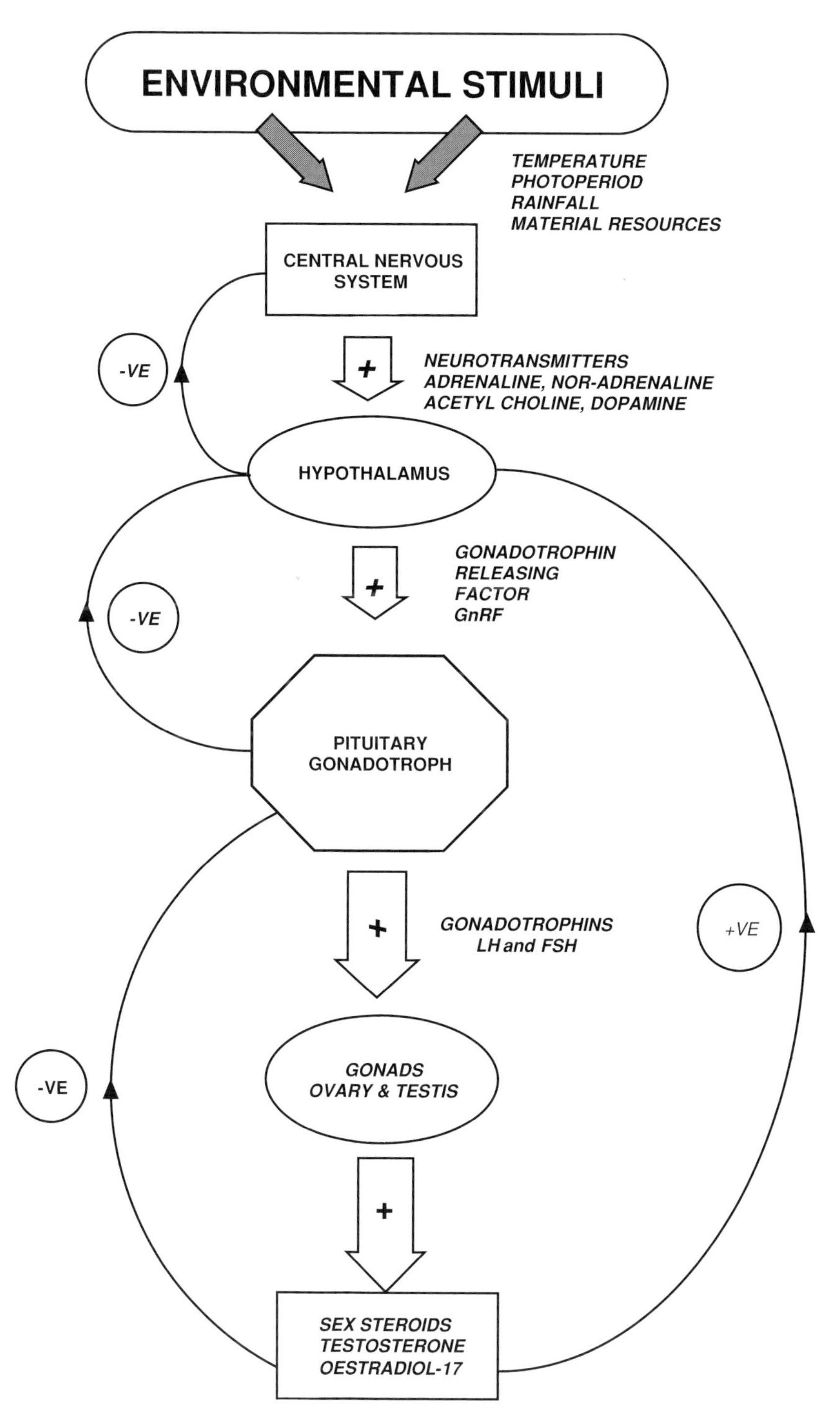

Figure 1.3. Schema illustrating the hierarchical nature of the hormonal control systems regulating the secretion of steroid hormones by the vertebrate gonads.

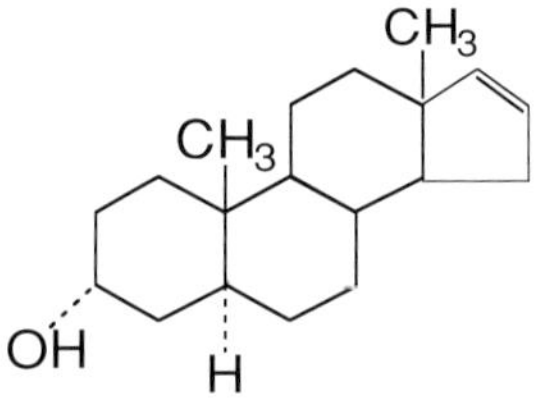

Figure 1.4. Structure of the steroid molecule, androstenol (5α-androst-16-en-3α-ol).

The rates of secretion of these hormones into the blood are in turn influenced markedly by changes in the external environment, and it is the hypothalamus of the brain, with its many specialised neurosecretory neurones, that is most involved in transducing environmental cues such as temperature and photoperiod into hormonal cues which help maintain homeostasis. The pineal gland in the centre of the brain in mammals also produces the hormone melatonin, which is secreted with a marked diurnal–nocturnal rhythm, and has a strong influence on reproductive processes in many mammals (Reiter, 1978; Reiter and Follett, 1980; Tang *et al.*, 1996).

Pheromones are also chemicals produced by animals that are released into the surrounding air and water and communicate information between different individuals in a population, particularly in relation to sexual and social status. These are often steroid molecules and a fascinating example of how plants and animals may co-evolve interlocking strategies is provided by the steroid molecule androstenol (5α-androst-16en-3α-ol) shown in Figure 1.4. Truffles are fungi that have long been known for their aromatic properties and these are prized in cooking, especially in France and Italy. The subterranean truffles were traditionally located with the aid of a sow or 'truie' as seen in Figure 1.5. Nowadays dogs (and even portable gas chromatographs) are used to locate the valuable truffles, but the mystery of why female pigs were particularly susceptible to the smell of the truffles became apparent when the identity of the mating pheromone of the boar was discovered by Claus *et al.* (1981). This is also androstenol and the male secretes it in foam around the mouth when trying to mate. The sow, on smelling it, adopts the lordosis posture and allows the male to mount her. One can only surmise that the sow in the forest assumes that a handsome boar is buried for some reason underground and, on unearthing the truffle, shows her disgust by trampling on the truffle and thus releasing its spores into the atmosphere. In this way, the truffle is using a mammal and its reproductive signalling system to complete its own amazing reproductive cycle. Nor does the story end here. Gower and Ruparelia (1993) have recently carried out tests that suggest that the musky-smelling androstenol functions as a mild aphrodisiac in humans.

Figure 1.5. Searching for truffles in France, using the ancient method with 'la truie', or a sow.

There are periods of an animal's life, however, when the internal state is not maintained constant but varies systematically. The most important of these is during the period of growth from the juvenile to the adult state, but there are also other periods – such as during the process of reproduction – where there may be important changes in the *milieu intérieur*, especially that of the female, engendered by the presence of an embryo in viviparous vertebrates (see Hytten, 1976). Pathological states are often associated with dramatic changes of the *milieu intérieur* and, in some of these, a new homeostatic régime is established by an apparent 'resetting' of the upper and lower set-points. Fever is a good example of this: the body temperature is maintained homeostatically, but at a higher set-point than normal, owing to changes in the ionic composition and osmolality of the cerebrospinal fluid (CSF) (Myers *et al.*, 1971; Turlejska and Baker, 1986).

2

Stress: the concept and the reality

We talk very glibly these days of stress, especially the 'stress of modern-day living', and in the scientific literature one often finds mention of things such as 'temperature stress', 'water stress' and 'the stress of reproduction'. It has proven extraordinarily difficult, however, to agree on a common definition of the term, let alone develop methods for measuring the incidence and intensity of stress in animals.

There is general agreement among biologists that 'stress' is an important ecological factor, often contributing to the extinction of rare and endangered species (Bradshaw, 1996, 1999). There are thus a number of important questions that need to be addressed concerning stress, including:

What is stress?
How do we define it and measure its effects?
How does one measure its incidence and severity?
What effects does stress have on an animal's ability to maintain homeostasis?
Does prolonged stress reduce fitness?
Are threatened and endangered species more susceptible to the effects of stress than are other species?
Can we use instances of stress physiology to gauge the level of susceptibility of animal species to environmental change and their likelihood of extinction?

Definitions

It was Walter Cannon again who, as early as 1914, first developed the idea that organisms react to unfavourable situations in terms of highly integrated metabolic activities. Cannon's concept of stress was derived by simple analogy from Newtonian physics where an imposed force (stress) produces a deformation (strain) in an object, with the strain being proportional to the stress. He further developed this mechanistic analogy by introducing the concept of a 'critical stress level' in biological systems, producing a 'breaking strain' in the appropriate homeostatic mechanism, which then failed to counter the stress (Cannon, 1929).

By far the greatest contribution to the study of stress, however, was made by Hans Selye who, in his monumental paper, defined stress as 'a state of non-specific tension in living organisms', thereby engendering the current problem of how best to identify its presence and also measure its intensity (Selye, 1946). Selye (1936) clearly identified what he termed the 'stress triad' of adrenocortical enlargement, atrophy of the thymus and lymphoid tissues, and ulceration of the digestive tract, which may be observed in experimental animals subjected to a wide variety of nociceptive (i.e. harmful) agents of high biological intensity. He then formulated his famous 'General Adaptation Syndrome', comprising three stages:

(i) a stage of '*alarm*',
(ii) a stage of '*adaptation*' and
(iii) a stage of '*exhaustion*' (Selye, 1946).

Unfortunately, Selye came to regard the 'stress triad' as an extreme manifestation of a normal non-specific 'stress response' necessary for the maintenance of the specific homeostatic states identified by Cannon. He later described this non-specific response as a 'state of stress', as exemplified in his definition above, and believed that a minimal level of stress was essential for normal existence ('eustress') and distinguished this from potentially harmful levels of stress ('distress') leading to pathological states (Selye, 1976).

Selye's approach to the concept of stress has not escaped criticism, because of its inherent vagueness and imprecision, and the essence of such criticism is that the term 'stress' should rather be applied instead to the environmental factors that elicit homeostatic adjustments as defined by Cannon (Mason, 1975; Levitt, 1980; Hoffman *et al.*, 1993).

Stress is thus defined by Brett (1958) and Koehn and Bayne (1989) as:

> any factor that inhibits the growth and reproduction of individuals in a population

and by Sibley and Calow (1989) as:

> ...an environmental condition that, when first applied, impairs Darwinian fitness.

Stress and stressors

I personally think that Selye's approach is the better and that those environmental factors inducing stress in organisms should be called *'stressors'* and that the term 'stress' should be restricted to the effect, rather than the cause. I would thus not consider abnormally high temperatures, or a severe shortage of water, as a stress, but as a stressor with the potential to induce stress in an animal. Whether it does so or does not, however, depends on how effective the animal's homeostatic mechanisms are in combating and controlling the impact of the stressor. We thus need to know more about the animal's defences against environmental perturbations; in the face of a given stressor, it should be possible to determine whether stress is engendered or not by what happens to the animal's *milieu intérieur.*

A stressor that leads to a change in the animal's internal state, from whatever might be considered to be optimal, is one that we can say has produced a stress state in the animal. Take the case of a mammal that is short of water. The normal reaction would be for the animal to experience thirst and search for water to drink. If this is not available then there would be a slight elevation of the osmotic pressure of the blood that would lead to the secretion of increased amounts of antidiuretic hormone (ADH) from the pituitary gland. This, in turn, would increase the rate of water reabsorption in the collecting ducts of the renal medulla and, thus, increase the concentration of the urine being produced and decrease its volume. In this way the animal protects its body water and avoids the potentially deleterious effects of dehydration by conserving water that would otherwise be lost via the kidneys.

There is no way that one would interpret this as an animal experiencing stress. It is exposed to a stressor (the shortage of water) to which it is responding adequately by the use of its normal homeostatic mechanisms. If, however, the animal starts to experience dehydration, despite secreting maximal amounts of ADH in the blood, and its kidney producing urine of the maximum concentration possible for that species, we can consider that it is experiencing stress. Dehydration is simply loss of body water and it leads to an increase in the osmotic pressure of the remaining body fluids, which may be easily measured with an osmometer.

Our definition of stress thus encompasses two aspects: a maximal stimulation of regulatory mechanisms (in this case ADH from the pituitary) opposing the stressor, which are none the less inadequate to prevent a significant change in state of the *milieu intérieur.* By using this approach it should be possible to identify the existence of stressful states in biological organisms through their reaction to specific stressors, and also measure the intensity of the stress induced by the extent of the subsequent physiological reaction. This approach assumes that stressors will modify the internal state of an organism in some characteristic way, which can then be used to measure the intensity of the stress.

An operational definition of stress

A few years ago I proposed such an operational definition of stress as:

> ...the physiological resultant of demands that exceed an organism's regulatory capacities (Bradshaw, 1986)

in an attempt to dissociate potentially harmful changes of state from beneficial regulatory responses. By an 'operational definition' I mean one that can be used to identify the presence of stress and measure its intensity, without necessarily defining its nature. By this definition, a 'stressed' animal is one that shows a significant deviation of the *milieu intérieur* from whatever state is considered to be optimal, and this change is accompanied by a maximal activation of regulatory responses that have, none the less, been surpassed.

Elevated concentrations of corticosteroids in the plasma may thus be an indication of the presence of a stressor, but cannot be taken as proof that an animal is experiencing stress. The level or intensity of the stress experienced by the animal can only be gauged from the extent of the deviation of the *milieu intérieur* from its normal homeostatic state. Measurements of plasma concentrations of other hormones involved in the maintenance of homeostasis, such as ADH in the example above, are also of great value but can only be interpreted in the light of a thorough understanding of the normal boundaries of the *milieu intérieur* (e.g. plasma electrolytes). One uses hormone concentrations therefore as indices of activated regulatory responses in the organism, but stress is only invoked if the *milieu intérieur* is significantly perturbed, indicating that normal homeostatic processes are no longer operating effectively (Bradshaw, 1992a, 1996).

An example

A good example of stress in a field population of marsupials comes from the study of Jones *et al.* (1990) on the quokka (*Setonix brachyurus*), a small wallaby restricted to Rottnest Island, some 20 km off the coast of Perth in Western Australia (see Figure 2.1). In this study the effects of the availability of surface water on the animals' ability to maintain homeostasis during the long, hot, dry summer were observed, along with changes in circulating concentrations of lysine vasopressin, which is the antidiuretic hormone in these marsupials (Chauvet *et al.*, 1983).

Figure 2.1. The marsupial quokka, *Setonix brachyurus*, once widespread on the mainland of Western Australia but now found in any numbers only on Rottnest Island. (Illustration by L. Burch with permission from *Quokkas in the Southern Forests* (1984). Fauna Series of Forests Dept of WA (now CALM).)

The quokka was originally a forest-dwelling animal found commonly on the mainland of Australia but is now restricted to a semi-arid and highly degraded coastal habitat totally lacking in fresh water (Bradshaw, 1983). Measurements of body condition (Miller and Bradshaw, 1979) show a significant negative correlation with circulating concentrations of adrenal corticosteroids, as seen

Table 2.1. *Water and electrolyte balance of quokkas on Rottnest Island*

Data expressed as mean ± S.E.

Parameter	West End	Lake Baghdad	Significance
Body mass (kg)	2.09 ± 0.15	2.66 ± 0.13	$p < 0.01$
Condition index	5.33 ± 0.26	6.45 ± 0.25	$p < 0.01$
Lysine vasopressin ($pg\ ml^{-1}$)	89.2 ± 19.5	35.6 ± 15.8	$p < 0.05$
Total body water (ml)	1492 ± 18.8	1888 ± 18.6	$p < 0.001$
Urine production ($ml\ kg^{-1}\ d^{-1}$)	20.35 ± 2.75	58.18 ± 13.79	$p < 0.02$
Urine osmolality ($mOsm\ kg^{-1}$)	1253.1 ± 44.7	968.7 ± 101.3	$p < 0.02$
Plasma osmolality ($mOsm\ kg^{-1}$)	301.3 ± 6.0	279.8 ± 2.4	$p < 0.005$
U/P_{osm}	4.16	3.46	$p < 0.02$

From Jones *et al.* (1990).

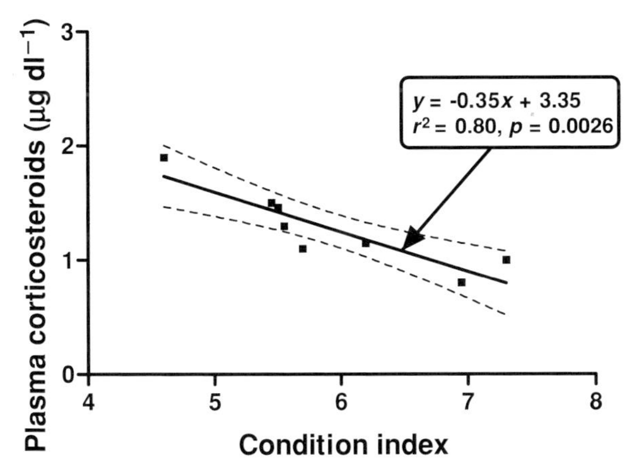

Figure 2.2. Significant negative correlation between condition index and the circulating level of adrenal corticosteroids in the Rottnest Island quokka, *Setonix brachyurus*, during late summer and autumn (modified from Miller and Bradshaw, 1979).

in Figure 2.2, indicating the presence of stressors. Debate still rages over the precise role played by glucocorticoids such as cortisol in the stress response; the discovery of the anti-inflammatory effects of glucocorticoids confounded Selye's erroneous belief that these steroids enhanced the response to stress (Munck *et al.*, 1984; Sapolsky *et al.*, 2000).

The data presented in Table 2.1 show the variation in concentration of ADH between two sub-populations of quokkas on Rottnest Island, one with access to brackish water (Lake Baghdad) and the other deprived of any source of free water (West End). The Lake Baghdad animals show an increased water turnover (measured with tritiated water; see Chapter 4) compared with the West End animals and the latter were in significantly poorer condition, with elevated plasma osmolalities. Their total body water content (TBW) was also significantly lower, indicating that they were dehydrated when compared with the Lake Baghdad animals.

The significantly higher urine to plasma osmotic ratio (U/P_{osm}) of the West End quokkas is a clear measure of the extent to which water-conserving mechanisms have been fully deployed, and ADH levels in the plasma are amongst the highest reported from free-ranging marsupials (Bradshaw, 1990). Our conclusion is thus that the West End quokkas are experiencing stress as a result of water deprivation in late summer on Rottnest Island and these data highlight the importance of even small quantities of brackish water for the maintenance of good condition in this species over the long, hot, Mediterranean summer in Western Australia.

More examples of stress in natural populations of animals will be considered in detail in Chapter 5, as well as recent theoretical papers impinging on stress and its measurement such as those of Wingfield (1994), McEwen (1998a,b) and Bradshaw (1999).

3

Basic methods used in ecophysiological studies

Animal welfare and the ethics of experimentation

Recent years have seen a welcome change in the attitude of some scientists towards animals, especially towards those species that are routinely used for experimental purposes. In the not too distant past, experiments were carried out where animals were subjected to considerable distress and it is largely due to public attitudes that such experimentation is now banned and all procedures are carefully scrutinised by Animal Ethics Committees before being approved. Copies of the Australian National Health and Medical Research Council (NH&MRC) Code of Practice may be had by visiting the following website: *http://www.health.gov.au/nhmrc/research/awc/code.htm*

Experiments in which animals are subjected to stress are without value scientifically as, often, the animal's appropriate responses are masked or overwhelmed by the stress to which they are exposed. One can never be sure that animals housed in laboratory situations are behaving 'normally', although the laboratory is the only place where hypotheses can be properly tested. It is thus extremely important that every effort is made to ensure that animals held in laboratories are given an environment sufficiently diverse for them to display their normal behaviour and physiology. A classical case in point is that of reptiles, which were thought for many years to be 'cold-blooded' because, in laboratory situations, their body temperature closely parallels that of the surrounding environment. If given access to a source of heat, however – such as an infra-red lamp – most lizards will bask under this and rapidly elevate their body temperature to between 34 °C and 37 °C, which is their 'thermal preferendum'. One of the great advantages of ecophysiological studies over classical laboratory investigations is that, for the most part, they are carried out with animals living in their natural environment, where the intervention of the experimenter is kept to an absolute minimum.

Capture methods: trapping and marking animals

A variety of trapping methods is available for the capture and recapture of vertebrate animals, depending on their size and general mobility. The most frequently used live traps are 'Elliots' and 'Tomahawks'. These are collapsible metal traps (either aluminium sheet or wire grid) that are activated by a spring mechanism, usually a treadle that closes the door to the trap once it is depressed by the animal's weight. Large versions of this system are usually called 'box traps' and have a door that closes when the animal reaches, and touches, a bait suspended from a hook at the rear of the cage (see Figure 3.1). Traps are commonly baited with what is called 'universal bait', a mixture of rolled oats, peanut butter, honey and raisins, which many animals find attractive. In some cases anchovies or fish meal are added and this is very effective in attracting carnivores such as the marsupial chuditch, *Dasyurus geoffroii*. When trapping in winter it is important to cover the traps with a hessian bag to avoid hypothermia in caught animals. Traps may be cleared throughout the night if this does not cause undue disturbance, otherwise they should always be cleared at first light. Larger animals such as rock wallabies – which are very prone to injure themselves in typical metal traps – can be caught successfully with the specially designed 'Bromilow' trap (Kinnear *et al.*, 1988), which has an internal netting structure in which the animal is held once it has entered the trap. Larger netting structures with a swinging gate may also be used to trap more than a single animal that are attracted to bait distributed internally. Aniseed oil has also been found to be very effective in attracting marsupials to such multi-trap structures.

Another very effective method for trapping small vertebrates such as lizards and many small mammals is to use pitfall traps, often in association with a fenceline. Pitfall traps are easily constructed from PVC piping (150 mm diameter, 400 mm length) set into the ground flush with the surface. These are closed with a metal lid when not in use and small animals readily fall into them, especially if directed towards them by a small fenceline of 5–10 m in length, which crosses the trap. For larger animals (> 50 g), a 20 l plastic bucket is used instead of narrower PVC pipe. Pitfall traps also obviate the need for bait, and are thus the only means of trapping that is effective with nectarivores, such as the marsupial honey possum (*Tarsipes rostratus*), that are not attracted by universal bait. Interesting analyses of the effectiveness of pitfall and other traps are given in Laurance (1992), Parmenter *et al.* (1998), Friend (1984) and Friend *et al.* (1989). Small birds and bats are normally captured with

Figure 3.1. Two types of live trap used routinely in animal studies. Above is the aluminium Elliot trap used for small vertebrates (20–250 g) and below is a larger Tomahawk trap used for larger species (250–750 g).

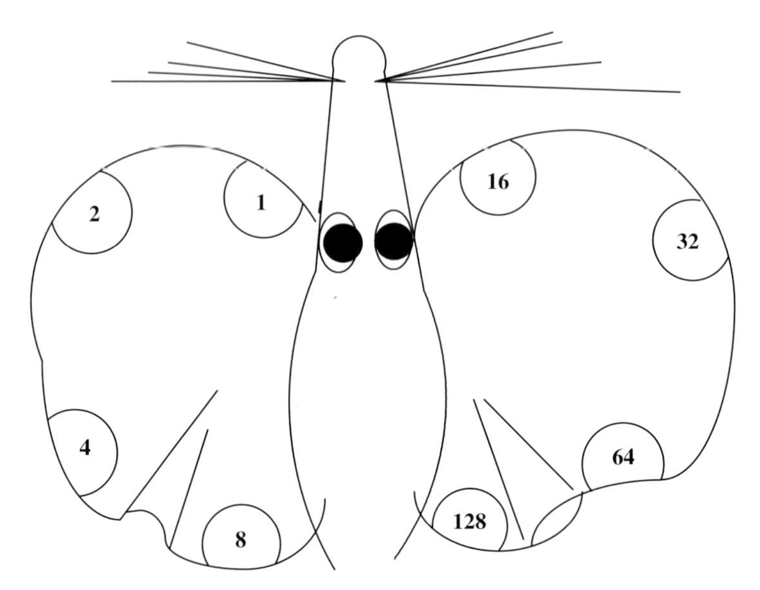

Figure 3.2. Diagram of the binomial code used for marking, by ear-notching, the marsupial honey possum, *Tarsipes rostratus*, and suitable for other very small mammals.

fine mist nets that are suspended between metal or wooden poles. Larger birds may be captured with cannon nets, which are fired above a flock of birds when foraging on the ground. A very useful handbook, giving details of census methods for a wide variety of terrestrial vertebrates, is that edited by Davis (1982).

Standard methods for marking animals individually include metal or plastic leg bands for birds (a bird-bander's licence is needed for this) and small numbered metal ear tags with mammals (National's 'Jiffy' wing bands from National Band & Tag Co., Kentucky, USA, are very good). Toe-clipping should be avoided with small mammals but this seems to cause lizards little distress and toes are often lost naturally. Snakes can be marked by scale-clipping and there are now a number of freeze-branding techniques available for marking amphibians (see Klewen and Winter, 1987). Very small animals such as the honey possum, with a mass of 6–10 g, are marked by ear clipping using the code shown in Figure 3.2. Every effort should be made to ensure that marking is carried out efficiently and without imposing undue stress on the animal. There is little point in releasing an animal that is likely to die as a result of the stress of a badly managed capture and marking procedure.

Specialised capture techniques

Many large mammals cannot be captured effectively in traps, and darting is now an established and reliable method for their capture. Dart guns are available commercially, which propel darts filled with an anaesthetic mixture that rapidly immobilises the animal (e.g. the PneuDart 270™ (manufactured by Pneu-Dart Inc.; see *http://www.pneudart.com/*), which uses an 0.22 blank to propel an 0.5–5.0 ml syringe from a second barrel). The syringe contains a small charge that explodes by inertia on impact and serves to inject the dose rapidly into the muscle of the animal (Capchur™ is another brand based on an 0.410 shotgun firing explosive darts, and DEEB manufacture a Black Wolf dartgun powered by CO_2). The most commonly used anaesthetic combination is a mixture of Rompun™ (xylazine) and Ketalar™ (ketamine hydrochloride), which is effective in a large range of mammals. The darted animal usually lies down within 4–8 min of being hit and can be approached and handled within 10–13 min. It is important not to chase darted animals as they may injure themselves in the chase and they can also 'work off' the effects of the anaesthetic. Once processing of the darted animal is complete, an antidote is usually administered (Reversin™ or yohimbine is commonly used) to offset the effects of the anaesthetic as quickly as possible. Darted animals should always be kept under observation until they have recovered and are able to move again in a co-ordinated manner. If it is hot, move the animal into the shade or cover with a bag. In very large mammals, such as elephants, the drug M999 (Etorphine™), often combined with xylazine, is used in dosages of 12–19 mg for both females and bulls (0.004 mg kg^{-1}). Revival is effected by injecting M5050™ (diprenorphine). A compressed-air dart gun with a range of 40 m (Daninject) has been used very effectively by one of my former students, Juliet King, working with elephants in Kenya. Mixtures of ketamine and diazepam and of tiletamine and zolazepam (Zoletil 100™) at a combined dosage of 1 mg kg^{-1} has been found to be very effective in immobilising grey seals and southern elephant seals by Baker *et al.* (1988, 1990).

Meristic and body mass measurements

There are a number of basic measurements that are usually taken when animals are captured: body mass (measured with an accurate balance, battery-powered for small animals or Persola™ and Salter™ spring balances for larger individuals), and usually a number of size measurements. Measurement of pes length (length of the foot, measured with vernier callipers) is usual in mammals, snout–vent length in reptiles, and also head or crown length in small mammals. Carapace 'length' is routinely measured in tortoises; this is really a curve, owing to the convex shape of the carapace. It is very important to know the extent of the error inherent in such measurements and it is essential, when first commencing a research programme, to repeat the measurement a number of times and then calculate the coefficient of variation (CV) of the parameter (the CV is expressed as a percentage and calculated as 100 × SD/mean, where SD is standard deviation). CVs of less than 10% are desirable when taking important body measurements and practice is often needed before such precision can be achieved, especially under field conditions. Pes length, although commonly used, is often difficult to measure with high repeatability, owing to the fact that the foot in mammals is made up of a number of separate bones and can vary in shape quite readily. In an effort to overcome this, Bakker and Main (1980) introduced a leg measurement, which they called the 'short leg'. This is effectively the length of the tibia, measured with large callipers, from the base of the patella to the calcanial ridge; although practice is needed to apply this measurement, the level of precision that can be achieved is much greater than with the pes. Body mass and size measurements are often used to estimate 'condition indices' which relate mass to body size, in an attempt to relate the mass of a given individual to theoretical or empirical models (Reist, 1985; Bradshaw and De'ath, 1991; Krebs and Singelton, 1993; Jakob *et al.*, 1996) Such condition indices have come under criticism because they are normally ratios (e.g. body mass in grams : pes length in millimetres, or some transformation of body mass such as the cube root) and ratios are difficult to treat statistically as they are not normally distributed (Atchley and Anderson, 1978). Packard and Boardman (1987, 1988) launched the attack with two papers arguing that the use of ratios should, at all cost, be avoided in ecological studies. Dodson (1978) defended their use, with appropriate care, and Magnusson (1989) responded in a paper recommending the use of an analysis of covariance (ANCOVAR) to overcome the problem. Tracy and Sugar (1989) then joined the debate, highlighting situations where the ANCOVAR would be inappropriate. The problem of non-normally distributed ratios at the basis of condition indices can be avoided by using the residuals from a regression of body mass (or some derivative) on body length (Bonnet and Naulleau, 1994), which are normally distributed around the least-mean-squares line of best fit. Even this stratagem, however, has its detractors, and Kotiaho (1999) has pointed out that residuals usually correlate slightly with body mass itself (larger animals have larger absolute residuals than small ones) and thus may contribute a systematic (although very small) error to the estimation of body condition. Despite these statistical concerns, condition indices are used widely by ecologists and ecophysiologists alike because, whatever their limitations, they provide precious information on the prior history of the animal that cannot be acquired in any other way. To know that an animal may have lost 30% of its expected body mass, given its material size, is vital information, even if the estimate may have an error of ±5%.

Measuring the extent to which an animal's limbs are symmetrical is also of some interest and there is a number of studies suggesting that animals that experience some degree of stress during their developmental period end up by being slightly asymmetric, i.e. with right and left sides not equal (Parsons, 1990). The study of what is now called 'fluctuating asymmetry' (FA) has received considerable attention in recent years as, if true, it provides a very simple way of determining whether particular individuals in a population evidence the effects of developmental stress (Palmer and Stobexk, 1986; Sarre and Dearn, 1991; Sarre *et al.*, 1994). Conservation biologists have been particularly interested in such measures as they could provide evidence of inbreeding and loss of genetic diversity in endangered species (Gilligan *et al.*, 2000). There are studies that both support and refute the claims of 'fluctuating asymmetry' as an epigenetic measure of stress and a recent analysis of the literature by Simmons *et al.* (1999) and Tomkins and Simmons (2003) suggests that this field may have become something of a 'bandwagon'. Nevertheless, it is of interest to know the extent to which individuals in a population show evidence of bilateral asymmetry and this has certainly been found to be significantly elevated in an insular population of the rock wallaby *Petrogale lateralis* on Barrow Island off the arid northwest coast of Western Australia (J. M. King and S. D. Bradshaw, in prep.). The level of genetic divergence in this population has recently been analysed by Eldridge *et al.* (1999) and found to be the lowest of any mammal yet studied.

Mark-and-recapture estimates of population size

There is a very large literature devoted to attempts to estimate the size of vertebrate populations, and the inherent difficulties of the methodology are well covered in a number of ecological texts such as those of Caughley (1977), Caughley and Sinclair (1994), Caughley and Gunn (1996) and Krebs (1999, 2001). The procedures involved are very simple: animals are marked in some way, released into their habitat and then recaptured at some later date. Basic methods for estimating the population size are based on the 'dilution principle', which is widely employed in the biological sciences. Fluid spaces such as the extracellular fluid volume (ECFV) are estimated, for example, by injecting a known amount of a substance that distributes itself normally in this space (for the ECFV, the plant polysaccharide inulin or sodium thiocyanate (NaCNS) are used as they do not penetrate cell membranes). After adequate mixing, a blood sample is taken and the concentration of the molecule measured. If x is the amount injected and y the final concentration, then V, the volume of distribution (or ECFV in this case), is given by x/y. In the case of our population estimate, x becomes the number of animals marked and released and y is the proportion of marked to unmarked individuals at some time after adequate 'mixing'; V then becomes the estimated population size. This 'dilution principle' was first used by Lincoln in his simple Lincoln Index to estimate the size of waterfowl populations (Lincoln, 1930), and Bailey (1952) provided a means of estimating the error involved. There are many more sophisticated methods available now which rely on multiple samples over a designated period of time to estimate better the size of the population (see Leslie, 1952; Schnabel, 1938; Schumaker and Eschmeyer, 1943) and the best of these, such as the Schnabel, are based on the rate of increase of the proportion of marked to unmarked animals with a continuous trapping and marking régime. Some methods allow for separate estimates of rates of birth and death (or immigration and emigration from the population) and the stochastic method introduced by Jolly (1965) is far superior when following populations over longer periods of time, and when recapture rates may be low. The Achilles' heel of all such methods, however, is the problem of 'mixing' or, in population terms, the assumption that all animals are equally likely to be trapped. In practice, this is never the case; 'trap happy' and 'trap shy' animals are the bane of population ecologists! There are computer programs available (e.g. CAPTURE[2] and MARK[3]) that allow one to test this assumption and select a procedure that, to some extent, corrects for the vagaries of animal behaviour. In some cases, where recapture rates are extremely low and the animal very long-lived, none of these methods gives reliable estimates and one may have to rely on the KTBA estimate, i.e. 'known to be alive', which is simply the number of individuals that have been caught over a given time span. The western swamp tortoise, *Pseudemydura umbrina*, is a case in point. This threatened species had declined to extremely low numbers in the wild in Western Australia and a recovery programme was put in place to breed young in captivity and release them into the wild (Kuchling *et al.*, 1988, 1992) (see Chapter 9). Individuals of this species are, fortunately, very long-lived and in some cases individuals that were thought to be long dead were recaptured up to 20 years later. The only method for estimating the size of this population that has proved of any use is that of Manley and Parr (1968), which was designed for use with insects. In practice, no method gives reliable estimates of population size unless something like 70–80% of the population is marked, and this is one of the major problems that one needs to contend

[2] Software developed by Department of Fishery and Wildlife Biology, Colorado Coop. Fish and Wildlife Unit, Colorado State University: *http://www.cnr.colostate.edu/~gwhite/software.html*

[3] Program by Dr G. C. White, Colorado State University; downloadable from *http://www. cnr. colostate.edu/~gwhite/mark/mark.htm*

with in the field. Simple Excel programs for estimating population size are given in Appendix 1 for the Jolly–Seber method and Schnabel method, which are two of the easiest to apply and give results as reliable as those of some of the more complex procedures.

Radiotracking and spooling to evaluate individual movements

The development of small, reliable, frequency-modulated radiotransmitters that may be affixed to animals has revolutionised our understanding of the behaviour of many vertebrates in the field (see Harris *et al.* (1990) for a review). A great deal of information may be gained about an animal's movements and home range, defined by Burt (1943) as '... that area traversed by an individual in its normal activities of food gathering, mating and caring for young' from repetitive trapping, but there are limitations. Although traps are usually placed in regular grids, areas utilised by the animal may still be missed. The animal may also move surprising distances in one night and avoid the traps completely. Dickman *et al.* (1995) have recently documented long-range movements of a number of species of small mammals in arid regions of Australia. Species such as the sand-inland mouse *Pseudomys hermannsburgensis* were found to move distances of up to 14 km. Moro and Morris (2000) also used radiotelemetry very effectively to determine the home ranges of two competing rodents on Thevanard Island in the arid northwest of Western Australia. The tiny marsupial honey possum, *Tarsipes rostratus*, is a species that, from trapping data alone, was thought to be highly sedentary, moving as little as 20–30 m over a 12 month period (Wooller *et al.*, 1981; Garavanta *et al.*, 2000). A recent radiotelemetry study has revealed, however, that males move large distances between feeding areas (where they are routinely trapped) and daytime refuges, which may be as much as 250 m distant (Bradshaw and Bradshaw, 2002). Estimates of home range size in this species based on trapping alone thus vastly underestimate that of males, although they are reliable for females, which move far less.

There is a large variety of radiotransmitters available on the market: Biotrak™ transmitters from the UK have been used extensively, as have AVM™ from the United States. Sirtrak™ in New Zealand manufacture single-stage transmitters weighing as little as 0.9 g for use with very small mammals, and Titley Electronics in South Australia (*http://www.titley.com.au/*) have recently marketed a tiny collar-fitting transmitter weighing 0.7 g with a battery life in excess of three weeks (see Figure 3.3).

What is critical in all telemetry studies is that the weight and shape of the transmitter do not impede or stress the animal in any way. Typically, one should

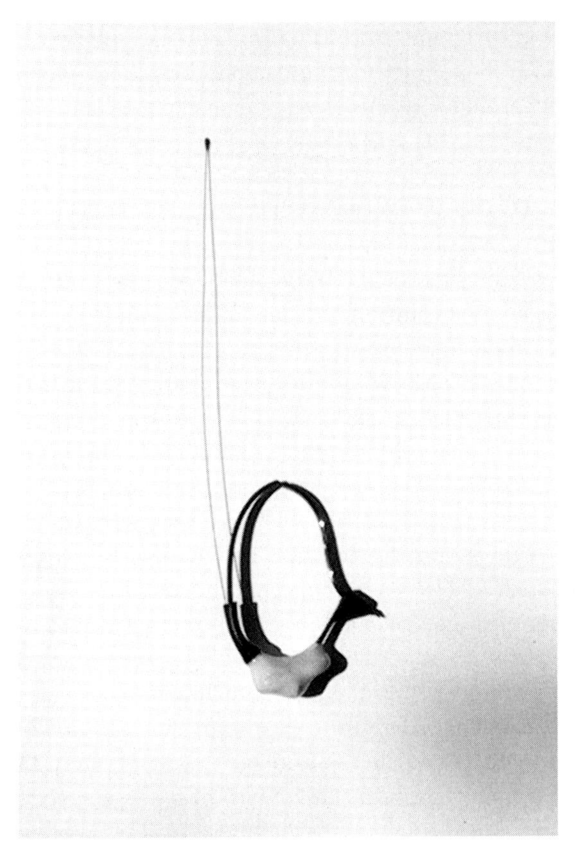

Figure 3.3. Lightweight (< 1 g) radiotransmitter used for tracking very small vertebrates. This radio has a range of 200–400 m and a battery life of *c.* one month.

try to affix a transmitter that weighs 5–8% of the animal's total mass, and this fraction should never exceed 10%. In some cases, transmitters may be inserted internally. In snakes, this can be done by placing them inside a small prey species that is then fed to the snake (see Saint Girons and Bradshaw (1981), for example, with the dugite, *Pseudonaja affinis*). The larger the animal, of course, the larger the transmitter that it can carry and some remarkable results have been achieved by French scientists radiotracking wandering albatrosses (*Diomedea exulans*) in the Antarctic region (Jouventin and Weimerskirch, 1990) as will be seen in Chapter 8. These birds are large enough to carry satellite-linked transmitters and 'fixes' may be acquired every 20 minutes; these are downloaded at the CNRS Centre d'Études Biologiques de Chizé in southwestern France. Similar satellite-linked transmitters are used routinely with elephants in Kenya and aid in tracking the migratory movements of these large herbivores. Global positioning system (GPS) collars from Lotek (Canada) are used and these also have a standard VHF beacon fitted for locating the elephant. The GPS in the collar takes a fix from satellites at regular intervals, selected by the experimenter (e.g. hourly), and stores the data in a data-logger for later downloading. The collars may also be purchased with a modem so that data may be downloaded to a laptop when flying or driving close to the elephant. Battery life at the moment is limited to one year, but it is hoped that improvements will soon increase this to two years.

Spooling is another very useful technique used with smaller mammals and reptiles. This technique also enables one to reconstruct the actual trajectory of a given individual's movements over time, which is very difficult with radiotelemetry. With radiotracking, successive fixes over short periods of time may be used to reconstruct an animal's movement path but there can be large errors involved in triangulation (Saltz, 1994) and autocorrelation of the data may also introduce an error if fixes are too close together in time (Swihart and Slade, 1985a,b). Tracking an animal constantly may also perturb it and lead to 'false' movements provoked by the observer. With 'spooling', a centre-unwinding cotton thread spool is attached to the animal and, as it moves, it leaves a trail of thread behind it (much like Theseus in the Minotaur's lair). The thread is then collected and the distance travelled per day calculated by weighing the thread, which has a precise mass : length ratio. The only problem with the technique is that, sometimes, the animal may run out of thread if it moves a large distance over 24 h, but this can be anticipated in larger individuals and one may adjust the thread accordingly. Some surprising results can be had with this method, such as finding that the small agamid lizard known as the mountain devil, *Moloch horridus*, will occasionally climb trees in search of ants, on which it feeds exclusively (P. C. Withers, pers. comm.).

Radiotelemetric acquisition of physiological data

Radiotelemetry is also used most effectively to gather information on vital physiological parameters, such as body temperature, heart rate, or respiration rate, of otherwise unrestrained animals. Small transmitters may be implanted or affixed to the body of the animal and be tuned for the transmission of selected parameters. Body or skin temperature is the most common parameter chosen with free-ranging animals, but heart and respiration rate can also be of great interest in attempting to measure the costs of locomotion in different vertebrates. McCarron *et al.* (2001) have recently published data on

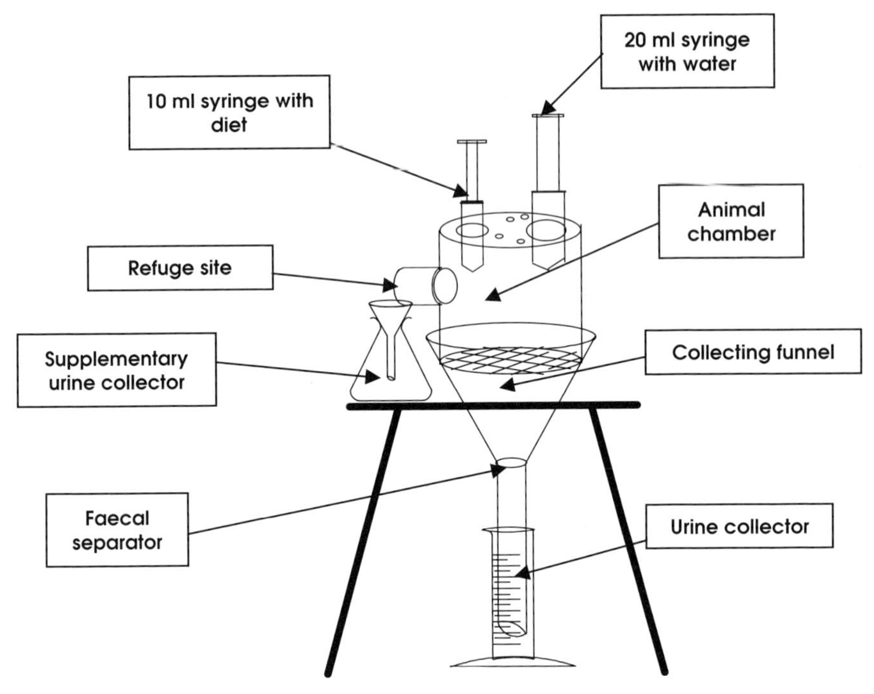

Figure 3.4. Diagram of purpose-built metabolism cage used to collect urine and faeces from the 10 g marsupial honey possum, *Tarsipes rostratus*.

free-ranging heart rate, body temperature and energy metabolism in eastern grey kangaroos and red kangaroos in arid southeast Australia and this paper gives an excellent idea of the scope of the technique.

Metabolism cages for the collection of urine and faeces

There are a number of occasions where it is highly desirable to collect urine and faeces from animals that have just been trapped in the field. From the volume and concentration of urine voided one can learn much about the state of hydration of the animal, and faeces, when properly analysed, can provide vital clues to the animal's diet. Metabolism cages are special cages designed to hold animals for short periods of time for this purpose. They can be purchased (for example, Nalgene™ make plastic cages suited for rats and mice) but are usually hand-made to suit the size and shape of the species being studied. Typically they consist of a holding cage positioned over a grid that collects the faeces but lets the urine pass, to be collected via a funnel. The design of the special metabolism cage that we use for the tiny marsupial honey possum is shown in Figure 3.4. Larger metal metabolism cages

Figure 3.5. Large metabolism cages used to collect urine and faeces from 1–3 kg marsupial possums, bettongs and wallabies.

used to collect urine and faeces from chuditch (*Dasyurus geoffroii*), woylies (*Bettongia penicillata*) and brushtail possums (*Trichosurus vulpecula*) are shown in Figure 3.5. These cages, of course, are not 100% accurate and urine is lost by evaporation and also in wetting the collecting surfaces. These problems can be partly overcome by spraying the collecting surfaces with a silicone spray each time before use and by pouring a small volume of Ondina™ oil (Shell) in the receptacle used to collect the urine. If urea concentrations are also to be measured later, a small amount of 1N HCl or other acid (1–2 ml) is also added to inhibit bacterial activity and loss of urea by breakdown to ammonia. Metabolism cages need to be calibrated before use to determine

Table 3.1. *Calibration of marsupial metabolism cages*

Cage no.	Volume added (ml)	Volume recovered (ml)	Recovery (%)
1	0.5	0	0
1	1	0.5	50
1	2	1.4	70
1	4	3.0	75
1	8	7.4	92
1	16	15.5	97
1	32	31.8	99
1	64	62.0	97
2	0.5	0.2	40
2	1	0.4	40
2	2	1.1	55
2	4	2.8	70
2	8	6.4	80
2	16	14.8	93
2	32	31.0	97
2	64	60.2	94
3	0.5	0.4	80
3	1	0.8	80
3	2	1.7	85
3	4	3.4	85
3	8	6.9	86
3	16	15.0	94
3	32	31.4	98
3	64	64.0	100

the extent of urinary losses in the collection process and this is done by adding known amounts of water to the cages and then assessing recovery. Some typical data from the large cages shown in Figure 3.5 are shown in Table 3.1 and, graphically, in Figure 3.6. These data show that, in general, losses are greater with small volumes, but cage no. 3 performed much better than nos. 1 and 2. When these data are graphed and a linear regression made, as in Figure 3.6, the cages turn out to be surprisingly accurate, despite their very simple design and construction. The equation of the combined regression for these three cages is $y = 0.982x - 0.556$ with a remarkable $r^2 = 0.999$, where y is the volume recovered and x the volume added. A 'perfect' regression equation would be $y = x$ but this equation shows that average recovery of the added water is close to 98%. We may now use this

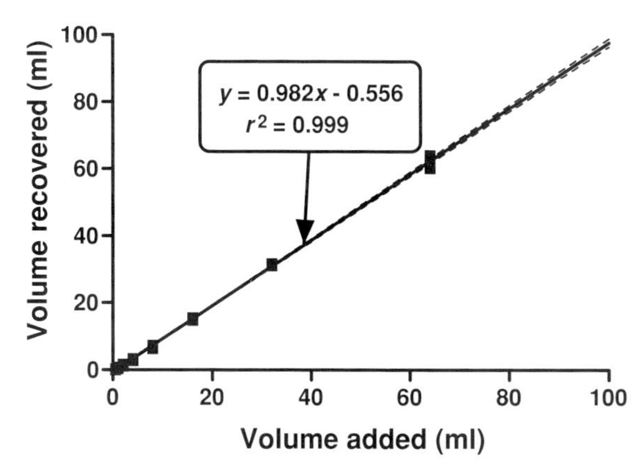

Figure 3.6. Urinary calibration curve for the metabolism cages shown in Figure 3.5, generated by dropping known volumes of water into the animal cage and then collecting as with urine. The equation of the regression is volume recovered = 0.982 (volume added) – 0.556 with $r^2 = 0.999$.

equation to correct for the actual volumes of urine recovered from our animals, where:

$$\text{Corrected volume} = (\text{volume collected} + 0.556)/0.982.$$

If, for example, we collect an actual volume of 12.8 ml of urine, then this would be corrected to a 'true' volume of 13.6 ml by the equation and 1 ml would correct to 1.59 ml.

Care needs to be taken in setting up the conditions by which the calibration curve is determined, as all of the data will subsequently be corrected by the use of this one equation. One needs to know whether the animal produces urine in large volumes at one time, or by small spurts over a longer period of time, and attempt to replicate this pattern when adding water. The calibration curve will also be different if water is added each time to a dry collecting surface, as opposed to one that has already been wetted by the addition of water. Care at an early stage can result in the collection of much more reliable and meaningful data sets.

Methods for the identification of food items

Although it is one of the most important aspects of an animal's biology, the quantification of dietary intake is, without doubt, one of the most difficult of tasks. Animal nutritionists know that, in grazing ruminants, large differences may exist between feed intakes of different individuals, which may translate into significant variations in productivity. Much effort has gone into devising methods for the estimation of feed intake of individual farm animals, including sophisticated methods involving the monitoring of jaw movements and the muscles controlling swallowing. These methods, however, cannot be applied to native animals that are not confined to paddocks and that are free to browse in habitats where the range of dietary items available to them may be enormous.

Various routine methods that are used to determine the diet of free-ranging native animals include the following.

- Direct observations of animals eating
- Identification of mouth contents
- *Post mortem* analyses of gut contents
- Impact of grazing on the species structure of the vegetation, measured with field exclosures (see, for example, Bell *et al.* (1987) with the Garden Island tammar, *Macropus eugenii*)
- Studies utilising marker substances, such as naturally occurring plant alkanes
- Faecal analysis

Of these, the last, faecal analysis, is the most utilised following its introduction in the 1960s by Glen Storr (Storr, 1961, 1963, 1964) working with the Rottnest Island quokka, *Setonix brachyurus*. This approach works best with strict herbivores and depends on the creation of a library or 'épidethèque' of epidermal fragments from all of the species of plant that occur in the animals' habitat. Microscopic slides are prepared from faecal pellets, from which an estimate may be made of the frequency with which given items are consumed. If the nutritive value of the plants is known, estimates of intake may then be made (Storr, 1964; Algar, 1986). To make slides, macerated faecal scats are first sieved, heated in a water bath in a chromic acid – nitric acid mixture and then neutralised and washed. Resultant fractions are stained, rewashed and then evenly distributed over labelled microscopic slides. Once air-dried, each sample slide is mounted in Aquamount™ and systematically scanned for food items under a dissecting microscope (×40).

A number of small mammals and bats feed on plant pollen and nectar and some species such as the honey possum, *Tarsipes rostratus*, are complete nectarivores. Identifying the plants on which these animals feed is done by means of the pollen, which is brushed from the animal immediately after its collection. Pollen grains can be stained and identified under the microscope by comparison with a collection of pollen grains representing the common species in the animal's habitat (Wooller *et al.*, 1983) (see Figures 3.7 and 3.8).

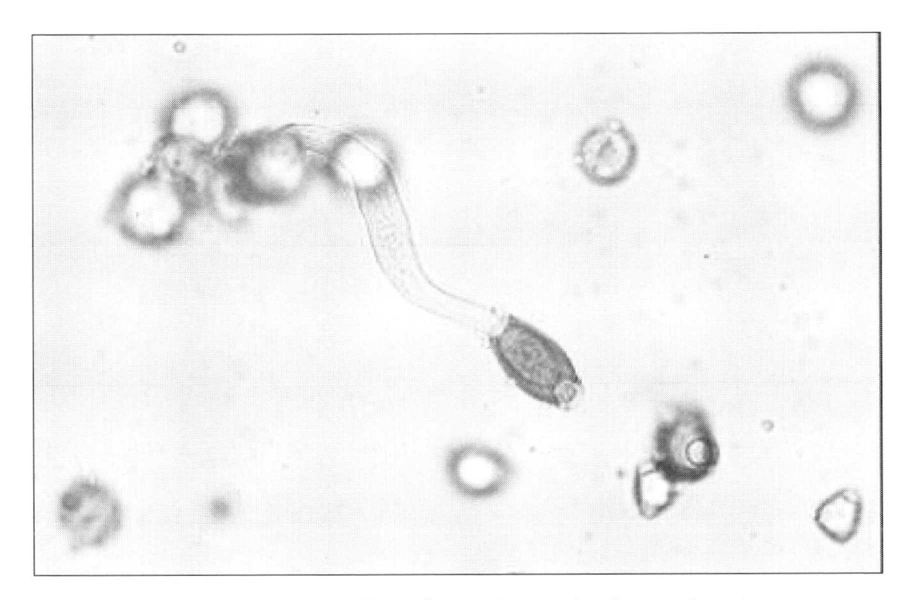

Figure 3.7. Microphotograph of a pollen grain germinating on the microscope slide (×100).

The use of radioisotopes to measure turnover rates of essential molecules and to estimate rates of food intake

By far the greatest advances in determining dietary intakes have been made, however, by using isotopes to measure the rates of turnover in the bodies of animals of essential molecules such as water, sodium and oxygen, all of which are linked in some way to the animal's diet (Bradshaw, 2000). Tritium (^{3}H), for example, is an isotope of hydrogen with two extra neutrons in the nucleus, which render it unstable and hence radioactive (see Figure 3.9). It decays with a half-life of approximately 12 years (i.e. half its radioactivity will be gone in 12 years), emitting weakly charged electrons in the process. Tritiated water, i.e. water in which one of the hydrogen atoms has been replaced by a tritium molecule (^{3}HHO), can be used to measure the rate of water turnover of an animal that is living undisturbed in its own habitat. The principle of the method is quite simple: the animal is injected (intramuscularly or intraperitoneally) with a small amount of tritiated water (1 mCi kg^{-1} = 37 MBq kg^{-1} is recommended) and a blood sample (the equilibration sample or ES) is taken 2–6 h later, depending on the size of the animal, by which time the isotope has fully equilibrated with the body water of the animal. The animal is then released and recaptured at some future time (usually 5–7 days later) and a second small blood sample taken, the recapture sample or RS. From

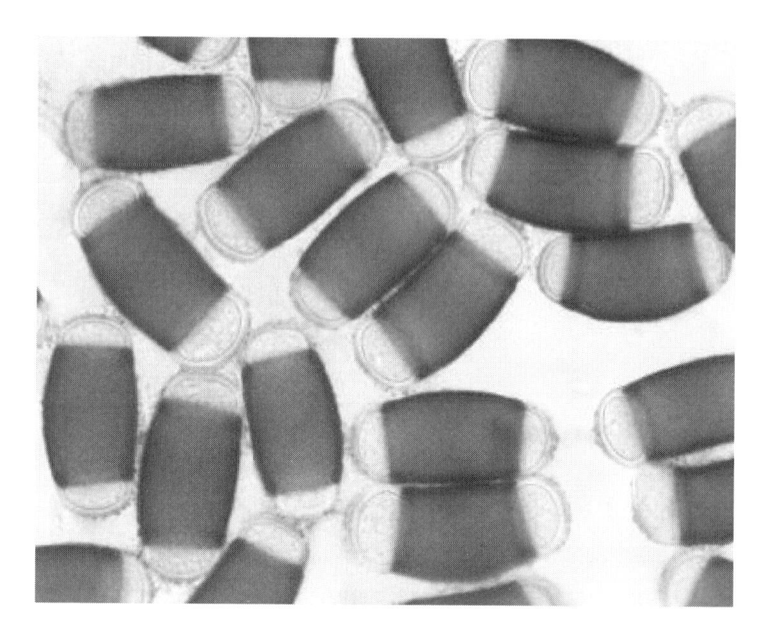

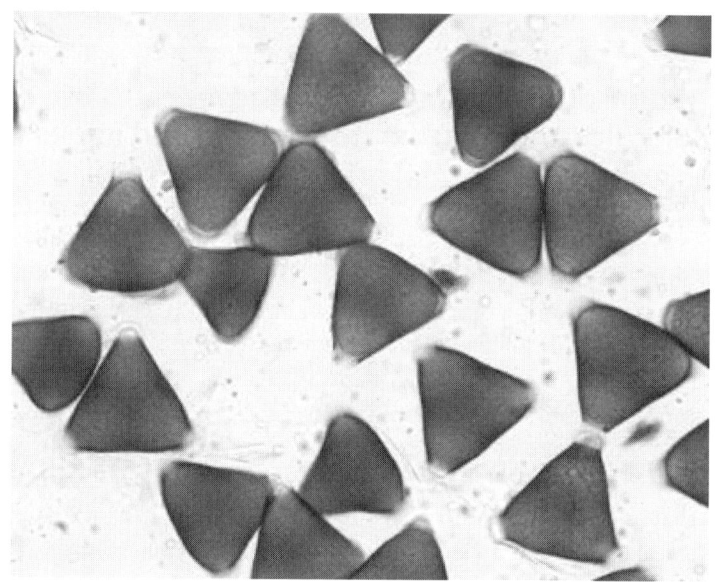

Figure 3.8. Microphotograph of pollen grains collected from the snout of the marsupial honey possum, *Tarsipes rostratus*. The species present are *Banksia ilicifolia* (above) and *Adenanthos meisnerii* (below) (×100).

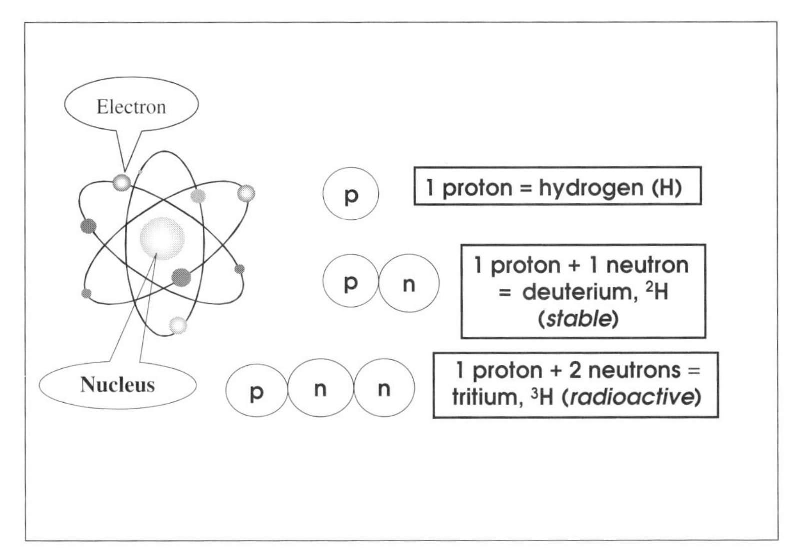

Figure 3.9. Diagram illustrating the atomic structure of the two isotopes of hydrogen: ^{2}H (deuterium) and ^{3}H (tritium).

the decline in specific activity (SA) of the injected isotope in the second blood sample (in this case the ratio of tritium isotope to hydrogen, or ^{3}H : ^{2}H) one may calculate the fractional turnover (k) of the water pool. If one knows the volume of this pool (i.e. the total body water content of the animal), which may be estimated accurately from the initial dilution of the amount of tritium injected, then the fractional turnover may be converted to a volume of water per day. This may then be corrected for metabolic water production (0.658 ml water per litre of CO_2 expired), knowing the body size and approximate metabolic rate of the animal, and this then is the volume of water procured from the diet, on condition that the animal drank no free water over the period of measurement. If the water content of the actual diet can be assessed, then the water turnover data give a direct estimate of food intake (see Nagy (1987b) for a detailed treatment of the topic).

The fractional turnover (k) of the labelled pool of body water is calculated as follows, where ES is the equilibration sample at time t_1 and RS is the recapture sample at time t_2 (see Figure 3.10):

$$\frac{k = \ln(\text{ES}) - \ln(\text{RS})}{t_2 - t_1}.$$

The product $k \times$ (TBW) then gives the rate of water influx in millilitres per unit time, where TBW is the total body water content of the animal in millilitres.

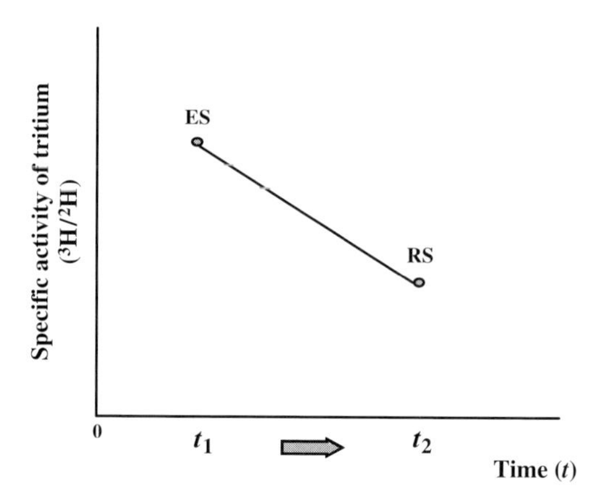

Figure 3.10. Graph illustrating the decline in specific activity (SA) of injected tritium as a result of water turnover over time $t_1 \rightarrow t_2$ in a vertebrate animal.

Counting of the radioactivity in blood samples is done either by centrifuging the blood to obtain plasma, or by distilling water (containing the tritium) from the blood sample (details of the method for microdistillation of blood samples are given in Nagy, 1983). A small sample (20–50 μl) of either plasma or water is then added to 5 ml of a scintillant (this is a solution containing an organic phosphorescing compound, e.g. Emulsifier Safe™ or Ultima Gold™ from Packard) in a plastic vial and counted in a liquid scintillation spectrometer (LSS). These instruments count the radioactivity in each sample by measuring the frequency and intensity of pulses of light produced when electrons emitted by the decaying tritium interact with the phosphor to emit a photon of light. The light is captured by two sensitive photomultiplier tubes operating on a 'coincidence circuit'. Only if both tubes record a photon within a time interval of 5 ms is it registered as a 'count'. Otherwise it is treated as 'noise' and not counted and by this stratagem the fidelity of the counter is enhanced considerably. It is important to note that no radioactive counter is 100% efficient and only a portion of the disintegrations occurring in the isotope are captured as 'counts' by the machine. This efficiency can be very high when the energy given off by the isotope is high (e.g. of the order of 96% in the case of ^{14}C, which has a peak emission energy of 156 keV) but is much lower and of the order of 30–40% in the case of tritium, which has a peak emission energy of only 18.6 keV. The counts per minute or cpm recorded by the LSS thus need to be converted to disintegrations per minute or dpm by correcting for the efficiency of counting, which is normally determined automatically for each sample. The dpm thus equal $100 \times \text{cpm}/E$, where E is the percentage efficiency. The error of counting is proportional to the number of counts accumulated and a minimum of 10 000 counts should be accumulated to ensure an error of 1% or less. Remember also, when counting plasma, that this is not pure water and plasma proteins account for about 6% of the overall volume. Plasma counts thus need to be corrected by 100/94 to reflect plasma water.

The water turnover method has been used extensively by Ken Nagy and his colleagues at UCLA to estimate the food intakes of a variety of desert animals that do not have access to free water (Nagy *et al.*, 1976; Nagy and Costa, 1980; Nagy, 1982a, 1988a; Nagy and Peterson, 1988). An example of how water turnover data may be used to estimate food intake, which varies from 6.3 to 61.7 g d^{-1} in the large scincid lizard *Tiliqua rugosa*, based on a third-year class exercise at the University of Western Australia, is given in Appendix 2. More details on methodology will be given in Chapters 4 and 5.

A second isotope, sodium-22, which is a potent γ-emitter, has also been used very effectively to estimate food intake in carnivorous animals via measurements of their daily sodium intake. The same assumptions apply to this isotope and the procedures followed are virtually identical with the collection of an equilibration and recapture sample over a suitable interval where the animal has had access to its habitual diet. The decline in SA of the injected sodium-22 again allows a measure of the animal's sodium influx per unit time. This figure needs to be corrected for assimilation as, normally, in vertebrates, retention of sodium from the diet is less than 100%, and usually falls between 80 and 90%. If the sodium content of the animal's diet is known – this is done by capturing and measuring the sodium concentration of selected prey items, usually determined by faecal analysis – then the food intake of the carnivore can be estimated. Measuring the specific activity of the sodium (i.e. the ratio of $^{22}Na : {}^{23}Na$) requires an additional technique, measuring the concentration of 'cold sodium' or ^{23}Na in the plasma sample, and this is usually done on 10 μl samples with a flame photometer. This extra step is not needed when working with tritium because, when working with pure water, the volume itself is a measure of the amount of hydrogen present, the 'cold isotope' in this case.

This sodium-22 method has been used extensively by Brian Green and his colleagues at the CSIRO in Australia to estimate food intake in a variety of animals including rabbits, dingoes, buffalo, chuditch, emus and penguins in the Antarctic (Williams and Green, 1982; Green and Eberhard, 1983; Williams and Ridpath, 1983; Herd, 1985; Green *et al.*, 1986, 1988, 1992; Green and Brothers, 1989). See also Gallagher *et al.* (1983), Gauthier and Thomas (1990) and Robertson and Newgrain (1992) for studies using sodium-22 on other vertebrates. The technique has even been used in migrating locusts

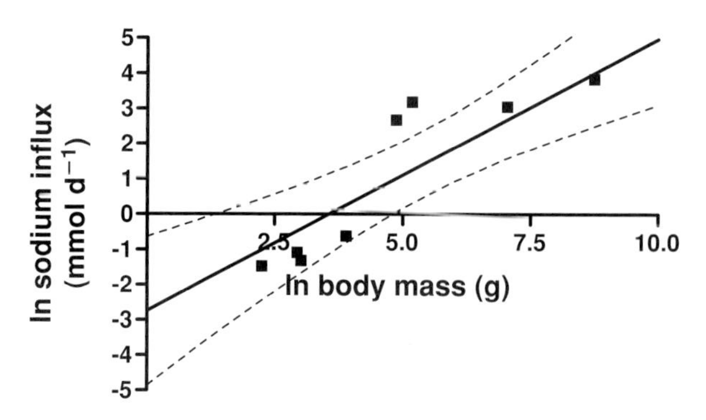

Figure 3.11. Relation between sodium influx, measured with sodium-22, and body mass in birds (modified from Goldstein and Bradshaw, 1998).

(Buscarlet, 1974). Goldstein and Bradshaw (1998) reviewed measurements of sodium flux in birds and found that they are tightly related to body mass, as shown in Figure 3.11. This highlights an important point when assessing measurements of food intake in wild animals: to what extent are such measurements simply a function of body size, or do they in fact reflect adaptive differences between species and the impact of environmental variables such as temperature or rainfall? Nagy (1994b) poses a similar question when considering the allometry of field metabolic rates in mammals. After adjusting for mass and infraclass effects, he found that residual variation is still substantial, ranging from 2- to 5-fold among marsupials and 6-fold among eutherians.

Table 3.2 presents some unpublished data from an honours thesis by a student in the School of Agriculture at the University of Western Australia, working with the threatened marsupial chuditch, *Dasyurus geoffroii* (S. Lacey, K. D. Morris, S. D. Bradshaw and R. Bencini, unpublished). These data were used to test the hypothesis that female chuditch have higher rates of water utilisation than males, which is clearly not the case in the two seasons measured. They also suggest that the food intake of females is greater than that of males in July, when they are carrying young, from their significantly higher influx of sodium ($p = 0.02$).

The doubly labelled water method for the measurement of field metabolic rate

The doubly labelled water (DLW)[4] method is one of the most powerful that has been developed over the past 25 years for the estimation of the rates of

[4] Doubly labelled water is water labelled with both tritium and oxygen-18.

Table 3.2. *Rates of water and sodium turnover in free-ranging chuditch* (Dasyurus geoffroii) *in the southwest of Western Australia*

Month	Sex	*n*	Body mass (g)	TBW (%)	Sodium pool (mmol kg^{-1})	Water influx (ml kg$^{-0.82}$ d^{-1})	Sodium influx (mmol kg^{-1} d^{-1})
April	F	5	860 ± 41	62.5 ± 2.6	28.3 ± 0.8	96.3 ± 9.2	4.4 ± 0.1
	M	3	1050 ± 58	64.4 ± 1.7	33.4 ± 0.8	133.7 ± 30.9	4.2 ± 0.6
July	F	2	825 ± 32	72.3 ± 0.3	50.7 ± 0.8	152.0 ± 31.1	6.6 ± 1.1
	M	3	1050 ± 96	70.1 ± 1.2	45.1 ± 1.1	143.1 ± 11.1	3.7 ± 0.4

From S. Lacey, K. D. Morris, S. D. Bradshaw and R. Bencini (unpublished data).

energy expenditure of free-ranging animals. It is based on the pioneering studies of Lifson and McClintock (1966), who first demonstrated that the oxygen of water is in isotopic equilibrium with that of expired carbon dioxide (CO_2), paving the way for measurements of the rate of CO_2 production. CO_2 and water combine to form carbonic acid and this reaction is favoured in animals by the presence of the enzyme carbonic anhydrase in the erythrocytes.

Carbonic anhydrase

$$CO_2 + H_2 \rightleftharpoons H_2CO_3 \longrightarrow H^+ + HCO_3^-.$$

When doubly labelled water is injected into an animal it interacts as follows:

$$\text{DLW} = {}^3\text{HH}^{18}\text{O} \rightarrow {}^3\text{HHC}^{18}\text{OO}_2 \rightarrow {}^3\text{H}^+ + \text{HC}^{18}\text{OO}_2^-.$$

Oxygen-18 is then lost from the body in the form of both water and carbon dioxide. The rate of decline of the specific activity of the injected ^{18}O is greater than that of the injected ^3H (which is only lost in the form of water) and the rate of CO_2 production is given by the difference in the slope between the two curves, as seen in Figure 3.12 where the alternative hydrogen isotope deuterium (^2H) is used in place of tritium.

Estimates of the errors involved in this methodology have been extensively studied by Nagy (1980, 1992; Nagy and Costa, 1980) and an excellent text detailing all aspects of the methodology is now available (Speakman, 1997). Comparisons with indirect calorimetric methods, and measurements of oxygen consumption, show that the method is accurate to within ±8% of classical laboratory methods for the measurement of energy expenditure. Nagy (1987b) and Nagy *et al.* (1999) have brought together a mass of data where the doubly labelled water method has been used to measure field metabolic rates (FMR) in birds, reptiles and mammals, from which rates of food intake may be estimated (see also Nagy (2001) for a recent review). Chapter 4 will discuss in detail some of the theoretical and practical aspects of working with doubly labelled water and also use the honey possum, *Tarsipes rostratus*, as an example of how reliable estimates may be obtained of food intake in free-ranging animals.

Methods of blood collection and the use of anaesthetics

Many ecophysiological techniques require that a small sample of blood be taken from an animal, and there are many methods available. Of paramount importance when taking a blood sample is the well-being of the animal, and the need to avoid inducing stress. Venepuncture (or phlebotomy) is the method

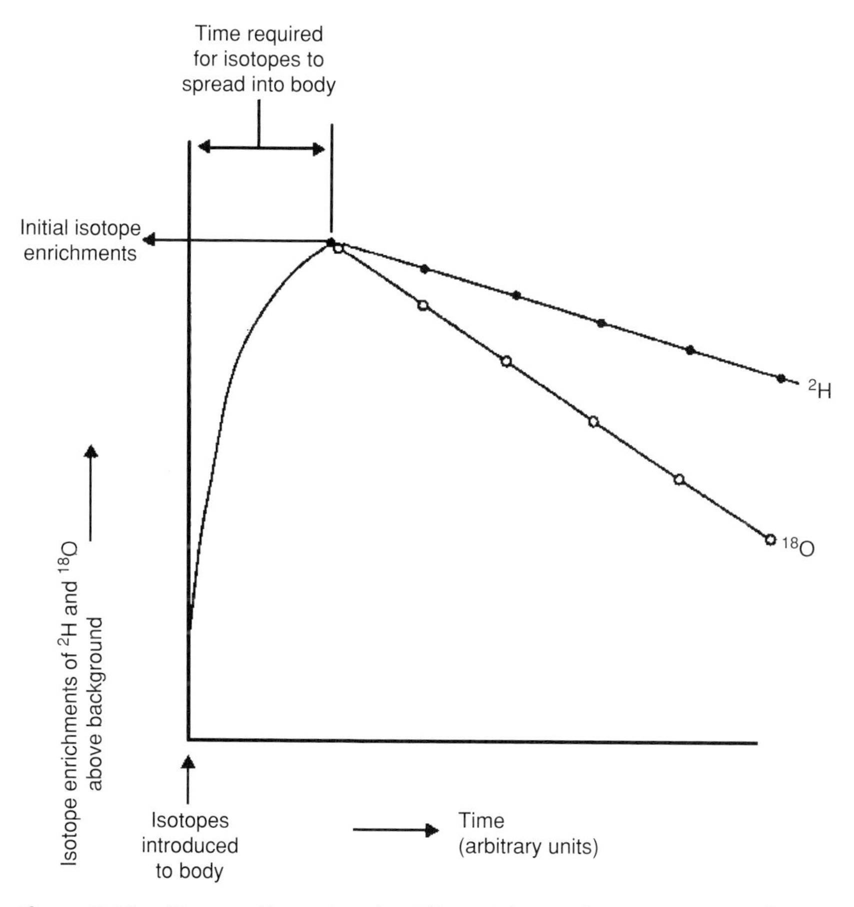

Figure 3.12. Diagram illustrating the differential rate of decay of tritium (^{3}H) and oxygen (^{18}O) isotopes when injected into an animal as doubly labelled water (from Speakman (1997) with permission).

of choice where a large superficial vein is accessible. Tail-vein punctures in kangaroos and wallabies are simple to carry out and cause the animal little distress. The animal is held in a calico or hessian bag, with just its tail extended and, with the head in the dark, the animal is much quieter and less likely to struggle. In macropod marsupials there are two large tail veins located laterally on the tail and the site is prepared by removing as much of the fur as possible with scissors. The site is then wetted with a local antiseptic (Cetavlon™ is ideal) and then shaved clean with a scalpel blade held at a very shallow angle so as to ensure that the skin is not cut. At this stage the vein usually becomes apparent and can be readily visualised by occluding it anteriorly (i.e. close to the body of the animal) either with the thumb or with a rubber tourniquet. A 2 ml syringe, fitted with an appropriately sized needle (e.g. 21G), is prepared and heparinised by flushing with a concentrated solution of sodium or ammonium heparin (Fisons Pty Ltd. or CSL, Australia) and then 'pumping' with air to ensure that only a film of this anticoagulant remains in the syringe. The vein is then punctured, keeping the bevel of the needle uppermost, and gentle suction applied as soon as blood is seen to enter the syringe. As the blood is being collected retrograde from the vein (i.e. blood is flowing from the tail to the body of the animal) the tourniquet can be released at this stage and this may increase the flow rate. When approximately 1 ml of blood has entered the syringe, the needle is removed in one flowing action while applying pressure to the puncture site with a sterile cotton swab. The site is then dressed, usually with a swab wetted with Cetavlon™ and tied around the tail with a single closure. The animal will easily remove this swab some time later when grooming itself. Collecting blood from an arm vein (saphenous) follows the same procedure and may be required in the case of animals that have obviously damaged tails and hence veins that are difficult to access (e.g. males that have been involved in fighting).

Bleeding small mammals is usually carried out by puncturing the infra-orbital sinus that is located in the eye socket. The technique, known as orbital sinus bleeding, was first published by two French workers in 1951 who described phlebotomy from the ophthalmic plexus in laboratory rats and guinea pigs (Halpern and Pacaud, 1951). These workers emphasised the ease with which blood may be obtained and the advantage of the technique in terms of lack of stress and trauma to the animal. The technique has since been applied widely, especially with small mammals such as rats and mice, and has replaced other more traumatic techniques such as tail cutting and cardiac puncture. The method is usually employed for the collection of relatively small volumes of blood (*c*. 0.1–0.3 ml). If larger volumes of blood are needed, conventional venepuncture techniques are normally used. Orbital sinus bleeding has also been found to be extremely effective in a number of lower vertebrates such as agamid lizards and frogs, but not in birds (MacLean *et al.*, 1973).

This technique has been evaluated quite extensively in the literature, especially as there was initial concern expressed about possible damage that may occur to the eye socket as a result of repetitive bleeding. Suber and Kodell (1985) compared collection from the orbital sinus with tail vein incision, and cardiac puncture in rats anaesthetised with carbon dioxide, and reported a significant problem with haemolysis in samples taken by cardiac puncture and tail vein incision. They went on to conclude that '... orbital puncture was the technique of choice with the least variance and greatest precision....'

Brookhyser *et al.* (1977) used orbital bleeding to collect quite large volumes of blood (1–3 ml) from the chinchilla and concluded that '. . . by monitoring food intake and body weight, this technique was shown to have no detrimental effects on the general health of the animals.'

Vanherck *et al.* (1998) compared the effects of orbital sinus bleeding by experienced and non-experienced technicians and emphasised the importance of training. They went on to note that

> The use of either a Pasteur pipette or a haematocrit capillary did not necessarily produce different results. Neither did puncturing the lateral versus the medial canthus of the orbit. By not applying chloramphenicol eye ointment in the conjunctival sac after puncture, the number of abnormalities in 'ocular discharge' and 'corneal alterations' in the punctured orbits was significantly decreased. Four punctures in the same orbit with 14-day intervals by a skilled animal technician did not cause a significant increase in abnormalities.

Izumi *et al.* (1993) compared plasma biochemical values when collected from the orbital sinus, heart and tail vein in mice, rats and hamsters and found small differences in enzyme concentrations, but reported no adverse impact of the technique.

Parmenter *et al.* (1998) instituted long-term studies to monitor population density and prevalence of infection in rodents, which constitute the reservoir for Sin Nombre virus (SNV). In this study, field techniques used in sampling small mammals for SNV infection were evaluated to determine whether trapping and handling protocols were having significant effects on future trappability, or mortality, of animals. Analyses were based on 3661 captures of 1513 individuals representing 21 species from three rodent families and two species of rabbit. They concluded that: '. . . the handling/bleeding procedures had no significant effects on recapture rates or mortality.' Dameron *et al.* (1992) also reported on characteristics of blood collected directly from the posterior vena cava and concluded that: 'For haematologic testing, there were no biologically significant differences between samples collected from the OVP and PVC.'

Although the use of this method is still opposed by some of the more conservative veterinarians, there is little doubt that, in the hands of a skilled practitioner, it is the most efficient and least stressful means of collecting blood from small animals. The method can be used either with or without anaesthesia, depending upon the species. It should always be remembered that applying an anaesthetic imposes a considerable physiological stress on the animal itself and, although it may relieve the experimenter's anxiety somewhat, it may impose an added and undue strain on the experimental animal. I have collected blood from the orbital sinus of rodents such as the North African merione (*Merionis shawii* and *M. libycus*), *Pseudomys albocinereus*, *P. nanus*, *Zyzomys argurus*, *Mus domesticus* and *Leggadina lakedownensis*, and marsupial species such as *Isoodon obesulus*, *I. auratus*, *Antechinus flavipes*, *Parantechinus apicalis*, *Tarsipes rostratus*, *Pseudantechinus macdonnelennis*, *Sminthopsis crassicaudata*, *S. griseoventer* and *Planigale maculata*. Of particular note is the 10 g honey possum, in which all previous attempts to bleed the animal by conventional means resulted in its death. Using the orbital sinus technique, many hundreds of individuals have been bled in the past few years without a single loss.

Anaesthetics are required whenever any procedure is likely to result in pain or obvious distress to the animal. Again, the choice of anaesthetic depends very much on the species concerned. Gaseous anaesthetics are preferred with mammals (e.g. halothane or Fluothane^{TM} mixed with oxygen) as their rate of administration can be readily controlled. A very interesting field-adapted method developed for use with the pebble-mound mouse, *Pseudomys chapmani*, is described in Anstee and Needham (1998). Barbiturate anaesthetics, which were once widely employed (e.g. Nembutal^{TM} or sodium pentobarbitone, Abbot), now tend to be avoided because of their side effects, but sodium pentothal (Pentothane^{TM}) administered intravenously can be very effective for complex surgery with marsupials (see, for example, McDonald and Bradshaw, 1993). Ketamine hydrochloride (Ketalar^{TM}) is often recommended for use with reptiles and amphibians (see, for example, Arena *et al.*, 1988) but it must be remembered that ketamine hydrochloride is a neuroleptic drug and not an anaesthetic *per se*, and its long-term usage is not recommended in mammals. I have found Brietal Sodium^{TM} (methohexitone sodium from Lilly Corp.) at a dosage of 10 mg kg^{-1} and administered intraperitoneally to be one of the most effective anaesthetics in reptiles. With frogs, anaesthesia may be induced with Brietal Sodium and then controlled by the percutaneous administration of MS222^{TM} (tricaine methane sulphonate, Sigma). This is most easily done by wetting a swab with MS222 and applying it to the 'pelvic patch' (McClanahan and Baldwin, 1969) of the frog or toad, from which it is readily absorbed. Alternatively, the frog may be placed in 0.3% MS222 for approximately 30 min if it reacts poorly to Brietal Sodium. What is often not appreciated is that the reaction to anaesthetics may be highly variable in lower vertebrates such as reptiles and amphibians. Whereas it is relatively easy to define dosage limits for mammals, these vary widely between different individuals in lower vertebrates. I have found often that one individual may require ten times the dosage of an anaesthetic that works well in another of the same species. This means that, when using anaesthetics, it is critical to commence with the lowest possible dose and increment this gradually so that losses do not eventuate.

Measurement of plasma constituents

Knowledge of the composition of the internal environment, or *milieu intérieur*, of an animal may be critical when studying changes that occur in its immediate environment. As discussed in Chapter 2, our ability to discern states of stress in animals depends very much on our capacity to measure changes in internal state, which may be inimical to the long-term survival of the individual. Foremost among such parameters are plasma concentrations of basic electrolytes, such as sodium, potassium and chloride, and the osmotic pressure, or osmolality, of the plasma that is the resultant of such constituents. A number of auto-analysers are available for carrying out such analyses routinely, but these are often expensive to purchase and maintain and often only available in hospitals and pathology laboratories where large batches of samples need to be processed daily.

Concentrations of sodium and potassium in very small samples of plasma or urine may be readily measured by emission spectroscopy in a flame photometer (FP), or by absorption in an atomic absorption spectrophotometer (AA). Measurements by AA require less sample volume (of the order of 1 μl), because of the greater precision of the instrument, but FPs are much simpler to operate and give reliable results with 10 μl samples. Interference between ions such as Na^+ and K^+ can be a problem at high concentrations in FPs and is avoided by diluting all samples in a 15 mmol l^{-1} solution of lithium chloride, which serves as an internal standard. A simple protocol for the simultaneous measurement of Na^+ and K^+ concentrations in plasma and/or urine is given in Appendix 3 for an IL FP.

Osmotic pressure, or osmolality, is a colligative property, which means that it depends upon the number of molecules present, and not their chemical nature. Various techniques are available for measuring the osmolality of biological fluids, which is expressed in mOsm kg^{-1} of water. Freezing-point depression is the oldest method, and various freezing-point depression osmometers are available on the market (e.g. Gonotek, Model Osmomat 030; Knaver). This method has the advantage that the sample is not destroyed in the process of measurement and may be recovered for further analysis. Vapour pressure osmometers (e.g. Wescor) are used widely to measure plasma and urinary osmolality and operate reliably with sample volumes in the 8–10 μl range. Plasma chloride concentrations are usually measured in 10 μl samples by amperometric titration using a Buchler–Cotlove™ or Radiometer™ chloridometer. Blood and plasma glucose concentrations may be measured with glucose oxidase kits (Sigma) but there are many small electronic instruments used by diabetics for monitoring their blood glucose levels that are ideal for small animal studies (e.g. the ExacTech^{TM} blood glucose monitor produced by the Italian firm MediSense Inc.). Only one small drop of blood is needed and the measurements can be readily performed within minutes in the field.

Concentrations of urea in plasma and urine are usually measured in 5–10 μl samples via a modification of the Conway microdiffusion method (Wolfe, 1992) whereby ammonia (NH_3) is liberated from the urea and measured by its interaction with Nessler's reagent. The method is old, but very reliable; a detailed protocol is given in Appendix 4.

Measurements of plasma hormone concentrations

There are many hormones that are of interest to ecophysiologists. Sex steroids such as progesterone, testosterone and oestradiol-17β are vital in controlling sexual behaviour and influencing the activity of sexual organs such as the oviduct and uterus in vertebrates. Other steroid hormones, such as aldosterone, cortisol and corticosterone secreted by the adrenal cortex, control such diverse processes as water, electrolyte and carbohydrate balance of the animal, the latter also being assisted by the polypeptide hormone glucagon and the catecholamine adrenaline (norepinephrine). Concentrations of other hormones, such as antidiuretic hormone (arginine vasopressin (AVP) in most eutherian mammals and arginine vasotocin (AVT) in lower vertebrates), are also of critical importance in controlling and modulating the water economy of animals.

Until relatively recently, the routine measurement of these hormones has only been possible in specialist laboratories but radioimmunoassays (RIAs) are now widely available, many in commercial kits, which allow ready assay of many of these hormones in small plasma samples. The principle on which these assays are based comes from the work of Roslyn Yallow who, while studying the effects of insulin in diabetics noticed that, over time, the injections became less and less effective (Yallow and Berson, 1960). She surmised that the patients might be developing antibodies to the insulin, a suggestion that was met with derision by her colleagues who argued that antibodies could not be produced against such a 'small' molecule. She had her revenge when accepting the Nobel Prize for her work, which involved developing an RIA for insulin using these antibodies, by reading from the rejection letter that she received many years earlier from the prestigious American journal *Science*.

The basic principle of RIAs is treated in detail in a number of excellent texts (see Ekins, 1974; Chard, 1978) and the following is only a brief outline. Antibodies are protein molecules that have the capacity of binding specifically

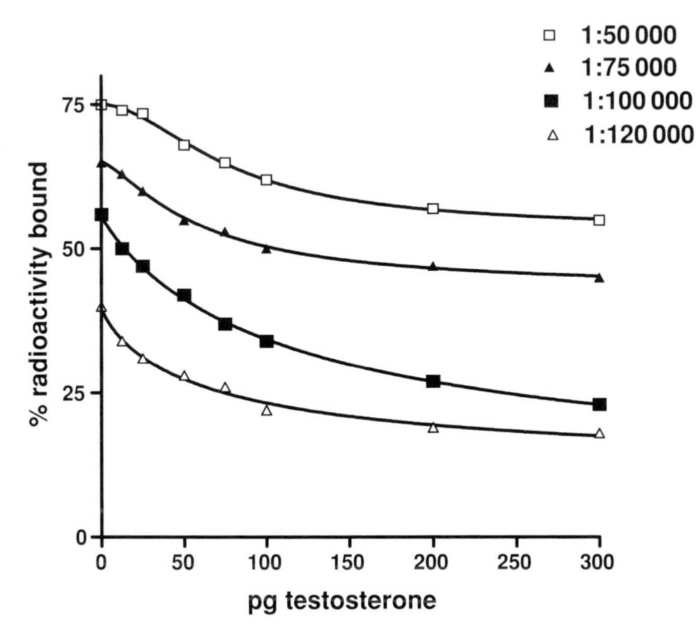

Figure 3.13. Successive dilutions of a progesterone antibody showing how an appropriate dilution is chosen to establish the standard curve. The 1 : 100 000 dilution binds in excess of 50% of the label and gives the maximum displacement of the label over the range of standards (0–300 pg) and is thus the antibody dilution of choice.

with another molecule, the antigen or ligand that was used to provoke their formation. Antibodies are usually raised in mammals such as the rabbit and they are produced by injecting the animal with a solution containing the antigen that one wishes to measure. Often, the antigen is combined with a protein, such as bovine serum albumin (BSA) or thyroglobulin, in order to increase the size of the molecule being used, so as to challenge the animal's immune system. Producing antibodies is something of a hit-and-miss procedure and one never knows until one collects blood from the animal whether a 'useful' antibody has been produced. This is an antibody that will bind the antigen with high and unique specificity and thus be useful in an RIA. Once such an antibody (AB) has been produced to a hormone such as progesterone, aldosterone, or arginine vasotocin, the assay is then established by firstly 'titrating' the AB to determine the optimum concentration for its usage. Serial dilutions of the AB are incubated with a range of concentrations of the antigen to be measured and a series of 'displacement curves' produced, as shown in Figure 3.13. Of these, the dilution at 1 : 100 000 is chosen because it shows maximum displacement over the concentration range of the hormone that we can expect in our blood samples. Hormone molecules are measured in the RIA by assessing their ability to replace a radioactive form of the same molecule that is bound to the AB. The AB is first exposed to a solution containing a tracer amount of the hormone, labelled with a radioactive isotope such as tritium (^{3}H) or iodine-125 (^{125}I). The amount added is calculated, knowing the binding characteristics of the AB, to occupy about 50% of the available binding sites. If we now add to this a solution containing the hormone that we wish to measure, these non-radioactive hormone molecules will 'compete' for binding sites on the AB. This competitive interaction can be described by the Law of Mass Action as follows:

$$\mathrm{AB} + \mathrm{Ag}^* \rightarrow \mathrm{AB{-}Ag}^* + \mathrm{Ag}^*,$$

where AB is the antibody and Ag* the radioactively labelled antigen (e.g. ^{3}H-cortisol).

When unlabelled antigen (i.e. cortisol) is added to this equilibrium, then some of the unlabelled cortisol will compete for, and replace, some of the labelled cortisol bound to the AB, as follows:

$$\mathrm{Ag} + \mathrm{AB{-}Ag}^* + \mathrm{Ag}^* \rightarrow \mathrm{AB{-}Ag}^* + \mathrm{AB{-}Ag} + \mathrm{Ag}^* + \mathrm{Ag}.$$

It is this 'competitive displacement' of the labelled (Ag*) by the unlabelled antigen (Ag) that forms the basis of the RIA. All one needs to do now is to measure the decrease in the amount of labelled antigen that is bound to the AB and this is a measure of the amount of unlabelled hormone that has been added to the system. In practice, it is usually easier to measure the amount of radioactive label displaced from the AB and a number of simple systems are available for separating 'free' and 'bound' fractions – dextran-coated charcoal and ammonium sulphate precipitation being among the most common.

The process of binding between an AB and an antigen is a reversible one and can be best represented by the following equilibrium equation:

$$\mathrm{AB} + \mathrm{Ag} \underset{k_2}{\overset{k_1}{\rightleftharpoons}} \mathrm{Ab{-}Ag},$$

where k_1 and k_2 are the rate constants of the two reactions. The Affinity Constant,

$$K_\mathrm{a} = \frac{[\mathrm{AB{-}Ag}]}{[\mathrm{Ab}] + [\mathrm{Ag}]} = \frac{k_1}{k_2} \quad \text{expressed in mol l}^{-1},$$

and the Dissociation Constants, $K_\mathrm{d} = 1/K_\mathrm{a}$.

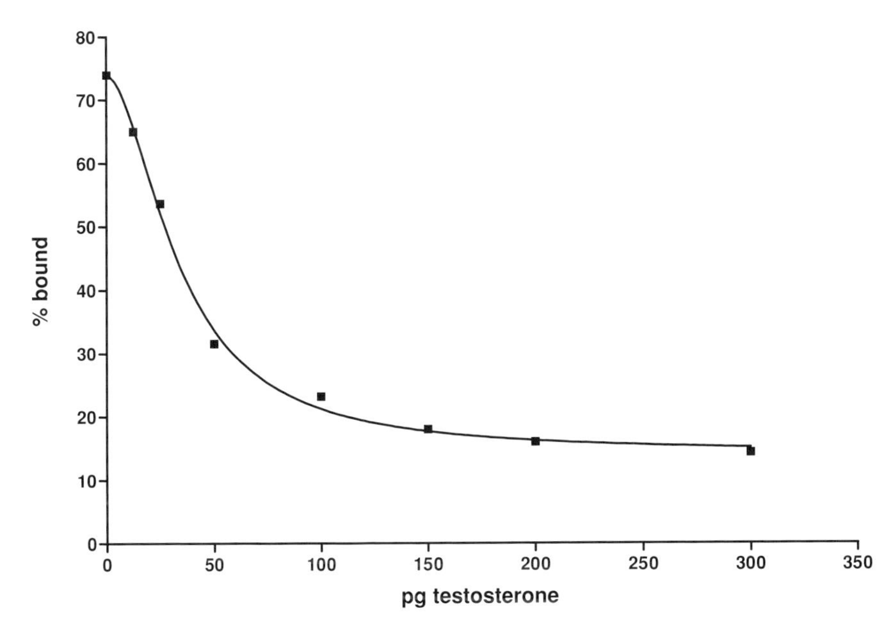

Figure 3.14. Standard displacement curve used for the radioimmunoassay (RIA) of the steroid hormone testosterone. Specific binding to the antibody (1 : 80 dilution) decreases from 75% to approximately 20% with the addition of increasing amounts of testosterone and the useful part of the curve extends from approximately 10 to 100 pg.

An example of a protocol for the measurement of a steroid hormone in plasma samples, in this case testosterone, is given in Appendix 5. A typical standard curve for testosterone, using an AB in a 1 : 80 dilution, is shown in Figure 3.14. Note that maximum binding is close to 75% and that this falls progressively to a minimum of about 16% as more and more cold testosterone is added. Not all of the curve can be used, however, for the assay, as it becomes very flat, has minimal discriminating ability above 150 pg. Best results with this AB will be obtained with samples containing testosterone in amounts ranging from 10 to 150 pg (i.e. 1–150 ng ml^{-1} if a plasma volume of 10 μl is being assayed). Figure 3.15 shows another procedure that may be needed when attempting to determine how effective the AB is in discriminating between closely related hormones. This graph shows the binding characteristics of an AB to a number of closely related neurohypophysial peptides, arginine vasotocin (AVT), mesotocin, phenypressin and lysine-vasopressin (LVP). As may be seen by the displacement curves, the AB is quite specific for AVT, LVP and phenypressin, but shows no cross-reactivity with mesotocin over a considerable dose range.

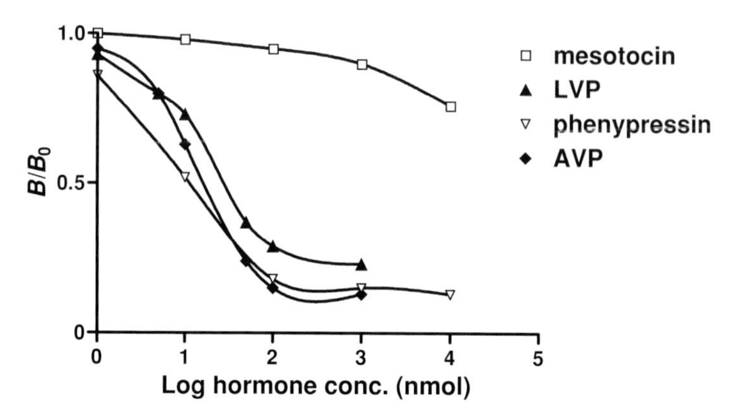

Figure 3.15. Competitive binding data for an antibody raised to the neurohypophysial peptide arginine vasopressin (AVP). The antibody cross-reacts substantially with both lysine vasopressin (LVP) and phenypressin (Phe^2-Arg^2-vasopressin), but shows no cross-reactivity with mesotocin.

There is a fairly long history of attempts to find the best curve to 'fit' the binding data from RIAs; Rodbard and Ekins have done most to explore mathematically the nature of the binding changes that occur in an RIA (see Ekins, 1974; Rodbard, 1978; Rodbard *et al.*, 1978). There is now general agreement that the four-parameter logistic equation (4PLE) most accurately describes the shape of the curve; it is calculated as follows.

$$y = \frac{A - D}{1 + (x/C)^B} + D,$$

where A is the upper asymptote of the curve, D is the lower asymptote of the curve, C is the value of x at the mid-point of the curve, and B is the slope of the curve. Curve-fitting computer programs such as Prism^{TM} (Grahpad, US) will rapidly fit a curve to RIA binding data and also allow the automatic calculation of unknowns.

One topic that can be confusing is when attempting to calculate or describe how accurately the assay measures the actual hormonal concentrations in the plasma. By 'accuracy' we mean how closely the measures approach the 'true' concentration, and this is normally assessed by repetitive measurement of an 'internal standard' of known concentration. Such an internal standard should be run with every assay so that day-to-day variations in the performance of the assay can be monitored. Typically, an internal standard is made by adding a known amount of the steroid to be assayed to a volume of 'stripped' plasma, i.e. a plasma pool from which all the endogenous steroid has been 'stripped' by heat incubation with charcoal (see Appendix 6 for details). Aliquots of this internal standard are then included in each assay of the standards and plasma samples and provide an appropriate quality control for the assay. These values can also be used to calculate the 'precision' of the assay (usually referred to as 'sensitivity' in the literature), which is defined as the smallest amount of hormone that can be significantly differentiated from zero. There are two basic methods for calculating the sensitivity of an assay, The first is that of Frankel *et al.* (1967), where sensitivity (S) is calculated as:

$$S = \frac{ts}{\sqrt{n}},$$

where s is the SD for the chosen standard, n the number of replicates and t the value of Student's t for $p = 0.05$ or 0.01 for $n - 1$ degrees of freedom.

The other approach is to calculate sensitivity from the standard deviation (SD) of replicates of the lowest standard in the standard curve. If, for example, ten replicates of the same 10 pg standard average $72.3 \pm 1.6\%$ bound, then the precision of the assay can be estimated by converting the error to an actual amount, i.e. 10 ± 0.22 pg. If we accept two standard deviations as our measure of precision then these data suggest that the assay is capable of discriminating an amount of 0.44 pg of the steroid from zero with 95% probability.

In some cases it may be necessary to extract and partially purify the plasma hormones before they are assayed, and some form of chromatography is the procedure used. Paper and thin-layer chromatographic methods were supplanted by column liquid chromatography in the 1970s, and high-pressure liquid chromatography (HPLC) is now used routinely to separate many hormones. Simple glass columns packed with a suitable compound, such as silica, celite or Lipidex™ (Packard) as the 'stable' phase, are very effective for separating the major sex steroids (progesterone, oestradiol-17β, testosterone and dihydrotestosterone (DHT)) into discrete fractions. Fairclough *et al.* (1977) detail procedures for the chromatographic separation of the major sex steroids, using Lipidex™, which is a modified form of Sephadex (hydroxyalkoxypropyl-Sephadex), as the stable phase and various concentrations of n-hexane and chloroform as the mobile phase. Bradshaw *et al.* (1991) used this method to measure seasonal changes in sex steroid concentration in two species of agamid lizard in the arid Pilbara region of Western Australia over a period of years, showing that reproduction in the two closely related species was cued by quite different climatic conditions.

Measurement of urinary and faecal steroids

The past decade has seen the exciting development of a number of RIAs that measure concentrations of steroid hormones excreted in the faeces and urine of an animal. Steroids are lipid molecules that have a very low solubility in water and therefore cannot be readily eliminated in an unmodified form in the urine. Hormones such as progesterone are metabolised in the liver, firstly by the addition of extra hydroxyl groups, and then conjugated with glucuronic acid to form glucuronides, whereas oestrogens are conjugated to form sulphates, both of which are water-soluble and may then be excreted in the urine. Early assays focused on measuring rates of excretion of conjugated steroids in the urine but, starting in the late 1970s, papers were published showing that many steroids are processed and excreted in the unconjugated form (i.e. native) through the bile and, ultimately, the faeces (Czekala and Lasley, 1977; Bishop and Hall, 1991; Lasley and Kirkpatrick, 1991; Wasser *et al.*, 1991). The advantages of measuring faecal hormones are obvious, when one considers that the animal does not need to be captured or bled, repeatedly in some instances, to obtain samples. The steroids in faeces, that have been collected over a period of 24 h therefore represent the pooled fractions of plasma steroids, and provide an integrated measure of physiological status over a 24 h period (Goymann *et al.*, 1999). Faecal steroid analysis has become a valuable method of monitoring hormonal changes in certain species, especially endangered mammals (Schwarzenberger *et al.*, 1996b; Spanner *et al.*, 1997; Stead-Richardson *et al.*, 2001) and is of considerable practical use to wildlife managers (Brown *et al.*, 1997). One of the major advantages is the avoidance of stress-induced changes in hormone concentrations, as faecal concentrations represent the situation some 24 h before the animals are captured (depending upon the rate of alimentary passage of the food; see Harper and Austad, 2001). In species that do not breed well in captivity, the ability to evaluate fertility non-invasively, as well as to document the occurrence of ovulation, through simple collection procedures, improves breeding efforts. Endocrine information is important, not only for characterising the female ovarian cycle and timing of reproductive events, but also for assessing the practical value of using external markers, such as behaviour or vaginal cytology, as reliable indicators of a female's reproductive status (Schwarzenberger *et al.*, 1996a), as well as to examine the influence of stress (Cavigelli, 1999). Faecal steroid analyses thus offer the potential of addressing many problems in reproductive, stress and conservation biology and the field is a burgeoning one at the present time. A sample protocol for the measurement of testosterone in faecal samples is included as Appendix 7 and a generalised faecal glucocorticoid assay for use in a diverse array of mammalian and avian species has recently been described by Wasser *et al.* (2000). A very useful study has just been published by Hiebert *et al.* (2000) detailing methods for measuring levels of corticosterone (B) in cloacal fluid collected from hummingbirds. The method has been well validated and corrects cloacal concentrations of B by reference to simultaneous measurements of creatinine concentrations in the fluid.

Measuring kidney function from urine collections

We saw earlier in this chapter how urine samples may be collected quantitatively from small to medium-sized mammals in the field by the use of suitably designed metabolism cages. As well as providing data on the composition of the urine so voided (electrolyte concentrations, osmolality, etc.), these samples can be used to provide a great deal of information about the animal's kidney function. To do this one needs to know the rate of urine production over a 24 h period, coupled with measurements of the concentrations of selected osmolytes in both plasma and urine. Twenty-four hours is normally too long a period to maintain an animal without food or water in the confines of a metabolism cage and, in practice, they are held overnight for a period of 10–12 h. Studies with a number of species of marsupial have shown that the rate of urine production estimated from 8, 12 and 24 h collection periods varies little and the volume of urine collected over a 10–12 h period may thus be extrapolated to a 24 h collection with confidence. By also measuring the osmolality (osmotic pressure) of the urine voided, and that of the plasma, osmolar and free-water clearances may also be calculated as follows:

$$V = C_{\mathrm{osm}} + C_{\mathrm{H_2O}},$$

where V = rate of urine production in ml kg^{-1} d^{-1},
C_{osm} = osmolar clearance in ml kg^{-1} d^{-1},
$C_{\mathrm{H_2O}}$ = free-water clearance in ml kg^{-1} d^{-1}.

The osmolar clearance is the volume of plasma 'cleared' of osmolytes per unit time, but is more usually visualised as that portion of the urine volume that is isosmotic with (equal in concentration to) the plasma. The free-water clearance can then be thought of as the amount of distilled (i.e. osmolyte-free) water either added to or subtracted from the urine to make the urine either dilute (hypo-osmotic) or concentrated (hyperosmotic) as the case may be. $C_{\mathrm{H_2O}}$ is normally negative in mammals since the urine is virtually always hyperosmotic, unless they happen to be water-loaded. If the urine voided were isosmotic then $C_{\mathrm{H_2O}}$ would be equal to zero and be positive if the urine were hypo-osmotic (i.e. more dilute than the plasma).

C_{osm} is calculated from the osmotic pressure of the urine and plasma as follows:

$$C_{\mathrm{osm}} = \frac{U_{\mathrm{osm}}}{P_{\mathrm{osm}}} \times V,$$

and from this it follows that:

$$C_{H_2O} = V - C_{osm}.$$

Thus, with the collection of some fairly simple data, we gain a very good idea of the animals' water economy during summer. Rates of water turnover and total body water (TBW) may be compared between sexes and between species and evidence of dehydration can be found by looking for any decrease in TBW below normal values associated with an increase in plasma osmolality. The osmolality of the urine is a good index of how water-deprived, or otherwise, is the animal, and the U/P_{osm} quotient of the urine is a direct index of the concentrating activity of the kidney. Clearances of selected osmolytes can also be calculated, as the clearance of a given volume of plasma of any osmolyte 'x' can be simply calculated from the formula:

$$C_x = \frac{U_x V}{P_x},$$

where C_x = the volume of plasma cleared of osmolyte x in ml kg^{-1} d^{-1},
U_x = urinary concentration of x in mmol l^{-1},
P_x = plasma concentration of x in mmol l^{-1},
V = rate of urine production in ml kg^{-1} d^{-1}.

Clearances of sodium, potassium, urea, etc., may thus be calculated, knowing plasma and urinary concentrations and V, the rate of urine production. Some actual data from a study of two species of marsupials, the woylie (*Bettongia penicillata*) and the chuditch (*Dasyurus geoffroii*) in late summer in the Perup Nature Reserve in southwestern Western Australia are shown in Table 3.3. Note that, although rates of urine production in the chuditch are much lower than those in the woylie, the urine of the former is more highly concentrated, with an U/P_{osm} quotient of 9.7 compared with 2.5, and the clearance data show that this is due to the higher osmolar clearance and more negative free-water clearance of the carnivorous chuditch. One would predict from these data that urea concentrations would be higher in the urine of the chuditch than in that of the herbivorous woylie.

Measurement of the glomerular filtration rate in the field

Although the collection of voided urine over a limited time span can give one a very useful 'snapshot' of an animal's state of hydration and level of activity of the kidney concentrating mechanism, more detail is needed when attempting to assess overall kidney function. For this, the rate of filtration of urine from the

Table 3.3. *Kidney function in woylies and chuditch*

Animal no.	Body mass (kg)	Urine vol. coll. (ml)	Urine vol. corr.[a] (ml)	Coll. time (h)	V (ml kg^{-1} d^{-1})	Urine osm.	Plasma osm.	U/P_{osm}	C_{osm} (ml kg^{-1} d^{-1})	C_{H_2O} (ml kg^{-1} d^{-1})
Woylies										
mj1487	1.17	38.00	52.91	8.78	123.61	298	279.5	1.07	131.79	−8.18
mj1547	1.37	3.60	7.04	6.33	19.48	687	296.5	2.32	45.14	−25.66
w4941	1.12	5.60	9.71	9.17	22.68	986	268.0	3.68	83.45	−60.77
w4942	1.19	10.00	15.57	8.33	37.71	691	282.0	2.45	92.38	−54.67
w4928	1.44	17.00	24.91	9.00	46.12	793	268.5	2.95	136.22	−90.10
Mean	**1.26**	**14.84**	**22.03**	**8.32**	**49.92**	**691**	**278.9**	**2.49**	**97.80**	**−47.88**
S.D.	**0.14**	**13.93**	**18.57**	**1.16**	**42.61**	**251**	**11.7**	**0.96**	**37.55**	**31.88**
S.E.	**0.06**	**6.23**	**8.31**	**0.52**	**19.05**	**112**	**5.2**	**0.43**	**16.79**	**14.26**
Chuditch										
w5311	0.94	5.80	9.97	9.68	26.31	3112	307.0	10.14	266.65	−240.35
w5762	0.56	2.10	5.04	6.43	33.59	3927	310.5	12.65	424.86	−391.26
w5764	0.7	0.80	3.31	6.37	17.81	2574	307.0	8.38	149.32	−131.51
w5768	0.54	3.00	6.24	7.07	39.25	2133	287.0	7.43	291.70	−252.45
w5775	0.92	0.80	3.31	6.40	13.48	3030	302.9	10.00	134.84	−121.36
Mean	**0.73**	**2.50**	**5.57**	**7.19**	**26.09**	**2955**	**302.9**	**9.72**	**253.47**	**−227.39**
S.D.	**0.19**	**2.07**	**2.76**	**1.42**	**10.69**	**670**	**9.3**	**1.99**	**118.25**	**109.64**
S.E.	**0.09**	**0.92**	**1.23**	**0.64**	**4.78**	**300**	**4.1**	**0.89**	**52.88**	**49.03**

[a] Urine volume corrected for losses.

plasma through the glomeruli in the kidney, known as the glomerular filtration rate (GFR) needs to be measured. In the laboratory, this is done by operating surgically on the animal and collecting urine directly from the ureters while infusing inulin at a constant rate into the animal's bloodstream. This procedure is usually terminal and thus quite unsuited for ecophysiological studies, where the primary aim is to return the animal unharmed to its environment. Another approach that has been used quite successfully with a number of species of wallaby is to measure the GFR, not by timed urine collection, but by measuring the rate of disappearance of injected inulin from the plasma. Inulin is a plant polysaccharide that is not processed by the kidney, i.e. it is neither secreted nor reabsorbed in the kidney tubule. The inulin, therefore, that is collected in the urine can only have been produced from the process of filtration, and the amount collected is a measure of its rate (Smith, 1951). Similarly, measuring the rate of disappearance of radioactively labelled inulin from the plasma will also allow one to calculate the GFR by using the equations originally developed by Sapirstein *et al.* (1955) to measure creatinine clearance in the dog.

The clearance method for GFR was first validated in the marsupial possum *Trichosurus vulpecula* by Reid and McDonald (1967) and then adapted by Bakker and Bradshaw (1983) to study renal function in the spectacled hare wallaby (*Lagorchestes conspicillatus*) on Barrow Island off the northwest coast of Western Australia. The animals were held quietly in hessian sacks with just their tails exposed, and cannulated with a Bardic Intracath™ that was inserted through a tail vein and then introduced well into the vena cava. They were then injected intravenously with a bolus containing a mixture of [^{14}C]inulin and [^{3}H]para-aminohippuric acid (PAH), and 1 ml samples of blood were collected at 5, 10, 20, 40, 80 and 120 min. PAH is both filtered and secreted in the kidney and provides a measure of renal blood flow from which one may then also calculate the fraction of the blood flowing through the kidney that is filtered per unit time (the filtration fraction, FF). Rates of decline of the two isotopes in the plasma of a hare wallaby are shown in Figure 3.16, with the disappearance equations fitted to each curve. Bradshaw *et al.* (2001) have recently used this approach to compare the renal function and levels of antidiuretic hormone in two desert-dwelling wallabies in the Pilbara region of Western Australia; the results of this study will be discussed further in Chapter 6.

The GFR has also been measured by David Goldstein in free-flying birds by means of implanted miniature osmotic pumps (Minipumps™). Details of the methodology are given in Goldstein (1993) and Goldstein and Rothschild (1993).

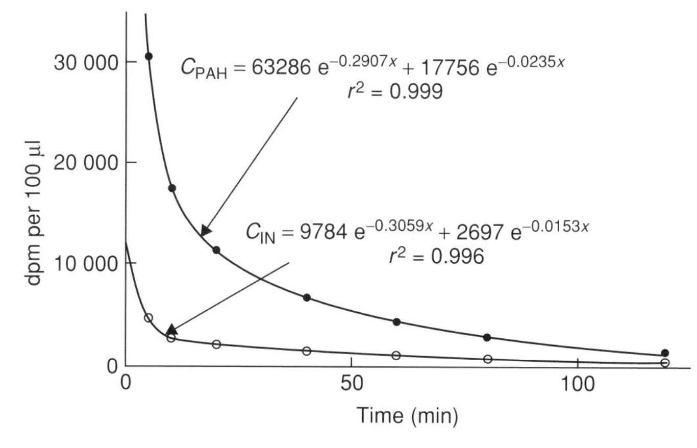

Figure 3.16. Disappearance curves over a period of 120 minutes for ^{3}H-PAH and ^{14}C-inulin injected intravenously into the spectacled hare wallaby, *Lagorchestes conspicillatus*. The curves have been fitted by using the method of Sapirstein *et al.* (1955) which enables the estimation of volumes of distribution as well as rates of clearance of the isotopes (adapted from Bradshaw *et al.*, 2001).

Kidney morphology in mammals

Considerable insight can be gathered into the functional capacities of the kidneys of a given mammalian species through an examination of its gross morphology. The kidney is not always 'kidney-shaped' in mammals (check out the shape of a cow's kidney next time you are at the butcher's) and they may also be unipyramidal (as in rodents and marsupials) or multipyramidal (as in humans), depending upon the number of medullary 'pyramids' that form the concentrating part of the kidney. The greater part of the blood supply to the kidney flows through the outer layer, or cortex, of the kidney, which is red as a consequence. The white inner medulla has few blood vessels (only the vasa rectae) and is composed primarily of urinary and lymphatic tubules. The relative size of the medullary region has long been known to correlate with the overall concentrating capacity of the kidney. Sperber (1944) was the first to develop a measure of renal medullary thickness, or RMT, which he applied as an index to predict the maximal concentration of the urine produced by a given species' kidney. Other workers have subsequently proposed slightly modified renal morphological indices (Heisinger and Breitenbach, 1969; Brownfield and Wunder, 1976; Greenwald, 1989; Beuchat, 1990), which generally allow reasonable predictions as to a given species' water-conserving abilities (Purohit, 1974). See also Bankir and de Rouffignac (1985) for an extremely insightful paper on kidney structure and function. General reviews may be found in Braun (1985) and Dantzler (1989). The length of the tubules leaving the juxtamedullary nephroi, and hence the overall length of the medullary pyramid, is also a good indication of concentrating ability, as longer nephric tubules allow the establishment of a higher standing osmotic gradient in the medullary interstitium. The midsagittal section of the kidney of the desert rodent *Leggadina lakedownensis*, shown in Figure 1.2, is a good example of a kidney with a high concentrating capacity; the medullary pyramid can be seen actually exiting from the kidney and extending down into the ureter itself. The maximum urinary concentration recorded for this species was 6330 mOsm kg^{-1} (Moro and Bradshaw, 1999; Moro, 2000) which places it among the highest concentrators of the desert rodents. MacMillen and Lee (1967) in an early paper reported urinary concentrations as high as 10 000 mOsm kg^{-1} in Australian desert rodents of the genus *Notomys*, and Diaz *et al.* (2001) have recently reported a maximal value of 9015 mOsm kg^{-1} for the South American desert marsupial mouse *Thylamus pusilla*.

The comparative method

Although this is a book on ecophysiology, the discipline is in essence a comparative one, as we wish all the time to compare and contrast the physiological and ecological performance of different vertebrate species in a wide range of environments and habitats. It is thus somewhat surprising that it is only in the past decade or so that biologists have come to realise that much of the comparative work done in the past could be flawed, or the conclusions misleading to some extent. The reason for this is quite simple. When comparing, say, the BMR or rate of water loss of a group of species from contrasting habitats, workers in the past (myself included) have always assumed that the actual data points are statistically independent and accordingly applied routine parametric statistical tests (*t*-tests, ANOVAR, etc.) in assessing the significance of differences between means. A moment's reflection, however, is sufficient to realise that species that are more closely related to one another in an evolutionary sense might be expected to be more similar in morphology, behaviour and physiology, than more distantly related ones. In fact the whole theory of Darwinian evolution that is based on the concept of modification through descent would predict this.

Joe Felsenstein seems to be the person who first took this problem seriously and attempted to develop methods that would allow one to correct comparative data for phylogenetic biases (Felsenstein, 1985). The whole approach and current methodology is explained in great detail in the excellent book by Harvey and Pagel (1991), who argue that closely related species tend to be phenotypically similar to each other as a consequence of at least three different biological processes: phylogenetic niche conservatism, phylogenetic time lags, and similar adaptive responses. Certainly, no biologist working today in the comparative field can afford to ignore the fact that it is statistically inappropriate to ignore the factor of evolutionary relatedness when comparing data from different species. As Felsenstein (1985) points out

> Comparative biologists may understandably feel frustrated upon being told that they need to know about the phylogenies of their groups in great detail, when this is not something that they had much interest in knowing. Nevertheless phylogenies are fundamental to comparative biology; there is no doing it without taking them into account.

The difficult part is to know or reconstruct the 'correct' phylogeny and in many cases the animals are known so poorly that it is impossible to do this in any meaningful way. Assumptions also must be made concerning rates of evolutionary change and branch lengths within a phylogeny, and reconstructing the data for nodal values (i.e. presumed ancestors) can be problematic (see Felsenstein, 1988). Nevertheless, where the phylogeny of a group is known with any certainty, the method has proven very useful as seen, for example, in the recent analyses by Williams *et al.* (1995) and Brooker and Withers (1994). An example of how to correct physiological data for two characters in a single phylogeny from four species for their implied phylogenetic relationship, using the 'independent contrasts method' of Felsenstein (1985) is given in Appendix 8.

Harvey and Pagel (1991, p. 170) summarise very clearly the need for an informed approach to the treatment of comparative data as follows:

> The branching structure of phylogenies ensures that species are not independent for statistical purposes. Various comparative methods differ in how they estimate and manage this non-independence. Some methods discard information in an attempt to create a set of independent points, while others which make use of all the variation in the phylogeny are to be preferred on logical grounds. These methods employ independent comparisons either to assess differences among species or higher nodes, or to assess the direction of evolutionary change. Evolutionary models implicitly or explicitly underpin all methods. The validity of a method depends upon whether the model on which it is based accurately describes the evolutionary processes that have generated diversity.

4

Turnover methodology: theory and practice

Basic assumptions

Many of the insights into the lives of animals that have been gained through the application of the ecophysiological approach come from turnover studies, using stable and radioactive isotopes. It is thus important not only to be able to implement such studies, but also to understand the strengths and limitations of the technology. Details of the basic turnover equation are given in Chapter 3 and Appendix 9, and the following outlines the various assumptions on which the method is based, along with mathematical procedures that have been introduced by various authors to cope with situational realities. There are six basic assumptions, most of which are violated to some extent in practice. These are discussed in great depth by Speakman (1997) and by Nagy (1980, 1983), to which reference should be made for specific details.

1. Rates of CO_2 production and water loss are constant

Clearly, this is never the case in reality, as animals feed and consume water over short periods of time, and changes in rates of activity will affect rates of gas exchange. Attempts to track varying rates of CO_2 production, by multiple sampling, are usually self-defeating as they involve capturing and bleeding the animal a number of times and the time interval is often too short to allow discrimination of slowly changing isotope levels. Rates of CO_2 production measured by the doubly labelled water (DLW) method are always averages over a given time span and are thus difficult to correlate with specific behaviours, or levels of activity. Speakman and Racey (1988) and Speakman (1997) recommend keeping sampling periods close to 24 h, or modules thereof, thus reducing variations due to diurnal patterns of behaviour.

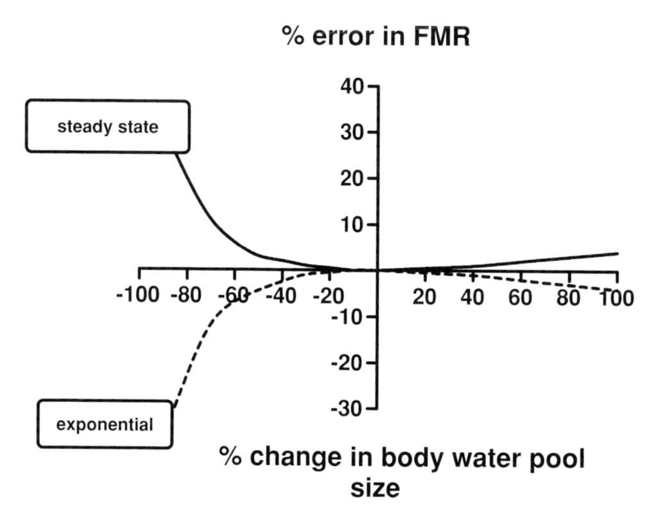

Figure 4.1. Graph illustrating the magnitude of the error when measuring field metabolic rate (FMR) if changes in the body water pool over the time of measurement are assumed to be linear, or exponential (adapted from Nagy (1980)).

2. Constant pool size

The isotope, once injected, is distributed within a pool (the volume of distribution) that is assumed to remain unchanged over the period of measurement. In reality, the total body water content or sodium pool of an animal will change slightly in size over time, but rarely by much during the usual time periods used to measure turnover (from 1 to 7 days). The safest way to correct for any such alteration in the volume of distribution of the isotope is to measure the pool size at both the beginning and the end of the experimental period, by making a second injection and collecting a second equilibration sample. In practice, however, the costs in terms of isotope, and the stress imposed upon the animal by the taking of yet another blood sample, often militate against this and changes in pool size are usually assumed to reflect changes in body mass. Corrections are given for either linear or exponential changes in the volume of the pool over time by Lifson and McClintock (1966). Figure 4.1 shows the percentage error in estimates of the rate of CO_2 production with linear change in the size of the body water pool if no correction is made (steady state) or if the change is assumed to be exponential.

3. Absence of isotopic fractionation

The isotope, once injected, is assumed to behave in an identical manner to the common isotope. In fact, small differences in physical behaviour occur, such as in the rate of evaporation from liquids. This means that isotopic fractionation occurs, such that gases leaving the body do not have an identical isotopic composition to the body pool from which they emanate. Lifson *et al.* (1955) were already aware of this problem when developing the DLW method, and proposed a series of correction factors to account for the fractionation of hydrogen and oxygen in gaseous water relative to liquid water, and for the fractionation of oxygen in gaseous CO_2, relative to oxygen in liquid water. The basic equation for the measurement of the rate of CO_2 production,

$$rCO_2 = (N/2)(k_o - k_d), \quad 4.1$$

was modified by Lifson and McClintock (1966) with the introduction of two small correction factors giving the following:

$$rCO_2 = (N/2.08)(k_o - k_d) - 0.015k_d N, \quad 4.2$$

where N = TBW content,
k_o = fractional turnover of oxygen isotope (oxygen-18), and
k_d = fractional turnover of hydrogen isotope (deuterium or tritium).

This equation, or modifications of it (see, for example, Nagy and Costa, 1980; Nagy, 1983), has been widely used by most workers employing the DLW method (see Appendix 9 for details).

A number of workers (e.g. Coward and Cole, 1991) have attempted to refine this correction for isotopic fractionation but the task is rendered somewhat difficult by the fact that losses via the formation of liquids (e.g. urine and saliva) are not fractionated, whereas gaseous losses may be either kinetically or equilibrium fractionated. The corrections of Lifson *et al.* (1955) only apply to equilibrium fractionation factors, and then only at 25 °C, and they also made the assumption that only 50% of the total water loss is fractionated, the other half being in the form of urine.

Equation 4.3 was revised by Schoeller *et al.* (1986) for a body temperature of 37 °C rather than 25 °C as follows:

$$rCO_2 = (N/2.078)(k_o - k_d) - 0.0123k_d N. \quad 4.3$$

Tiebout and Nagy (1991) took the most extreme estimate of tritium fractionation ($f_t = 0.84$) and assumed that 20% of the losses were fractionated, giving the following:

$$rCO_2 = (N/2.078)(k_o - k_d) - 0.0144k_d N, \quad 4.4$$

which differs from Equation 4.2 only in the number of decimals included.

Speakman (1997) offers two equations that combine the kinetic and equilibrium fractionation factors *in vivo* at 37 °C, assuming these two processes contribute in a ratio of 3 : 1 to the fractionated losses, with the assumption of 25% fractionated water loss:

$$rCO_2 = (N/2.078)(k_o - k_d) - 0.0062k_d N, \text{ using deuterium;} \quad 4.5$$

$$rCO_2 = (N/2.078)(k_o - k_t) - 0.0084k_t N, \text{ using tritium.} \quad 4.6$$

These would appear to be the most appropriate corrections to use in single-pool studies.

Several studies have compared the estimates of rates of CO_2 production from DLW studies with those measured by indirect calorimetry and show that correcting for fractionation effects improves the agreement between the two methods. The first study to do this was that of Lifson *et al.* (1955), who found that the DLW method overestimated energy expenditure in rats by 10%, but this fell to an underestimate of 3% when fractionation corrections were applied. Tiebout and Nagy (1991), working with the tropical hummingbird *Amazillia saucerottei*, reduced the overestimate by the DLW method from 15% to 10% by applying Equation 4.4 above, a difference similar to that reported by Schoeller (1993), working with humans.

Bradshaw and Bradshaw (1999), working with the nectarivorous honey possum, *Tarsipes rostratus*, found that mean estimates of the field metabolic rate (FMR) of this 10 g marsupial fell from 30.9 ± 3.1 to 28.6 ± 2.9 kJ d^{-1} when Equation 4.4 (Tiebout and Nagy, 1991) was replaced by Equation 4.6 (Speakman, 1997), the difference not being statistically significant ($t_{50} = 0.531$).

4. Isotopes turn over in the same pool, which is equal to the body water pool (TBW)

All of the equations given to date for calculating rates of CO_2 production using the DLW method assume that only a single pool is involved in which the hydrogen and oxygen isotopes are distributed. In reality, of course, this is a simplification, as the method assumes that the oxygen isotope is in exchange equilibrium with the oxygen in dissolved CO_2 and bicarbonate and the oxygen-18 must therefore extend into the body water pool as well as the bicarbonate and dissolved CO_2 pools. Schoeller *et al.* (1980) estimated the volume of this second pool to be 0.29% of the body water pool in humans and then went on to develop an alternative two-pool model for the measurement of rates of CO_2 production (Schoeller *et al.*, 1986). The basic equation used (ignoring fractionation) is as follows:

$$rCO_2 = {}^1/_2(k_o N_o - k_d N_d), \qquad 4.7$$

where N_o and N_d are the volumes of the oxygen and hydrogen isotope pools, respectively. Speakman and Racey (1988) recommend the use of the single-pool model for small animals (< 5 kg for mammals and < 1 kg for birds and reptiles) and the two-pool model for larger animals, but, in fact, the two-pool model has to date only been employed in studies of energy expenditure in humans (Speakman, 1997).

A related problem involves estimating the size of the body water pool, which is calculated by isotopic dilution, either of the hydrogen isotope (tritium or deuterium) or of oxygen-18. Speakman (1997) reviews estimates of the body water pool by isotopic dilution, compared with desiccation, and concludes that tritium slightly overestimates this volume (mean quotient of hydrogen/desiccation space $= 1.0448 \pm 0.01$ for mammals, 1.0473 ± 0.177 for birds). Oxygen-18, on the other hand, gives a better fit (1.0104 ± 0.0077 for birds and mammals) and is the isotope of choice for estimating body water content. An extreme case of the error that can arise when using tritium to estimate body water content is the hydrogen exchange that can occur in ruminant or ruminant-like mammals. Bakker and Main (1980) found that tritium significantly overestimated body water content in the quokka wallaby, *Setonix brachyurus*, and other small ruminant-like macropodid marsupials and proposed the use of a correction factor based on desiccation studies in these species.

5. Absence of isotope recycling

Entry of labelled or unlabelled carbon dioxide and water via the skin would introduce a potential error into the estimates of CO_2 production by the DLW method and a number of workers have tried to gauge this in both laboratory and field experiments. Isotope recycling is likely to occur in animals that spend time in confined spaces, such as burrows, where there can be a buildup of CO_2 and water vapour. Bradshaw *et al.* (1987b) attempted to measure metabolic water production (MWP) in *Lacerta viridis* lizards by maintaining them for five days without food in an underground cellar at 15 °C. The calculated rates of water turnover were far higher, however, than could be accounted for by MWP and indicated that the high relative humidity in the cellar had resulted in significant water exchange across the skin. A similar elevation of rates of water turnover by high levels of unlabelled water vapour was observed in pocket gophers (*Thomomys bottae*) by Gettinger (1983) and underlines the fact that, so far as water turnover is

Table 4.1. *Magnitude of error in doubly labelled water estimates of rates of CO_2 production (FMR) when isotopic recycling occurs*

Absolute humidity mg H_2O l^{-1} air	3.8	6.8	10.4	16.8	19.8
FMR by DLW ml g^{-1} h^{-1}	4.38 ± 0.26	4.22 ± 0.26	4.16 ± 0.26	3.96 ± 0.18	3.99 ± 0.43
FMR by gravimetric method ml g^{-1} h^{-1}	2.42 ± 0.13	2.48 ± 0.16	2.57 ± 0.09	2.75 ± 0.13	3.43 ± 0.43
Error (%)	81.3 ± 8.2	70.3 ± 10.9	62.0 ± 13.0	43.8 ± 7.0	18.2 ± 17.5

Modified from Nagy (1980).

concerned, the method works best with animals living in dry rather than humid environments.

Nagy (1980) examined the effects of unlabelled water and CO_2 on the accuracy of the DLW in kangaroo rats and found that high ambient levels of CO_2 could result in overestimates of the rate of CO_2 production of up to 80%, as seen in Table 4.1. Surprisingly, the magnitude of the effect varied negatively with increasing humidity.

Poppitt *et al.* (1993) proffer a word of caution in suggesting that many published estimates of increased energy expenditure in brooding birds and lactating mammals with young in nests may in fact reflect the inherent error due to surface fluxes of CO_2 that can occur in burrows and enclosed nests.

6. Background concentrations of isotopes are constant

This is not a factor normally taken into account by researchers in the field, even though it can be readily demonstrated that DLW calculations are quite sensitive to background oxygen-18 concentrations. Ideally, a number of samples of blood are taken from uninjected animals and used to determine the overall background concentration for the DLW calculations in the experimental individuals. This assumes two things: that all individuals share the same background concentration of oxygen-18, and that this does not vary significantly over the turnover period. Seasonal variations in oxygen-18 backgrounds from 0.1990 to 0.2006 atom% have been reported in the European robin (*Erithacus rubecula*) by Tatner (1990), probably reflecting variations in background concentrations in the birds' food. Individual differences may also be significant and the importance of obtaining a representative sample for, particularly, the

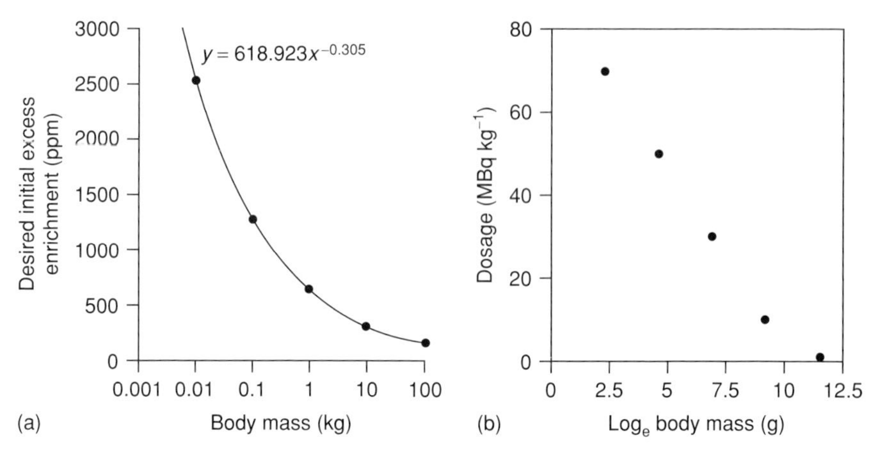

Figure 4.2. Graph illustrating the effect of body size and level of metabolism on the dosages of oxygen-18 and tritium needed for the measurement of FMR in birds and mammals using the doubly labelled water method (from Speakman (1997) with permission).

oxygen-18 background cannot be overemphasised in DLW studies. In a recent DLW study of aspic vipers (*Vipera aspis*) in France, an extremely low average ^{18}O background of 0.1741 was recorded from six individuals (X. Bonnet, O. Lourdais and S.D. Bradshaw, in prep.).

Injection solutions

Given that metabolic rate varies as a function of body mass in animals, isotopic dosages need to be tailored to meet the rate of metabolism of the species concerned. In general, the smaller the animal the faster the rate of washout of the injected isotopes and the shorter the time interval over which one may measure turnover rates. Nagy (1983) recommends a dosage of tritium of 1 mCi kg^{-1} (= 37 MBq kg^{-1}) and 3 ml kg^{-1} of 90–99% enriched oxygen-18. In practice, however, these dosages need to be adjusted for the body mass and expected field metabolic rate of the animal (or biological half-life of each isotope). Figure 4.2 (from Speakman, 1997) gives a useful guide to the amounts that need to be injected in birds and mammals. Bradshaw and Bradshaw (1999), working with the 10 g honey possum (*Tarsipes rostratus*), found that both tritium and oxygen-18 isotopes needed to be injected at 4–5 times these rates to have any chance of measuring their residuals after 24–36 h.

Analytical methods

Blood samples from animals are first centrifuged to provide plasma that, ideally, is then distilled to provide pure water for subsequent analysis. Tritium concentrations are measured by liquid scintillation counting, as already discussed in Chapter 3, and oxygen-18 enrichment is measured either by proton activation techniques or by isotope ratio mass spectrometry. Mass spectrometry is generally used with larger animals, where lower oxygen-18 enrichment is employed and a pre-injection sample can be taken from each individual for the measurement of background isotope concentrations. Enrichment concentrations of 0.01–0.05 atom% (100–500 ppm) are quite adequate for animals weighing more than 5 kg. Oxygen-18 is converted to carbon dioxide for analysis by mass spectrometry; the various methods for this are fully reviewed by Wong and Klein (1987).

Higher enrichment concentrations of 0.20–0.50 atom% (2000–5000 ppm) are used with smaller animals (< 1 kg) with the subsequent analysis of oxygen-18 by proton activation techniques. The most widely used of these is the method described by Wood *et al.* (1975) based on the $^{18}O(p,n)^{18}F$ prompt nuclear reaction, which has a resonance peak at 2.643 MeV. Distilled water samples in glass capillaries are spun in a metal target wheel and exposed to a proton beam from a 38 MeV cyclotron, which is degraded to approximately 7 MeV on exiting into air. The oxygen-18 is converted to radioactive fluorine-18 with the emission of a neutron and the samples are counted for γ emission after the decay of short-lived fission products. An alternative method, developed by Cohen *et al.* (1984) and fully described in Bradshaw *et al.* (1987a), uses a 3 MeV Van der Graaff linear accelerator to generate a weaker proton beam and exploits the $^{18}O(p,\alpha_0)^{15}N$ prompt nuclear reaction, which has a broad resonance peak at 840 keV. This method has the advantage of converting the oxygen-18 to nitrogen-15 and emitting alpha particles in the process. These have a zero background and are readily counted with a Rutherford backscattering (RBS) detector, a typical profile being shown in Figure 4.3. Sample preparation with this method involves the formation of an oxide of Ta_2O_5 from 25–50 µl of the distilled water sample in a microcell by anodic oxidation (see Figure 4.4). A 9 mm section of tantalum metal is then mounted in a target stick and exposed to the proton beam in a vacuum for 2–3 min, which is sufficient time for the generation of adequate counts of alpha particles (see Figure 4.5 for a typical calibration curve with oxygen-18 standards).

The two methods were compared in a double-blind trial with a series of samples that were divided and analysed by myself at the Lucas Heights nuclear facility in Sydney, Australia, and by Ken Nagy at UCLA using the method of Wood *et al.* (1975). Table 4.2 is from Bradshaw *et al.* (1987a) and the following statistical analysis confirms the excellent agreement between the two methods.

Table 4.2. *Results of a double-blind analysis of oxygen-18 enriched water samples by two separate proton activation analysis techniques*

Statistic	Wood *et al.* (1975)	Bradshaw *et al.* (1987a)
Means	0.6568 atom%	0.6525 atom%
Standard deviation	0.2493	0.2547
Standard error	0.08815	0.09004
n	8	8

From Bradshaw *et al.* (1987a).

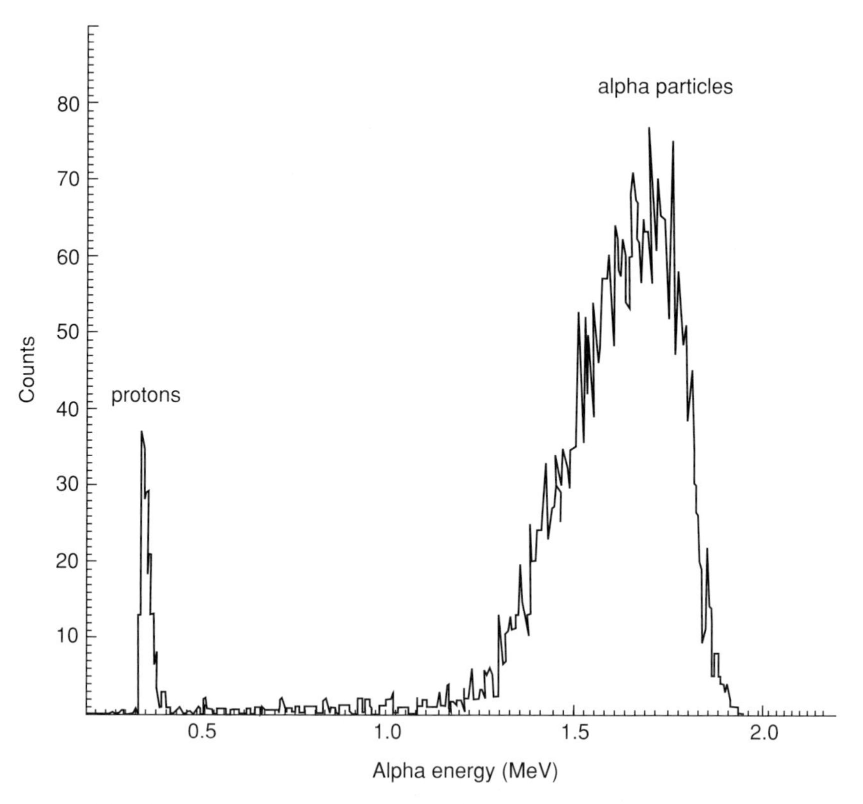

Figure 4.3. Energy profile of α-particles emitted by the prompt nuclear reaction $^{18}O(p,\alpha_0)^{15}N$ on bombarding a Ta_2O_5 target enriched with oxygen-18 with 840 keV protons in a Van der Graaf accelerator.

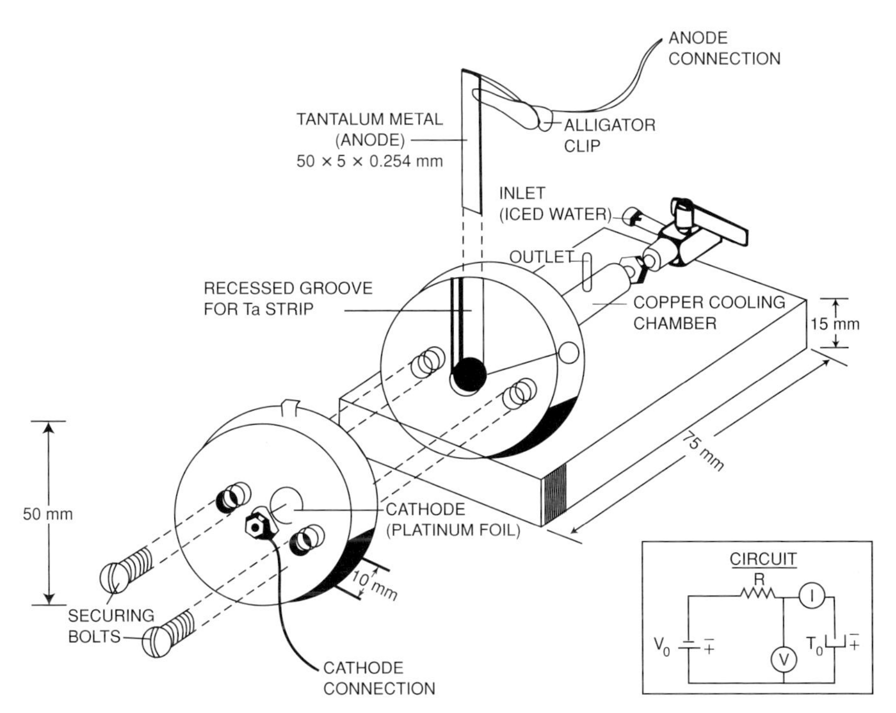

Figure 4.4. Exploded diagram of micro-cell designed for the anodic oxidation of tantalum metal strips with 50–100 µl samples of ^{18}O-enriched water, distilled from blood samples (see Bradshaw *et al.* (1987a) for details).

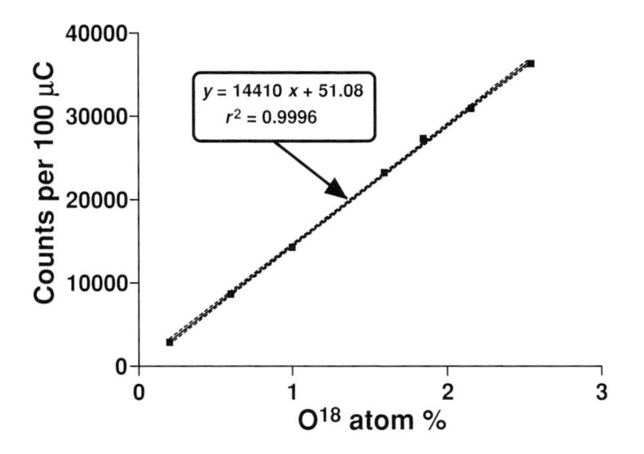

Figure 4.5. Calibration curve for measurement of oxygen-18 enrichment in biological micro-samples by the prompt nuclear reaction $^{18}O(p,\alpha_0)^{15}N$ at a resonance peak of 840 keV protons.

A 2-tailed t-test between the means returns $t = 0.3516$ with $p = 0.7355$, and a paired t-test gives $t = 0.3383$ with $p = 0.9735$. An F-test on the variances gives $F_{1,7} = 1.043$ with $p = 0.4783$. A correlation analysis between the two sets of samples returns the following linear regression:

$$y = 1.003 \pm 0.03791x - 0.005822 \pm 0.02491$$

with a coefficient of determination (r^2) of 0.9901 and $p < 0.0001$. Checking the data for normality by the method of Kolmogorov–Smirnov (KS) gives a distance of 0.16 with $p > 0.10$.

In practice, the analytical method of choice is determined by a number of factors: the size of the animal, and hence the cost of the oxygen-18 isotope, and the proximity to a laboratory that regularly carries out isotopic analyses. Speakman (1997) gives some useful addresses in his excellent book devoted to the DLW method.

Some useful rules of thumb

In all DLW studies there are a number of basic steps and procedures that need to be respected if reliable data are to emerge at the end of the study. Some of these are given below.

1. Carry out an equilibration trial with the species first, if possible, to determine the most appropriate time for collecting an equilibration sample from which body water content of the animal will be estimated. Speakman (1997) gives the relation: equilibration time in hours = 0.360 × ln body mass in kilograms + 2.555, based on published animal studies, which is a good starting point. Intramuscular injections of the isotope equilibrate more rapidly than intraperitoneal ones, and the intravascular route is of course the quickest.

2. Collect blood samples for preference. Although any body fluid should be suitable for DLW measurements, various workers have found significant differences between blood, urine and rumen samples from mammals (see, for example, Fancy *et al.*, 1986; Speakman, 1997) and I have also found urine samples to be unreliable in reptiles (S. D. Bradshaw, unpublished).

3. Use one injection syringe for all individuals. Calibrate this carefully after use and always use it to prepare a dilution standard, which consists of exactly the same volume of isotope solution injected into the animal, added to a known volume of water. Although one may argue that changing the syringe between animals reduces any chance of cross-infection, it introduces an undesirable additional error. If animal welfare concerns are critical, change the needle each time but retain the one syringe.

4. Never use all the injection solution. Always save enough to prepare a standard after all the animals are injected, as this is critical for the calculation of the results.

5. Distil water from plasma samples in preference to counting tritium concentrations in plasma. Plasma counts need to be corrected for the protein content (approximately 6%) and quenching can be a problem.

6. Always use dpm (disintegrations per minute) in your tritium calculations rather than cpm (counts per minute). Normally, with distillate, quenching is minimal and uniform but it is better procedure to correct for any quenching that may occur in your samples and all modern scintillation counters do this very effectively.

7. Store plasma samples in glass capillaries at 4 °C and then distil them as soon as possible after collection. We have found abnormal results with plasma samples stored frozen for periods of more than six months.

8. Respect your animals above all! Data from stressed animals are virtually useless and every effort should be made to handle and return your animals to their natural environment as soon as possible, and with as little disturbance and additional stress as possible.

Some other approaches to measuring the field metabolic rate

Although the DLW method is used widely in animal and human studies, other methods have been published for estimating energy expenditure of free-ranging animals. Perhaps the most promising of these is that developed in ruminant animals and based on the injection of ^{14}C-labelled bicarbonate, which is analysed either in the blood, or from the collection of expired CO_2 (see Corbett *et al.* (1971) and Whitelaw *et al.* (1972) for details of this interesting method). The DLW method has been found not to work well with insects, because their rates of water and oxygen turnover are very similar, and rates of elimination of other radioactive isotopes such as ^{137}Cs have been used as a means of estimating field metabolic rates (FMR) (see, for example, Taylor, 1978). An isotope of rubidium, ^{86}Rb, has recently been proposed as an alternative to ^{137}Cs by Peters (1996) working with goldfish (*Carassius auratus*) and toads (*Bufo terrestris*) and a series of radioisotopes was tested by Peters *et al.* (1995) in the desert iguana *Dipsosaurus dorsalis*, ^{86}Rb being found to be the most reliable predictor when compared with estimates of the FMR by the DLW method. Although one needs to balance the effects of exposing animals to more powerful γ-emitting isotopes, this approach may ultimately prove less invasive than the DLW method, which is based automatically on the collection of blood samples.

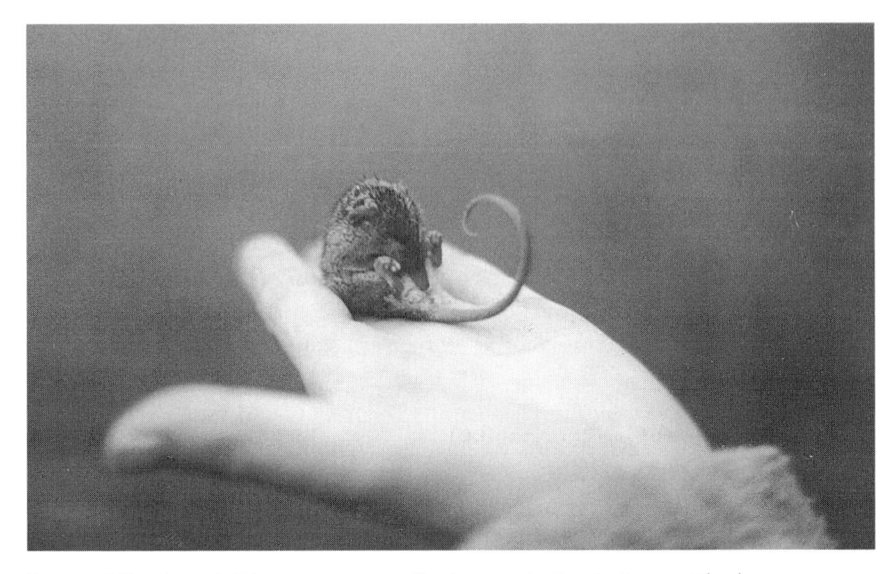

Figure 4.6. An adult honey possum, *Tarsipes rostratus*, in torpor. The honey possum is also known by its native aboriginal name of 'noolbenger'.

The honey possum, *Tarsipes rostratus*, as an example of the use of the DLW method for estimating the rate of food intake in the field

The tiny (6–10 g) marsupial honey possum is a challenging animal to study, both in the laboratory and in the field, as may be gathered from the photograph in Figure 4.6. It is too small to yield anything other than the tiniest of blood samples, and is so fragile that individuals can die within 10 s if held too tightly in the hand! None the less, it has been the focus of an ongoing ecophysiological study that has achieved accurate measurements of the daily nectar and pollen intake of free-ranging individuals in the field and related these to the animal's minimum requirements for nitrogen balance.

Smith and Lee (1984), in comparing reproductive strategies of Australian possums and gliders, speculated that diet is a major factor in shaping the diverse patterns observed in life-history traits amongst the different species. Their comparative analysis shows that honey possums have a reduced offspring production rate, when compared with both leaf-eating possums and similar-sized pygmy possums in relation to allometric expectations, as shown in Figure 4.7. They note that this pattern is also common to other sap-, gum- and nectar-feeding possums and gliders and suggest that these high-carbohydrate diets may be deficient in protein. The importance of determining the diet of the

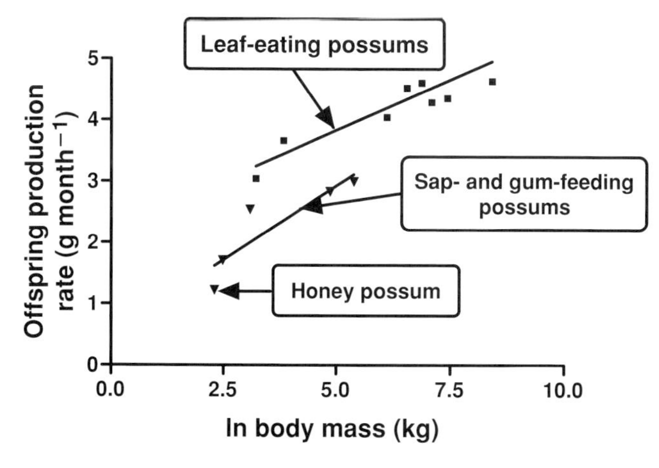

Figure 4.7. Regression of 'offspring production rate' in g per month of marsupial possums as a function of body mass. Note the relatively low reproductive rate of the honey possum (adapted from Smith and Lee (1984)).

honey possum, and its composition in terms of protein and carbohydrates, is thus clear.

Nagy *et al.* (1995) first measured the FMR of free-ranging honey possums in the Fitzgerald River National Park, on the south coast of Western Australia, and found that it averaged only 83% of allometric predictions from data on other marsupials. This was surprising, given that Withers *et al.* (1990) had earlier shown that the basal metabolic rate (BMR) of the honey possum is much higher than predicted, again from allometric equations. In part, this may be due to the fact that the published FMRs of the small insectivorous marsupials, such as *Antechinus stuartii* and *Sminthopsis crassicaudata* (Nagy *et al.*, 1978, 1988) are very high and this has a significant impact on the slope of the regression line, as seen in Figure 4.8. Although it is well known that the BMR of marsupials is some 30% lower than that of similar-sized eutherian mammals (the BMR in watts is 2.33 $BM^{0.75}$ in marsupials compared with 3.44 $BM^{0.75}$ in eutherians, where BM is body mass in kg), the same is not true of the FMR. Because the exponent of the marsupial regression is 0.59, compared with 0.75 in eutherian mammals (Nagy, 1987b; Nagy *et al.*, 1999) the FMR of very small marsupials is actually *higher* than that of similar-sized eutherians. Because of the combination of a high BMR and low FMR, the elevation of the field metabolism relative to basal rates in the honey possum is much lower than in other marsupials: 2.7 times in honey possums compared with 6.9 in dunnarts and 4.6 in brown antechinus. Obviously, small marsupial carnivores expend far more energy in tracking and capturing their insect prey

Table 4.3. *FMR and water influx rate of the honey possum* Tarsipes rostratus *in relation to allometric predictions*

	Field metabolic rate			Water influx rate	
Body mass	kJ d^{-1}	relative to predicted	relative to BMR	ml d^{-1}	relative to predicted
9.9 ± 3.4	34.4 ± 11.1	0.83 ± 0.26	2.71 ± 0.90	9.1 ± 9.2	0.92 ± 0.82

Modified from Nagy *et al.* (1995).

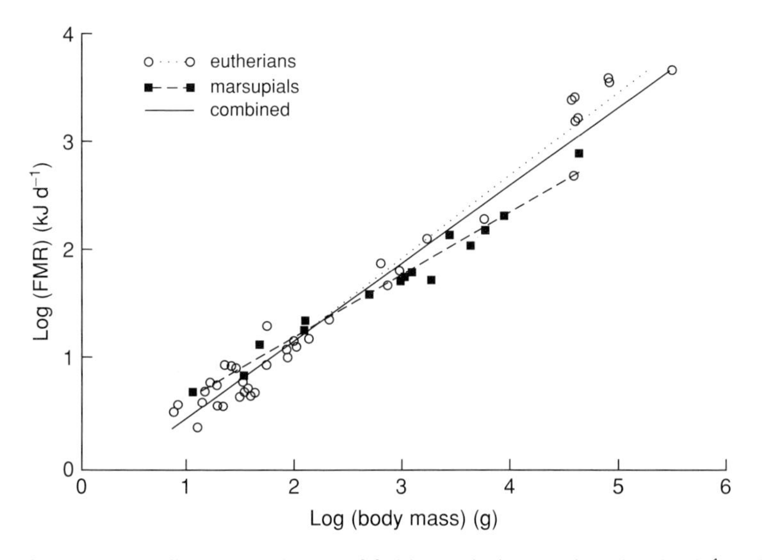

Figure 4.8. Allometric relation of field metabolic rate (FMR) in kJ d^{-1} with body mass in marsupial and eutherian mammals. The equation for the regression of eutherians is FMR = 4.63 $g^{-0.762}$ and for marsupials FMR = 10.8 $g^{-0.582}$ and the combined regression is FMR = 5.27 g $^{-0.723}$ (from Nagy, 1994, with permission).

than do honey possums in simply locating and feeding from large *Banksia* blossoms.

Table 4.3 summarises the FMR and water influx rate of the honey possum, both of which are lower than allometric expectations. Lest one think, however, that the overall energy expenditure of the honey possum is low, it is as well to remember that its FMR, at 30–35 kJ d^{-1} (Nagy *et al.*, 1995), is still quite high in absolute terms, and over 170% of that of a similar-sized eutherian mammal.

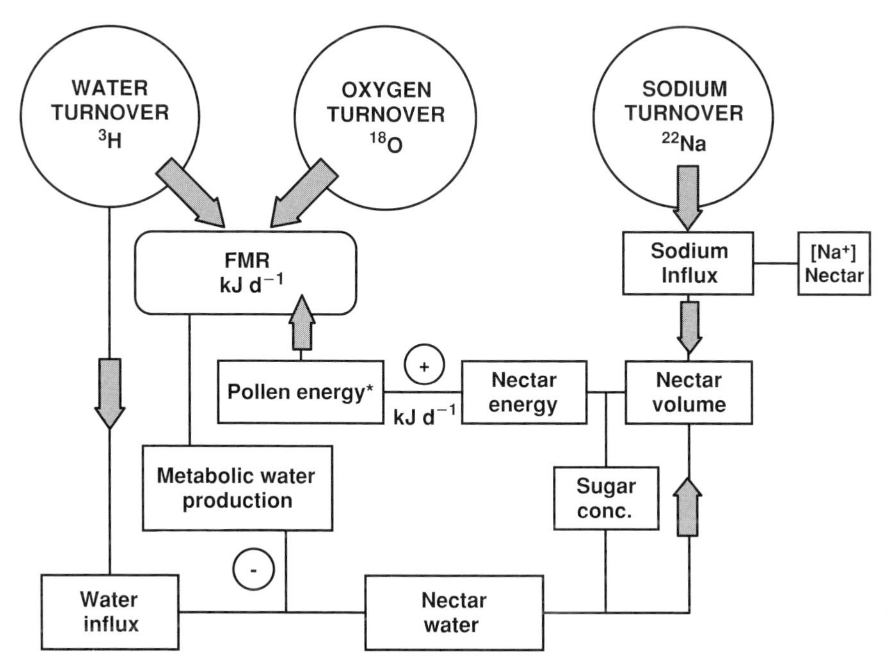

Figure 4.9. Schema illustrating the rationale used for the simultaneous measurement of rates of nectar and pollen intake in free-ranging marsupial honey possums, *Tarsipes rostratus* (from Bradshaw and Bradshaw (1999) with permission). Pollen energy (asterisked) is derived from subtraction of the nectar energy from the FMR, after correction for the energy digestibility of both pollen and nectar and energy losses in urine.

In order to quantify the daily nectar and pollen intakes of free-ranging honey possums, an experimental protocol was developed based on the utilisation of three isotopes: oxygen-18, tritium and sodium-22. This protocol is shown schematically in Figure 4.9. The FMR, in kilojoules per day, is determined by using oxygen-18 and tritium, and the turnover of tritium alone gives us the water influx in millilitres per day. Knowing the FMR, we may estimate the rate of metabolic water production (MWP), as we know that one litre of expired carbon dioxide produces 0.658 ml of water, and in calculating our FMR we assume, in view of their high carbohydrate diet, that 21 kJ of energy is produced per litre of CO_2. The MWP is then subtracted from the water influx to obtain the influx of water derived from the diet and from drinking free water if this occurs.

If we assume that honey possums do not drink free water and that their only source of dietary water is from nectar, then, knowing the sugar concentration of the nectar on which they are feeding, we may use the tritium data to estimate daily nectar intake. As already described in Chapter 3, we can determine the species of plant on which the honey possum is feeding by collecting pollen from the animal's snout, and then collect nectar from that plant to determine its sugar concentration. There is always the possibility that the honey possums may drink some free water, especially in mid-winter when all the vegetation is saturated, and if this were the case it would introduce an error and lead to an overestimate in the nectar intake. This is the reason for measuring simultaneously the rate of turnover of sodium, using the radioactive isotope sodium-22 as a tracer. If we assume that all of the sodium in a honey possum's diet comes from nectar, and none from the pollen it eats, then – knowing the sodium concentration of the nectar ingested – we may also estimate nectar intake, but in a manner completely independent of the estimate made with the use of tritium.

Some actual data from honey possums are shown in Table 4.4. as examples, giving the two estimates of nectar intake based on the two different isotopic turnovers. As may be seen, the agreement is fairly close and only in three cases (numbers 78M, 32M and 81F) does the nectar estimate based on the water turnover exceed that derived from the sodium turnover, i.e. we would appear to have instances of the possums ingesting free water. This may be real, in which case the sodium-based estimate is then used; otherwise the two estimates are simply averaged, as shown in the table.

Table 4.5 carries this analysis further and shows how pollen intake is estimated from these data. The estimate of nectar intake is converted to energy, knowing the calorific value of the carbohydrate content of the nectar (17.5 kJ g^{-1} dry matter) and then, as the only other source of energy in the possums' diet is pollen, the difference between the nectar energy and the FMR (after correction for the energy digestibility of both pollen and nectar and energy losses in the urine), must equal the energy derived from pollen (see again Figure 4.8). The energy content of the pollen was measured in a bomb calorimeter at 21.75 kJ g^{-1} and from this the actual pollen intake is calculated.

Table 4.6 gives actual data from the recent study of Bradshaw and Bradshaw (1999) and shows that honey possums in Scott National Park, in the extreme southwest corner of Western Australia, ingest daily close to 6 ml of nectar and 0.7 g of pollen, amounting to over 70% of the animal's body mass. Nectar accounts for approximately 60% of the possum's energy intake and pollen 40%.

These figures are based on a number of assumptions inherent in the model shown in Figure 4.8 and these are that:

1. nectar is the sole source of sodium in the diet;
2. drinking of free water is rare;
3. the possums are in energy balance.

Table 4.4. *Field metabolic rate (FMR) and rates of water and sodium influx used to estimate daily nectar intake in the honey possum*

No.	Sex	Mass (g)	H_2O in	Na^+ in	FMR ($kJ\ d^{-1}$)	H_2O met. ($ml\ d^{-1}$)	Nect. vol. H_2O	Nect. vol. Na^+	H_2O free ($ml\ d^{-1}$)	Nect. vol. ($ml\ d^{-1}$)
90	F	10.9	6.27	0.084	31.0	0.987	6.37	6.72	0	6.54
41	F	9.2	5.52	0.084	31.0	0.977	5.46	6.72	0	6.09
78	M	7.1	8.65	0.081	35.4	1.127	9.06	6.48	2.14	7.77
32	M	5.3	6.28	0.064	21.8	0.694	6.73	5.12	1.34	5.93
135	M	10.9	6.31	0.104	37.0	1.178	6.18	8.32	0	7.25
150	F	9.8	7.49	0.124	51.8	1.649	7.04	9.92	0	8.48
81	F	13.9	13.08	0.155	45.8	1.456	14.0	12.40	1.33	13.20
12	M	10.2	4.27	0.064	29.2	0.929	4.18	4.27	0	4.22
Mean ± S.E.	—	9.66 ± 0.91	7.23 ± 0.95	0.095 ± 0.001	35.37 ± 3.38	1.12 ± 0.10	7.37 ± 1.06	7.49 ± 0.93	0.60 ± 0.31	7.43 ± 0.94

Table 4.5. *Estimation of daily pollen intake by free-ranging honey possums in southwest Western Australia*

No.	Sex	Mass (g)	FMR (kJ d^{-1})	Nectar (ml d^{-1})	Nectar (kJ d^{-1})	Pollen (kJ d^{-1})	Pollen (g d^{-1})
90	F	10.9	31.0	6.54	19.47	11.53	0.52
41	F	9.2	31.0	6.09	18.12	12.88	0.59
78	M	7.1	35.4	7.77	23.12	12.28	0.56
32	M	5.3	21.4	5.93	17.63	4.17	0.19
135	M	10.9	37.0	7.25	21.57	15.43	0.70
150	F	9.8	51.8	8.48	25.23	26.58	1.21
81	F	13.9	45.8	13.2	39.28	6.47	0.29
12	M	10.2	29.2	4.22	14.77	14.43	0.66
Mean ± S.E.	—	9.66 ± 0.91	35.37 ± 3.38	7.43 ± 1.06	22.39 ± 2.68	12.97 ± 2.37	0.51 ± 0.10

From Bradshaw and Bradshaw (1999).

Table 4.6. *Estimated daily nectar and pollen intakes in free-ranging honey possums,* Tarsipes rostratus, *in the southwest of Western Australia*

Body mass (g)	FMR (kJ)	Nectar volume (ml d^{-1})	Nectar energy (kJ)	Pollen intake (mg d^{-1})	Pollen energy (kJ)
9.4 ± 1.0	34.3 ± 2.6	5.9 ± 0.6	19.8 ± 1.9	660 ± 156	14.5 ± 3.4
% of FMR			58		42

Modified from Bradshaw and Bradshaw (1999).

Bradshaw and Bradshaw (1999) attempted to leach sodium ions from pollen grains in the laboratory and found that approximately 95% of the sodium found in pollen is associated with the indigestible exine and is thus not available to the animal. Once known, this correction factor is easily applied to the calculations when estimating nectar intake from the turnover of ^{22}Na. Some intake of free water does occur from time to time, but only in winter and then it is not clear that the animals are deliberately drinking the water rather than being forced to imbibe it from very wet

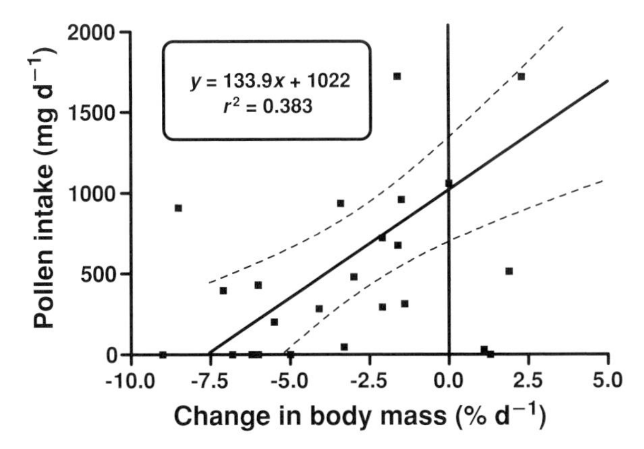

Figure 4.10. Correlation between the rate of change of body mass and the estimated daily intake of pollen of free-ranging marsupial honey possums (*Tarsipes rostratus*) (modified from Bradshaw and Bradshaw, 1999).

vegetation. Under these circumstances, the nectar intake is estimated from the turnover of ^{22}Na, rather than of tritium. The final assumption means that the method should only be used with possums that are maintaining body mass in the field and neither storing nor losing energy. In practice, no animal retains a completely stable body mass, even over short periods of time, and regressing pollen intake against daily changes in body mass enable one to estimate the amount needed for complete balance. Figure 4.10 shows that honey possums must ingest close to 1 g of pollen per day if they are to remain in balance.

This study shows that, even with very small animals and a most unusual diet, it is possible with a little ingenuity to devise means of measuring food intake in the field. We are perhaps a little further in responding to the specific hypothesis of Smith and Lee (1984) that stimulated this study: that the low reproductive rate of the honey possum results from a diet that is deficient in nitrogen. We have recently carried out a series of feeding trials in the laboratory and determined that the daily minimum nitrogen requirement (MNR) for balance is 89 mg kg$^{-0.75}$ (Bradshaw and Bradshaw, 2001). This value for the MNR is one of the lowest to be measured in any marsupial (Hume, 1999) and our field data show that honey possums are ingesting approximately ten times their MNR per day when feeding on 1 g of pollen. It thus seems unlikely that nitrogen is a factor limiting the reproduction of honey possums. We are currently measuring rates of protein turnover with ^{15}N-glycine, in breeding and non-breeding females, in an effort to determine what fraction of the nitrogen input is being diverted to the growing young.

5

Case studies of stress: incidence and intensity

We have already examined in Chapter 2 the question of stress, and how to recognise it. In this chapter some concrete examples will be discussed. The case of the marsupial quokka (*Setonix brachyurus*) has already been evoked as an illustration of an animal in mid-summer displaying a significantly modified *milieu intérieur* (dehydration with elevated plasma osmolality) associated with a maximal deployment of regulatory processes (very high circulating levels of the antidiuretic hormone lysine vasopressin; see Jones *et al.* (1990) for more details). This combination of a significant deviation of the *milieu intérieur* from whatever is considered to be the normal or ideal state, despite the activation of homeostatic regulatory processes, will be our *leitmotif* for identifying cases of stress, and also for measuring their intensity (Bradshaw, 1992a).

The silvereye

An interesting case is the silvereye (*Zosterops lateralis*), a small (10 g) Australian bird (see Figure 5.1) that has come to be the bane of vignerons in many parts of that country. These birds from time to time descend in their thousands to devastate grape crops in late summer, usually just before the harvest. The case is unusual for Australia, where most of the pests (such as the fox, cat and rabbit) are of European origin, and has necessitated a detailed study of this indigenous bird's ecophysiology (Rooke *et al.*, 1983, 1986; Rooke, 1984). Contrary to initial expectations, the silvereye has an extended lifespan, up to 15 years in some cases, and banding studies showed that they may move distances of over 400 km in search of suitable food supplies. In winter, silvereyes are insectivorous, feeding on a number of species including the potato tuber

Figure 5.1. The silvereye, *Zosterops lateralis*, feeding on nectar from the flowers of the marri eucalypt (*Eucalyptus callophylla*), which provides these small birds with their main source of energy and water during the dry summer and autumn months in Australia.

moth, which is a major pest in some agricultural areas. In summer, however, they become nectarivores, feeding principally on the flowers of the marri eucalypt tree (*Eucalyptus calophylla*), which normally flowers in late summer and autumn. It is also evident that the birds do not enter the vineyards and consume grapes every year, but usually every 4–5 years, which appear to be those particular years in which the marri trees have few or no flowers.

Measurements of the rate of water turnover of free-ranging birds, using tritium, showed that their water influx increases substantially in summer (from approximately 14 ml to 24 ml per day) but that the birds are still able to maintain their water balance (i.e. influx = outflux). In the autumn of a year in which the marri failed to flower, and the birds entered the vineyards in large numbers, Rooke *et al.* (1983) were able to show that rates of water influx fell well below rates of efflux and that the birds were consequently severely dehydrated (see Figure 5.2). Monthly measurements of circulating concentrations of corticosterone and plasma glucose in these birds throughout a full year give clear evidence of stress in the autumn period (February–April) with a significant elevation of plasma corticosteroids associated with a fall in blood glucose concentrations (see Figure 5.3).

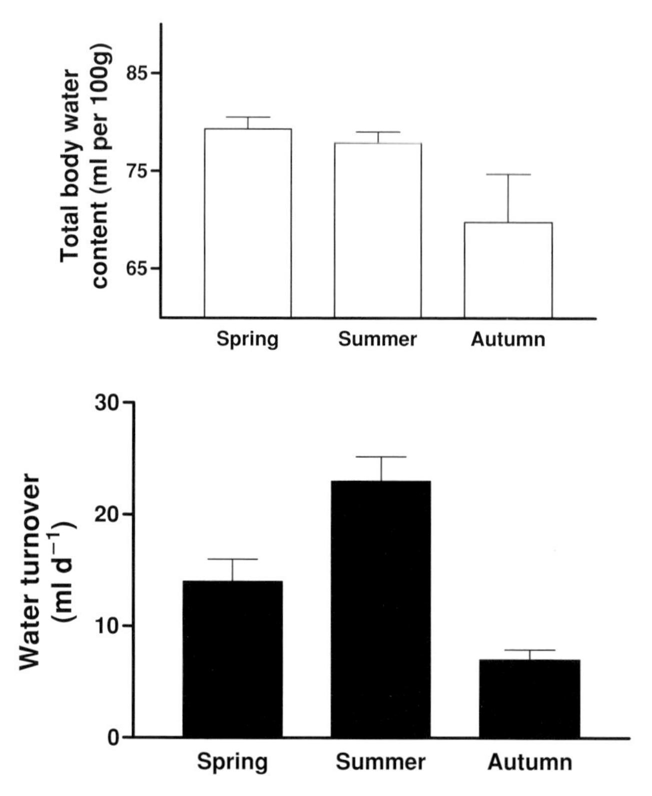

Figure 5.2. Seasonal changes in total body water content and the rate of water influx, measured with tritiated water, in silvereyes (*Zosterops lateralis*) in the southwest of Western Australia. Autumn coincided with the period when the birds were dehydrated and entered vineyards to feed on ripe grapes.

In the years prior to the European colonisation of Australia, one may speculate that the silvereyes would have experienced considerable mortality in those years when the marri failed to flower, as no other source of water and carbohydrate would have been available to them. The marri, in common with many other eucalypt trees, displays years in which 'masting' occurs, i.e. every plant, both small and large, flowers profusely and it is assumed that recruitment of seedlings is achieved, regardless of the number of insect seed predators. 'Mast' years in the marri appear to occur approximately every 4–5 years in the southwest of Australia and in-between years vary considerably in the production of blossoms. Rooke (1984) found that flower setting was positively correlated with mean maximum spring temperatures, and that nectar production by set flowers was negatively correlated with mean maximum summer temperatures, i.e. warm

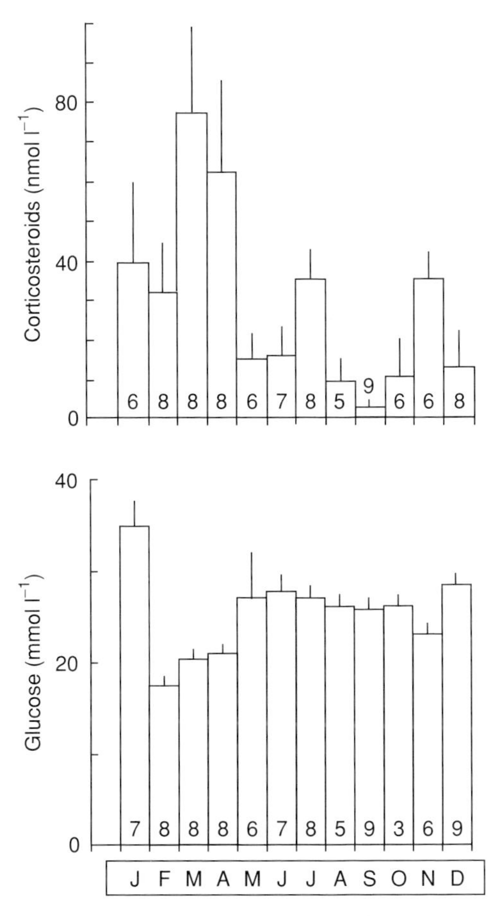

Figure 5.3. Seasonal changes in plasma corticosteroid concentrations and blood glucose concentrations of silvereyes (*Zosterops lateralis*) over a 12 month period in the southwest of Western Australia. Plasma corticosteroid concentrations were significantly elevated in March and April and glucose concentrations fell significantly in February, March and April. (Modified from Rooke *et al.,* 1986.)

springs and cool summers are needed for maximum nectar production by the marri.

With the advent of Europeans, and the planting of crops that fruit in summer and autumn (citrus crops, vines, etc.) the silvereyes were offered an alternative to dehydration and certain death in those years when nectar supply from the marri was deficient and it seems clear that populations of this species have increased markedly over the past 150 years. We thus have an interesting case of stress in an animal alleviated, rather than provoked, by humans, as is

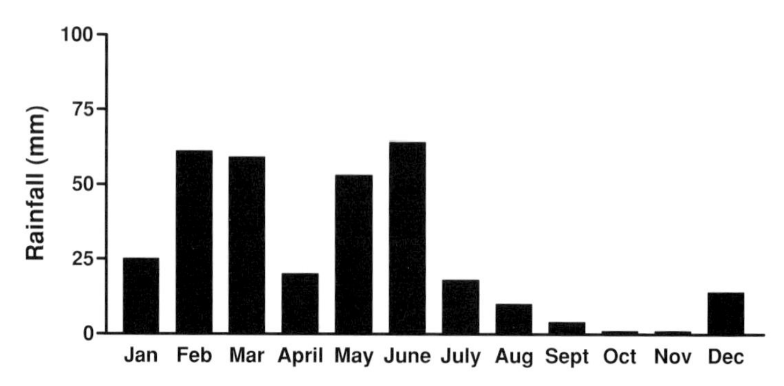

Figure 5.4. Mean monthly precipitation over the period 1968–88 for Barrow Island, which is situated 80 km off the arid Pilbara coast some 1400 km north of Perth, the capital of Western Australia. (Data courtesy of WAPET Pty Ltd.)

more usually the case. Of course, the problem is exacerbated by the clearing of marri trees to plant more vines, but the solution is in the hands of the vignerons, as alternative flowering food sources are readily exploited by the silvereyes.

Barrow Island macropods

Barrow Island, off the northwest coast of Western Australia, is one of Australia's most important nature reserves, being home to nine marsupial species, four of which are now extinct, or virtually so, on the adjacent continent. The island has an area of approximately 50 000 ha and lies some 80 km off the arid Pilbara coast, roughly 1400 km north of Perth, the capital of Western Australia. The environment is very arid (Stafford Smith and Morton, 1990) with unpredictable cyclonic rains falling in the summer period between December and March. The 'average' annual rainfall is approximately 200 mm, as seen in Figure 5.4, but it may vary from 5 to 750 mm in any one year. A four-year ecophysiological study of all the major vertebrate species on this island had, as one of its foci, the potential stress that these rare and endangered species might experience during long periods of drought. Rates of energy and material balance were measured during the dry and wet seasons, using the DLW method, and rates of water turnover are shown in Figure 5.5 with the figure for the hare wallaby of 27 ml $kg^{-0.82}$ d^{-1} being lower than that of the Arabian oryx (Williams *et al.*, 2001) and sharing with the Namib Desert rodent *Petromyscus collinus* (Withers *et al.*, 1980) the distinction of having the lowest rate yet measured for any mammal (see also Bakker and Bradshaw, 1989). Measurements were also made of the constancy of the *milieu intérieur*, and changes in kidney function and water balance were correlated with circulating

Table 5.1. *Genetic variation detected by microsatellites in Barrow Island and mainland black-footed rock wallabies* (Petrogale lateralis) *and euros* (Macropus robustus)

	Rock wallabies		Euros	
	Barrow Island	Exmouth	Barrow Island	Pilbara
Sample size	29	15	42	7
Number of loci	10	10	7	7
Polymorphic loci	10%	100%	71%	100%
Mean no. of alleles per locus	1.2 ± 0.2	3.4 ± 0.3	1.7 ± 0.49	6.7 ± 1.5
Average heterozygosity	0.053 ± 0.053	0.62 ± 0.032	0.18 ± 0.15	0.72 ± 0.09

From King (1999).

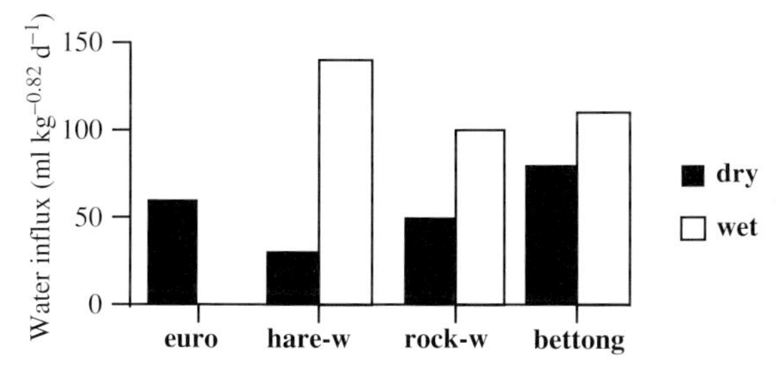

Figure 5.5. A comparison of the rates of water influx, measured with tritiated water, in Barrow Island macropods in wet and dry seasons. Euro, *Macropus robustus isabellinus*; hare-w, *Lagorchestes conspicillatus*; rock-w, *Petrogale lateralis*; bettong, *Bettongia lesueur*. (From J. M. King and S. D. Bradshaw, in prep.)

concentrations of adrenal and pituitary hormones (Bradshaw, 1992b, 1999; Bradshaw *et al.*, 1994, 2001).

Barrow Island was formed by rising sea levels some 11 000–13 000 years ago (Main, 1961) and the effects of this isolation are evident in some of the species now confined there. Levels of genetic diversity in the rock wallaby, *Petrogale lateralis*, are the lowest yet recorded for any mammal (Eldridge *et al.*, 1999) and are also considerably reduced in island compared with mainland populations of the euro kangaroo, *Macropus robustus* (see Table 5.1). Frequencies of fluctuating asymmetry (see Chapter 3) are also raised in island

Figure 5.6. A female of the dwarf form of the euro kangaroo, *Macropus robustus isabellinus*, found only on Barrow Island, some 80 km off the Pilbara coast of Western Australia. This female, which only weighs approximately 7 kg, has a large young in the pouch.

compared with mainland populations of the rock wallaby (J. M. King, unpublished) and the dwarf form of the euro kangaroo found on Barrow Island (*Macropus robustus isabellinus*) (see Figure 5.6) also exhibits a pronounced anaemia, which can be traced to an apparent deficit in red-cell enzyme function

Table 5.2. *Rates of water turnover and concentrations of antidiuretic hormone (LVP) in Barrow Island macropods during a drought year*

	LVP (pg/ml)	Water influx ($ml\ kg^{-0.82}\ d^{-1}$)	Water influx ($\%TBW\ d^{-1}$)
Euro	4.37 ± 1.4	59.0 ± 3.3	5.5 ± 0.5
Hare wallaby	14.65 ± 2.2	28.2 ± 6.1	3.5 ± 0.8
Rock wallaby	0.55 ± 0.4	54.8 ± 5.3	6.2 ± 0.7
Bettong	0.41 ± 0.3	79.6 ± 6.1	13.7 ± 0.9

From J. M. King and S. D. Bradshaw (unpublished).

(Billiards *et al.*, 1999). It seems likely, in the light of these data, that founder effects and/or genetic drift have contributed to this significant loss of genetic diversity in the 11 000 years that these populations of macropods have been isolated from their mainland equivalents. The anaemia shown by the euros is particularly interesting, as it must severely compromise their respiratory capacity, but it would appear that this potentially deleterious mutation has been retained in the population owing to the absence of predators on the island and associated selection pressure.

Evidence of stress was sought in all the macropodid species on Barrow Island (euro, hare wallaby, rock wallaby and burrowing bettong (*Bettongia lesueur*)), but was only detected in euros, and then only in 1994, which was the driest year on record. Concentrations of lysine vasopressin (LVP), which is the antidiuretic hormone in macropodid marsupials (Chauvet *et al.*, 1983), were elevated in hare wallabies during all dry seasons (see Table 5.2) but only in the euro in November 1994 was there evidence of maximal circulating concentrations of LVP associated with significant plasma hyperosmolality and raised concentrations of plasma cortisol (see Table 5.3). The euros thus displayed a significant perturbation of the *milieu intérieur* (increased plasma osmolality) that was not prevented by the maximal circulating concentrations of antidiuretic hormone. The fact that plasma cortisol concentrations were also elevated attests to the stressed condition of the euros at this time.

What is perhaps most significant in this study, however, is the fact that this was the only case of stress detected in the macropods over a four-year study period, which included the driest year ever recorded on Barrow Island (1994). This is a truly amazing testimony to the ability of these marsupials to maintain homeostasis under some of the most trying conditions of heat and total water deprivation.

Table 5.3. *Seasonal variation in physiological parameters of Barrow Island euros over a number of years with evidence of a stress syndrome in November 1994*

Asterisks indicate statistically significant differences. ND, not detectable.

	Dec 91	Dec 93	Nov 94	June 95	Sept 95	Sept 96	Nov 96
Plasma osmolality (mOsm kg^{-1})	261 ± 3	253 ± 4	286 ± 7*	274 ± 4	270 ± 3	274 ± 4	269 ± 4
LVP (pg ml^{-1})	ND	5.8 ± 4.2	16.7 ± 4.6*	2.1 ± 0.6	ND	ND	ND
Condition index (%)	99 ± 4	107 ± 4	102 ± 4	102 ± 6	96 ± 3	98 ± 4	110 ± 5
Cortisol (μg per 100 ml)	1.5 ± 0.25	1.2 ± 0.15	1.7 ± 0.22*	1.9 ± 0.4	1.0 ± 0.24	0.8 ± 0.22	0.7 ± 0.13
Corticosterone (μg per 100 ml)	0.6 ± 0.2	0.4 ± 0.1	0.5 ± 0.07	—	0.3 ± 0.1	—	—

From J. M. King and S. D. Bradshaw (unpublished).

Small marsupial carnivores

One of the most interesting and bizarre life-history stories to emerge in recent years is that shown by a number of small insectivorous marsupial species belonging to the family Dasyuridae. Most of the work has focused on one species, *Antechinus stuartii*, which John Barnett, as a PhD student, reported as exhibiting a complete male die-off following mating in the field (Barnett, 1973). *Antechinus stuartii* lives in high-rainfall forests in the southeast of Australia, where the seasons are quite predictable, but food is none the less not necessarily plentiful.

Barnett's discovery was quickly followed by papers describing the changes that occurred in the males during the breeding season, which led to their early demise (Lee *et al.*, 1977a,b, 1982). In contrast to the females in the population, males progressively lost body mass and condition, they lost fur (alopecia), and autopsies revealed severe gastric ulceration and blood-filled (haemorrhagic) adrenal glands. Other studies revealed that the immunological system of the males was also compromised (i.e. weakened), exposing them to a variety of infections, which were often the final cause of death.

Field studies of other species of the genus *Antechinus* revealed that this pattern of male die-off was not unique to *Antechinus stuartii* and, at one stage, appeared to be characteristic of all the species of the family Dasyuridae (Lee *et al.*, 1982; Lee and McDonald, 1985; Lee and Cockburn, 1985; Green *et al.*, 1989; Bradley, 1987). Further work, however, has shown that the larger 'native cats' do not share this rather one-sided life history and there have been various attempts to interpret the evolution of this unique case of mammalian semelparity[4] (see Braithwaite and Lee, 1979; Diamond, 1982; Lee and Cockburn, 1985; Dickman and Braithwaite, 1992; Dickman, 1993). Interestingly, a recent paper by Smith and Charnov (2001) speculates that the North American woodrats of the genus *Neotoma*, living in Death Valley, California, may also have been selected for semelparity during the Holocene, as the only means of persisting in this extreme environment.

The underlying mechanisms responsible for this unusual life-history pattern were disentangled in *Antechinus stuartii* in an important series of ecophysiological investigations by Tony Lee and his students (Lee *et al.*, 1977a,b; Bradley *et al.*, 1980) who measured changes in blood testosterone concentrations in males prior to and during the breeding season, and correlated these with changes in cortisol secretion from the adrenal gland. What was innovative about their study was that, as well as measuring total cortisol concentrations

[4] Semelparous is the term used to describe species that die after breeding only once; the name is derived from the unfortunate Semele, who succumbed after a night of passion with the Greek god Zeus.

in the plasma, they also used the technique known as equilibrium dialysis to determine the extent to which the hormone was bound to a specific carrier protein in the blood. Steroid hormones are normally transported bound to a carrier protein (transcortin or CBG for glucocorticoids, and sex-steroid binding protein (SSBG) for sex steroids such as testosterone), which protects them from metabolism in the liver and kidney, thus prolonging their biological half-life.

What they found in *Antechinus stuartii* was that plasma testosterone increased in the blood of males at the start of the winter breeding season in early June and reached very high concentrations by early August. Males of this species are very aggressive, forming male 'leks' (Cockburn and Lazenby-Cohen, 1992), and copulation with females is a long process, sometimes lasting as long as 24 h. Total cortisol concentrations in the plasma changed very little in males during the breeding season, suggesting, at first sight, that they were not stressed: plasma corticosteroids such as cortisol (F) and corticosterone (B) normally rise in stress situations. When, however, the ratio of free to bound steroid was measured by equilibrium dialysis, an enormous increase in the free component was observed in the latter part of the breeding season in the males. Normally something like 95% of the cortisol circulating in the blood is CBG-bound, and thus biologically inactive, and it is only the 5% free that exerts a physiological effect by binding with the specific hormone receptors in target tissues. This decrease in protein binding of the cortisol was found to be due to a direct effect of testosterone on the production and secretion of CBG from the liver, and the homeostatic feedback system that would normally have rectified this change failed to operate (McDonald *et al.*, 1986).

These changes are shown in Figure 5.7 and it is clear from these data that the males die as a result of the effects of abnormally high concentrations of free cortisol. Cortisol is a 'catabolic' steroid that breaks down muscle protein to form glucose (the process known as 'gluconeogenesis') during times of reduced food supply, and this explains the muscle wasting observed in *Antechinus stuartii*. High concentrations of steroid hormones (both cortisol and testosterone) also inhibit the proper functioning of the immune system (Grossman, 1984) and make the males more susceptible to respiratory and other infections. Prolonged secretions of glucocorticoids such as cortisol are also known to provoke gastric ulceration and degeneration of the adrenal cortex. In many ways, the whole pattern seen with male *Antechinus* is very reminiscent of Cushing's syndrome in humans, which is provoked by a hyper-secretion of cortisol from the adrenals (Peterson *et al.*, 1982).

Adrian Bradley has gone on to study other species in the family Dasyuridae and has found the same pattern of mating-induced mortality in the red-tailed phascogale, *Phascogale calura* (Bradley, 1987). He proposed an 'adaptive-stress

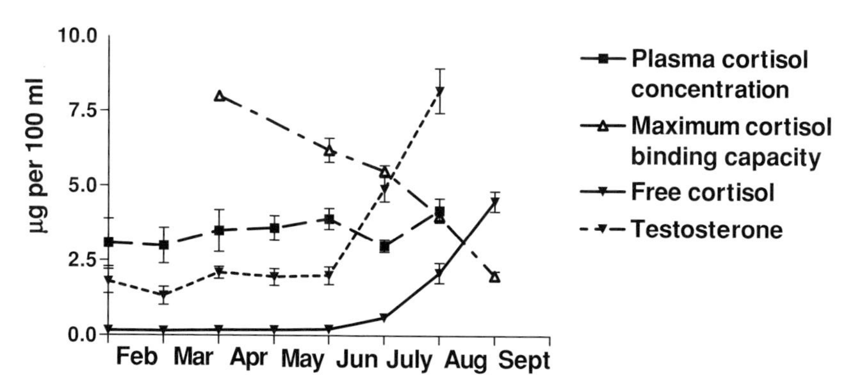

Figure 5.7. Changes in circulating levels of total cortisol and testosterone, along with changes in maximal cortisol binding capacity and free cortisol concentrations in male *Antechinus stuartii* during the breeding season. (Adapted from Bradley *et al.* (1980) and Lee and McDonald (1985).)

senescence' hypothesis to explain the evolution of this life-history strategy (Bradley, 1997). Chris Dickman has also studied patterns of post-mating mortality in a number of species of dasyurid and suggests that sperm competition may be the driving evolutionary force involved (Dickman and Braithwaite, 1992; Dickman, 1993).

Of particular interest is the species known as the dibbler (*Parantechinus apicalus*), shown in Figure 5.8, which has been studied both by Chris Dickman and by Harriet Mills during her PhD. This species was thought to be extinct on the Western Australian mainland but was first rediscovered in 1976 near the south coast town of Albany. The species was then found to occur in quite large numbers on two small offshore islands north of Perth, and Dickman and Braithwaite (1992) reported field evidence to suggest a similar male die-off to *Antechinus* in this separate genus. Follow up studies on Boullenger Island by the staff of the Department of Conservation and Land Management (CALM) failed to confirm this, however, and over a number of years males were found to live beyond their first breeding season.

Harriet Mills studied the ecophysiology of the dibbler on Boullenger Island and found evidence of a male die-off during one year when conditions of food supply appeared low. This suggests that the process may be facultative rather than obligatory in this species (Mills and Bencini, 2000). Another possible interpretation of the evolution of this novel life-history pattern is thus that it represents an attempt to divert food resources to the pregnant females when food is in critically short supply. In years when insect abundance is high, one would expect male dibblers to survive after breeding, but to be eliminated in

Figure 5.8. The dibbler, *Parantechinus apicalus.* This small dasyurid marsupial was thought to be extinct but relict populations have been found on the mainland of Western Australia and on Boullenger Island, a small sand and limestone island some 150 km north of Perth. (Photo courtesy of Harriet Mills.)

years when insects are scarce. This hypothesis remains to be tested, but the dibbler is obviously an ideal 'intermediate' species for its testing.

Thermoregulation in desert reptiles

By far the great majority of desert reptiles are ectotherms – regulating their body temperature during periods of activity between relatively narrow limits by behavioural means – utilising the sun as an external heat source (Pough, 1980, 1983; Avery, 1982). It is reasonable to ask, however, how precisely they regulate their body temperature, and whether they are ever forced to endure temperatures above what might be considered optimal (Huey and Kingsolver, 1993). In other words, do they ever show evidence of stress due to exposure to high environmental temperatures that are a characteristic of desert environments? If they did, from our operational definition of stress detailed in Chapter 2 we would expect to see a significant elevation of the body temperature of the animal above the preferred, despite the maximal deployment of high-temperature avoidance behaviour patterns. A recent paper by Hertz *et al.* (1993) details a promising new method for measuring thermoregulatory precision of lizards by comparing body temperatures of an animal in the field with the set of operant temperatures available in its immediate environment.

Table 5.4 collates published thermal information for two species of lizard, the desert iguana *Dipsosaurus dorsalis* and the Australian agamid *Ctenophorus nuchalis*, and contrasts the highest body temperature ever recorded in the field for each species with emergency and lethal temperatures. What is clear from this table is that maximal field temperatures fall below the panting threshold (PT) for both species and well below the critical thermal maximum (CTMax) and the lethal temperature (LT). Huey (1982) came to a similar conclusion in reviewing the thermal relations of lizards, snakes and turtles, and Table 5.5 provides clear evidence of the wide temperature safety margins that reptiles enjoy. Activity temperatures are, on average, 10–13 °C below the CTMax and the highest body temperatures ever recorded in the field for the various species sampled are some 6.5 °C less than their respective CTMax.

It is tempting to conclude from these data that desert lizards never experience 'thermal stress' – defined here as a body temperature elevated significantly above vital limits – in their natural environment, despite the operation of normal heat-avoidance or dissipation strategies. This may not be the case, however, as a close examination of the following examples shows.

van Berkum *et al.* (1986) studied thermoregulation in the tropical lizard *Ameiva festiva* (admittedly not a desert lizard) and made the point that:

> although these lizards are active foragers, their speed and duration of movements in the field fell far below the levels of speed and stamina that they achieved in the lab(oratory) when measured at temperatures they regularly experienced in the field.

They speculated that the phenotypic capacities of animals may not be shaped by routine activities, but instead by rare events that may be critical for an animal's survival, a point already made by Gans (1979) in developing his concept of 'excessive construction' (see also Bonnet *et al.*, 1999, for surprising data on the survival of blind tiger snakes on Carnac Island in Western Australia). This means that documenting routine thermoregulatory patterns may miss very rare environmental events when the animals are indeed exposed to excessive temperatures operating as a stressor which may, ultimately, influence fitness.

Table 5.4. *Known thermal profiles for two desert lizards in relation to the highest body temperatures recorded in the field*

Temperature	Definition	*Dipsosaurus dorsalis* (°C)	*Ctenophorus nuchalis* (°C)
Preferred body temperature (PBT)	Mean temperature selected in a thermal gradient	38.3[1]	36.2[2]
Maximum voluntary temperature (HBTS)	Body temperature at which lizard shuttles to shade	41.7[3]	?
Panting threshold (PT)	Body temperature at which mouth opens and ventilation rate increases	44.5[4]	44.1[5]
Critical thermal maximum (CTMax)	Body temperature at which animal loses righting reflex but from which recovery is possible	48.6[4]	48.5[6]
Lethal temperature (LT)	Lowest 'high' body temperature from which recovery is not possible	50.2[7]	49.0[2]
Highest body temperature recorded in field (T_bMax)	—	42.0[8]	43.8[2]

Sources: [1]de Witt, 1967; [2]Bradshaw and Main, 1968; [3]Berk and Heath, 1975; [4]Whitfield and Livezey, 1973; [5]Heatwole, 1970; [6]Heatwole, 1976; [7]McGinnis and Dickson, 1967; [8]de Witt, 1967.

A case in point was that recorded inadvertently by Bradshaw and Main (1968) at Port Hedland, in the arid Pilbara region of Western Australia, close to noon on a day with a shade temperature of 45.5 °C when, paradoxically, large numbers of small lizards (geckos and agamids) were seen elevated high above the ground in shrubs and bushes. The body temperatures of these lizards ranged from 41.0 to 46.0 °C (mean 44.3 °C) and a temperature profile recorded at the time confirmed that soil temperatures were in excess of 60 °C. The shade temperature in small clumps of *Triodia pungens* ranged from

Table 5.5. *Activity temperatures of reptiles in relation to critical temperatures*

Taxon	CTMax − T_b (°C)	CTMax − T_bMax (°C)
Turtles	13.7 ± 1.10	7.4 ± 1.24
Lizards (diurnal species)	10.4 ± 0.56	6.4 ± 1.03
Lizards (nocturnal/fossorial species)	13.0 ± 1.36	6.5 ± 1.17
Snakes	13.1 ± 0.66	6.8 ± 0.67

Modified from Huey (1982).

47.1 to 51.5 °C and, clearly, the coolest place at that time was as far above the ground as possible.

The panting threshold for one of the species concerned, the agamid *Diporophora bilineata*, is given by Heatwole (1976) as 31.3–40.4 °C, so we may conclude that in this particular case there is good reason to believe that these lizards were experiencing stress due to their abnormally elevated body temperatures. Other cases are equally anecdotal. Grenot (1967) mentions observing the large Saharan agamid lizard *Uromastix acanthinurus* panting at the mouth of its burrow in Béni-Abbès in Algeria, although the body temperature at this time was not known.

One of the most interesting lizards from this point of view is *Angolosaurus skoogi*, which inhabits slip-dune faces in the Namib desert (Seely *et al.*, 1988). This species is exposed to extremely high environmental temperatures when active and could be a candidate for Hamilton and Coetzee's (1969) 'maxithermy' hypothesis, based on insect thermoregulation. Pietruszka (1987), however, failed to confirm the thermal microhabitat predictions inherent in this hypothesis for *A. skoogi*. Nagy *et al.* (1991) measured the FMR of free-ranging individuals *in situ* and found that it was only 50% of that expected for a lizard of this size, although rates of water influx were similar to those of other desert lizards. Clarke and Nicolson (1994) have published details of the water and electrolyte balance of captive individuals of this interesting species, but essential data on thermoregulation are still lacking.

Another iguanid lizard with unusual thermoregulatory behaviour is *Sceloporus merriami*, which is constrained in its activity by having, at 32.2 °C, the lowest preferred body temperature of any North American desert iguanid (Grant and Dunham, 1988). A careful field study by Grant (1990) has established that average body temperatures of active animals are lower in the morning than in the afternoon (33.3 compared with 37.0 °C). Available operative temperatures in the environment frequented by this species are much broader in the morning (30–60 °C) than in the afternoon and early evening (36–38 °C) and the lizards are forced to accept higher body temperatures, higher rates of metabolism and water loss, and reduced locomotor performance if they are to remain active at all in the afternoon. Actual body temperatures did not exceed 40 °C, but this is well in excess of the PBT of 32.2 °C and very close to the 41 °C that Grant (1990) cites for 'lethal thermal stress' in this species. This would thus appear to be a case of a desert species that experiences stress on a daily basis.

The chuckwalla, *Sauromalus obesus*, is another interesting case because of some confusion in the literature as to its actual thermoregulatory capacities. McGinnis and Falkenstein (1971) reported a mean body temperature of 35.9 °C for lizards maintained in an enclosure, similar to the temperature reported for laboratory-maintained animals by Muchlinski *et al.* (1989), but lower than that reported by Zimmerman and Tracy (1989) and some 2.5–2.9 °C less than the mean reported by Muchlinski *et al.* (1990) in a later paper with instrumented free-ranging animals. Even more confusing is the report by Case (1976), who measured body temperatures as high as 43–46 °C from lizards wired with thermocouples.

Muchlinski *et al.* (1990) reported that chuckwallas ceased activity once their body temperature reached approximately 38 °C and then retreated into rock crevices. Temperatures there, however, could rise to quite high values in late afternoon and the maximum recorded was 42.9 °C, well below the 46 °C measured by Case (1976) and the CTMax for this species of 46.7 °C given by Cowles and Bogert (1944). Nagy (1973) notes that the chuckwalla abandons all diurnal activity, and hence thermoregulation, for long periods in the field when starved and in poor condition. Rates of CO_2 production measured with doubly labelled water typically fall in mid-summer to rates close to or even below 0.05 ml g^{-1} h^{-1}, which is lower than the basal rate for this species of 0.168 ml g^{-1} h^{-1}, predicted from the equation of Bennett and Dawson (1976) for iguanid lizards. Nagy (1972) documents changes during the drought year of 1970–71 in the Mojave Desert when marked individuals lost some 55% of their initial body mass and the specific activity of injected tritiated water remained constant (indicating no significant input of free water) for a period of ten months. These individuals thus experienced severe stress due to lack of water, but appeared to have avoided potential thermal stress during this prolonged drought by abandoning any attempt at thermoregulation.

There are other reports in the literature of a breakdown in thermoregulatory behaviour as a consequence of detrimental changes in the *milieu intérieur* of desert lizards. A combination of hypernatraemia and chronic dehydration in the field leads to a loss of thermoregulatory behaviour in the Australian agamid lizard *Ctenophorus (=Amphibolurus) ornatus* (Bradshaw, 1970) and Henzell (1972) found that dehydration in several desert species of the genus *Egernia* had a detrimental effect on the precision of their thermoregulation. Cogger (1974) found that even mild starvation was sufficient to lower the activity temperatures of the desert mallee dragon, *Ctenophorus fordii*, in central Australia, and Rice and Bradshaw (1980) found that a fall in body condition in *Ctenophorus nuchalis* was associated with a temporary loss of the ability of this lizard to alter skin reflectance with change in body temperature. Lee (1980), working with two tropical species of *Anolis*, found that well-nourished individuals thermoregulated more precisely than poorly nourished ones, although this was disputed by Henle (1992).

Thermal homeostasis and its maintenance thus appears to be conditional, at least in desert lizards, on the proper operation of other, more general, homeostatic processes controlling water, electrolyte, energy and probably nitrogen balance of the animal (Bradshaw, 1997a). Instances of desert lizards experiencing thermal stress appear to be rare in the literature, but this does not necessarily mean that they are without significance. Studying animals in the field during periods of climatic extremes – e.g. prolonged droughts – may be more revealing in the long run on the impact of natural selection on such populations than any number of studies carried out during so-called 'normal' years.

Breeding in Arctic birds

Many birds breed in the Arctic tundra, often after extensive migratory flights. On their arrival, they may be greeted with potentially stressful events including snow storms, high winds and freezing temperatures. Such stressors are likely to lead to increased rates of secretion of glucocorticoids which, initially, aid the birds in adapting to these conditions, but prolonged periods with elevated concentrations of corticosterone (B) may lead to a decrease in reproductive behaviour and interfere with the secretion of essential sex hormones (Silverin, 1986; Wingfield and Silverin, 1986). Elevated corticosterone concentrations in breeding birds have also been correlated with an increase in 'escape' behaviour (Astheimer *et al.*, 1992), which may lead to an abandonment of the reproductive effort. Prolonged exposure to high concentrations of catabolic steroids such as corticosterone can also lead to muscle wasting and downregulation of the immune system, with its attendant risks of adventitious infections (Grossman, 1984; Sapolsky *et al.*, 2000).

O'Reilly and Wingfield (2001) speculate in a recent paper that birds breeding in the high Arctic may have a reduced stress response when compared with those breeding in the lower Arctic, and thus be afforded some measure of protection from environmental extremes. In order to test this hypothesis they

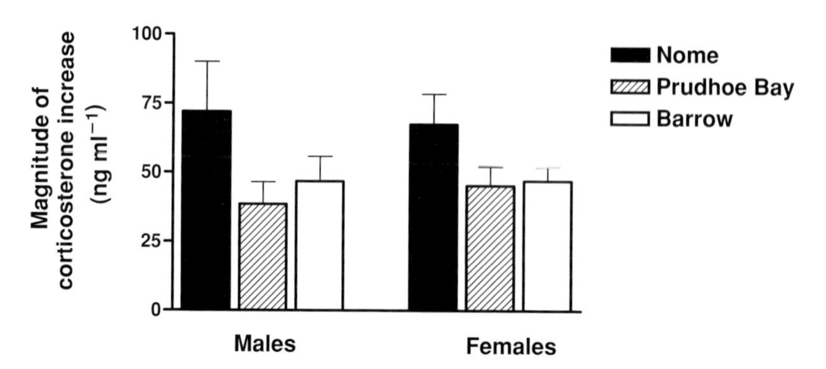

Figure 5.9. A comparison of the adrenocortical response, measured as an increase in plasma corticosterone (B) concentrations, following capture and handling of Arctic birds from northern and more southern latitudes. The response of the two northern populations (Prudhoe Bay and Barrow) is clearly lower than that of the birds from the more southern location of Nome. (Adapted from O'Reilly and Wingfield (2001).)

measured the elevation in plasma B concentrations following capture and routine handling (Wingfield, 1994) in three populations of semipalmated sandpipers (*Calidris melanotos*), two from northern sites in Alaska (Barrow and Prudhoe Bay) and one from Nome, which is some 6° further south. The results are shown in Figure 5.9 and it is clear that the adrenal response of both males and females from the two northern populations is attenuated when compared with that of the birds from Nome. O'Reilly and Wingfield (2001) also showed that individual birds most responsible for parental care within a given species had a lower stress response than birds with fewer or lesser responsibilities. Similar results were obtained by Wingfield *et al.* (1982) in an earlier study with females of the white-crowned sparrow, *Zonotrichia leucophrys*, which also breeds in the Arctic and in which the females are solely responsible for incubating the eggs. Astheimer *et al.* (1994) also found gender differences in the response of this species to injections of ACTH; the topic of adrenocortical responsiveness was reviewed by Astheimer *et al.* (1995).

Pravosudov *et al.* (2001) have recently studied the effects of variation in food availability on both basal and stress-induced concentrations of corticosterone in mountain chickadees (*Poecile gambeli*). They found that food-restricted birds had significantly higher baseline concentrations of B than those maintained on *ad libitum* food. These birds also increased their body mass significantly over the course of the experiment; B has been shown to enhance feeding activity and stimulate fat storage in other birds (Silverin, 1985; Gray *et al.*, 1990). These data suggest that birds wintering in temperate climates with harsh winters may have elevated baseline concentrations of circulating B over long periods of time and that this hormone assists them in coping with the demands of the environment.

John Wingfield has gone on to formalise much of this research into an integrated model of how Arctic and other birds cope with such unpredictable environmental events that are potentially capable of provoking stress (Wingfield, 2001). He applies McEwen's (1998a,b) concept of 'allostasis' (i.e. stability through change) and 'allostatic load' to the case of birds that are challenged with SLPFs, or 'short-lived perturbation factors'. These provoke the appearance of an ELHS, or 'emergency life history stage' (Wingfield *et al.*, 1998) that serves to enhance the bird's fitness in the face of the particular crisis and is mediated by adrenal corticosteroids (Wingfield and Romero, 2000). Three phases of adrenocortical activity are visualised:

1. Level A: baseline corticosteroid concentrations, which are essential for the maintenance carbohydrate and electrolyte homeostasis
2. Level B: daily and seasonal changes in corticosterone concentrations in relation to all allostasis
3. Level C: high transitory peaks that activate the ELHS and will provoke stress if the response is not adequate, or the animal is unable to escape from the stressor

This model fits well with the recent ideas of Sapolsky *et al.* (2000) on the rôle that glucocorticoid hormones play in helping an animal contend with stressors and the changing view that these steroids should be considered, at least at low concentrations, as anti-stress, rather than stress, hormones. Persistence of Level C events would signal the occurrence of stress in the Wingfield model, however, and sustained high circulating concentrations of glucocorticoids such as corticosterone would be detrimental, as noted above.

As mentioned already in Chapter 2, elevated concentrations of corticosteroids in the plasma very likely indicate the presence of a stressor, but cannot be taken on their own as proof that an animal is experiencing stress. The adrenocortical response is part of the animal's normal adaptive response to the presence of the stressor and its success or failure can only be judged by whether the animal is able to maintain homeostasis. The level or intensity of any stress experienced by the animal can thus be gauged from the extent of any deviation of the *milieu intérieur* from its normal homeostatic state.

Based on our operational definition of stress, the question of whether Arctic birds experience stress or not would appear to require further study of the concentrations of other regulatory hormones involved in homeostasis and concomitant measures of the birds' *milieu intérieur*. Concentrations of catecholamines, glucagon, insulin and arginine vasotocin, as well as pulse frequencies of gonadotrophin-releasing hormone (GnRH), are all things that one would like to know in such instances, coupled with details of FMR, total body water and fat content, blood glucose and urea concentrations and plasma electrolyte concentrations.

John Wingfield and his colleagues from the University of Washington in Seattle have made a promising start in this direction with their development of the 'challenge hypothesis' (Wingfield *et al.*, 1990), which seeks to relate patterns of hormonal secretion to environmental dictates, especially in Arctic birds. They argue that male–male aggression and the extent of parental care shown by birds are important factors in determining both the timing and concentrations of plasma testosterone (T) in breeding males. They predict that birds showing high levels of aggression and parental care will have short peaks of T production, in contrast to species where parental care is low and the profile of T will be lower and more sustained. They also contend that polygynous males are less responsive to environmental cues than monogamous males. These predictions are currently being tested by avian field endocrinologists and they provide a stimulus for research that is yet to be done and which evokes exciting possibilities.

Hibernation in mammals

An early paper by Suomalainen (1960) on hedgehogs in Finland proposed that the period of hibernation was one of severe stress for these mammals. The blood picture of hibernating animals was reminiscent of stressed animals, with decreased numbers of white cells (leucopoenia) and changes in the normal proportion of leucocyte types (neutrophilia, eosinopoenia and lymphopoenia). The adrenal glands were also enlarged, with reduced concentrations of cholesterol and cholesterol esters. The nuclei of cells in the supra-optic nucleus of the hypothalamus were found to be enlarged, and abundant neurosecretory material was evident in the pars nervosa, suggesting increased secretory activity of the neurohypophysis. Blood potassium concentrations fell and the sodium concentration of the serum was increased, suggesting a significant alteration in the *milieu intérieur*, although these changes are not common to all hibernating mammals (Reidesel, 1960). Some of these interesting early data from Suomalainen's (1960) study are summarised in Table 5.6.

Early workers concluded, on the basis of histological criteria, that the thyroid was non-functional during hibernation (see, for example, Popovic, 1960; Kayser, 1961; Gabe *et al.*, 1964; but see Lachiver, 1964). More recent work on mammalian hibernators, however, has done much to clarify the rôle played by hormones, and also gives an added perspective to the question of whether these animals are stressed or not. Demeneix and Henderson (1978a) measured

Table 5.6. *Changes in physiological parameters in hibernating and non-hibernating hedgehogs in Finland*

Condition	Leucocytes (per mm^3)	Percentage neutrophils	Adrenal mass (mg per 100 g)	Cholesterol in adrenals (mg%)	Cholesterol esters (mg%)
Active	18 167 ± 290	3.4 ± 1.2	51.7 ± 0.9	1024.6 ± 151	607.8 ± 124.2
Hibernating	4 100	47.5	62	204.2 ± 8.9	34.0 ± 4.5

Data summarised from Suomalainen (1960).

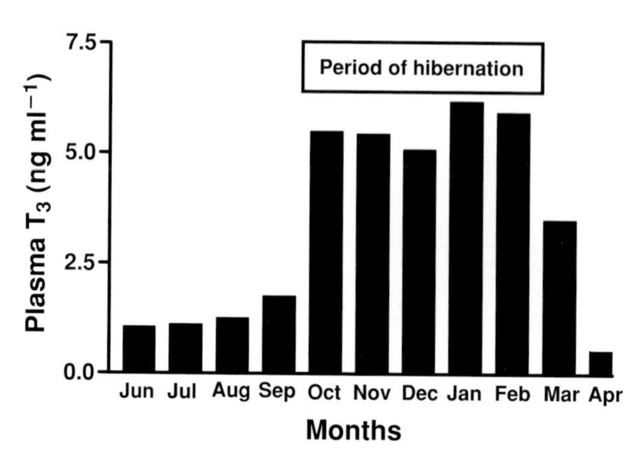

Figure 5.10. Circulating concentrations of tri-iodothyronine (T_3) in ground squirrels (*Spermophilus richardsoni*) prior to and during the period of winter hibernation (adapted from Demeneix and Henderson (1978a)). These were the first data to show conclusively that the thyroid gland of mammals is not inactive during the period of hibernation.

circulating concentrations of thyroxine (T_4) and tri-iodothyronine (T_3) in ground squirrels (*Spermophilus richardsoni*) and found that they were elevated rather than being depressed during hibernation (see Figure 5.10). Rates of clearance of these two hormones were also measured in active and hibernating animals by Demeneix and Henderson (1978b), who found that T_4 clearance rates were depressed during hibernation, but concentrations of free (unbound) T_4 were increased. Concentrations of tri-iodothyronine (T_3) and reverse T_3 were also elevated in hibernating individuals and these data suggest strongly that the thyroid gland is not inactive during this period, as was previously assumed (Hudson, 1981).

The question of whether or not hibernation is a period of stress for the animal is difficult to answer, and perhaps highlights a problem with our operational definition of stress proposed in Chapter 2. There are clearly profound changes in the composition of the *milieu intérieur* of hibernating mammals but, as Kayser (1961) pointed out many years ago, this altered state is none the less a regulated one (see also Lyman, 1978). Stress might only thus apply if the animal were unable to achieve or maintain this alternative *milieu intérieur* that is perhaps the norm for hibernation. Short bursts of spontaneous arousal occur throughout the period of hibernation, when the animal raises its body temperature and in some cases may even feed before dropping back again into torpor. Should one view these as attempts to avoid a state of stress induced by low environmental temperatures and deficient food supply?

I think not. Rather, we need to consider the possibility that an animal such as a hibernator has evolved two 'normal' states or *milieux intérieurs*: one the normothermic *milieu intérieur* and the other reverted to when environmental conditions militate against continued activity. Further research will hopefully enable one to clarify this question and refine our current understanding of what constitutes stress for an animal in its natural environment. Carey and Martin (1996) have recently measured heat-shock protein expression in hibernating thirteen-lined ground squirrels (*Spermophilus tridecemlineatus*) in an effort to see whether these stress proteins play a significant rôle in the process of hibernation. Carey *et al.* (2000) have also suggested that the intestinal tissues of hibernators may experience oxidative stress during the hibernation season and the question is clearly still an open one.

These various case studies show that some animals do experience stress at key points in their life cycle: silvereyes in years when their normal source of food fails; some macropods living on desert islands in exceptionally dry years; small male marsupial carnivores during their one and only breeding season; and, probably, Arctic birds. The field of stress physiology is still in its infancy, however, and future ecophysiological studies should do much to expand this list.

6

Survival in deserts

The desert environment is one traditionally viewed as being 'inhospitable' from an anthropomorphic (i.e. human) point of view. Many vertebrate animals, however, live and reproduce there successfully and the study of the many adaptations – morphological, physiological and behavioural – that make this a possibility has occupied comparative physiologists and ecophysiologists for decades. Every continent of the world contains a desert, including the Arctic and the Antarctic, which are classified as semi-arid deserts, and it is of interest to note, as shown in Figure 6.1, that the world's deserts include large areas of the ocean adjacent to each continent. The major physiographic and environmental characteristics of the world's deserts, both hot and cold, have been summarised by Bradshaw (1986) and interesting rainfall comparisons in terms of predictability and constancy (as defined by Colwell, 1974) are given by Low (1978), showing that the Australian deserts have the lowest probability of rainfall. Soil fertility is another important factor limiting productivity in arid situations; Figure 6.2 compares nitrogen and phosphorus concentrations in soils from various parts of the world, those of South Africa and Australia being by far the least fertile and therefore expected to have the lowest productivity (Stafford Smith and Morton, 1990).

A number of different vertebrates have been studied in their natural environment, in an effort to discern the nature of their adaptive solutions to the exigencies posed by high environmental temperatures and lack of free water. Some of these studies on desert frogs, lizards, birds and mammals will be reviewed in this chapter (Louw and Seely, 1982; Nagy, 1987a, 1988a, 1994a). Of particular interest is the question of whether these same adaptations have arisen through natural selection in these desert environments or whether, as

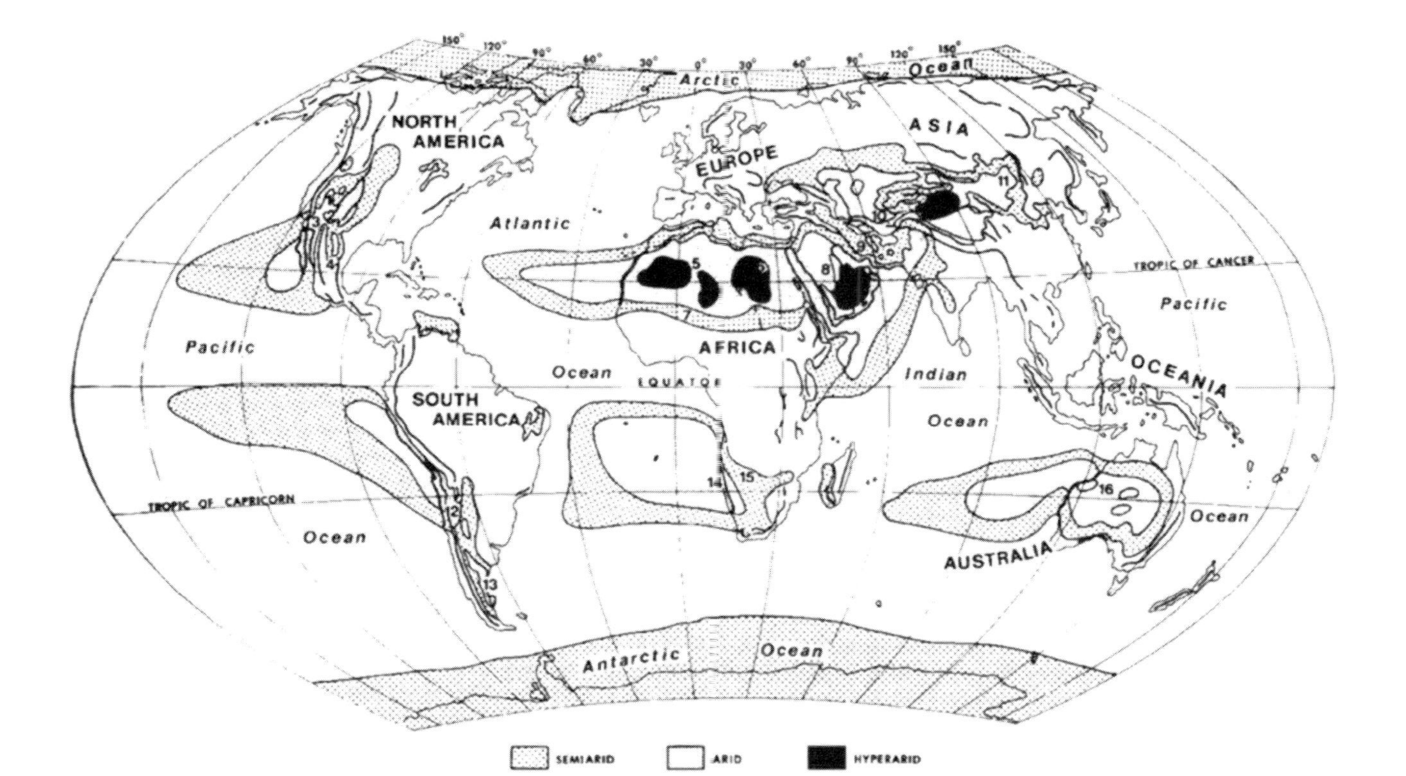

Figure 6.1. Distribution of the world's major deserts showing their concentration in subtropical latitudes and extensive oceanic extensions. (Adapted from Monod, 1973 and Bradshaw, 1986.)

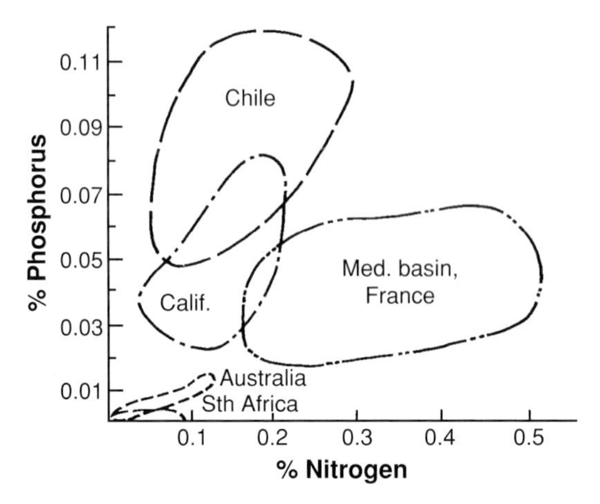

Figure 6.2. Percentages of phosphorus and nitrogen (total %) in the soils of five regions with Mediterranean climates. Modified from Rundel (1979) with information related to soils of the Mediterranean basin. Note the exceedingly depauperate soils that are characteristic of Australia and South Africa. (From de Castri (1981).)

appears to be the case in reptiles, they are 'pre-adaptations' or 'exaptations' (*sensu* Gould and Vrba, 1982) which underpin the success of these animals in such resource-limited environments (Bradshaw, 1988).

Frogs

Amphibians would, at first sight, appear to be the least likely of desert inhabitants, given their high rates of water loss from their moist skin and their almost universal need, at some stage, for free water to reproduce and complete their life history (Feder and Burggren, 1992). Despite these apparent constraints, deserts may contain quite diverse amphibian anuran faunae, and those of the United States and Australia have, to date, been the most studied (Shoemaker and Nagy, 1977; Shoemaker, 1988; Shoemaker *et al.*, 1992) although Warburg (1997) gives details on Israeli species. An early paper by Lee and Mercer (1967) described the formation of a 'cocoon' covering all but the nares of buried frogs (see Figure 6.3). Later studies showed that this is not, as first thought, a secretion of the skin but the result of the progressive buildup of layers of shed skin during the period of æstivation, as seen clearly in Figure 6.4. The frog normally ingests shed skin after each moult but this is clearly not possible when it is buried underground and in a state of suspended animation. The effect of this ever-thickening layer of dead skin is to increasingly waterproof the buried frog; rates of cutaneous water loss fall dramatically, and skin resistance increases, as the animal makes its cocoon

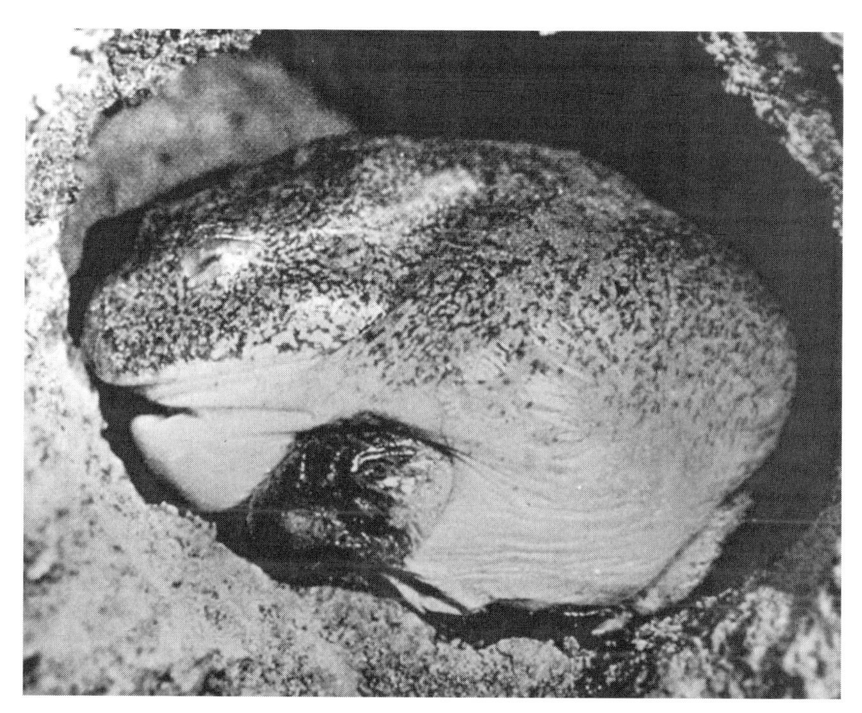

Figure 6.3. A desert frog buried in its cocoon. The species is the Australian water-holding frog, *Cyclorana platycephala*. (Photo courtesy of Phil Withers.)

(see Figure 6.5) (McClanahan, 1972; McClanahan *et al.*, 1976; Withers, 1998a).

Interestingly, not all desert frogs form cocoons and this adaptation is restricted to a few Australian genera (*Neobatrachus, Cyclorana,* and *Litoria (=Cyclorana?) alboguttata*), some African Ranidae (*Pyxicephalus*) and Hyperoliinae (*Leptopelis*) plus central and North American Hylidae (*Pternophyla* and *Smilicus*) and some South American Leptodactylidae (*Lepidobatrachus* and *Ceratophrys*) (Withers, 1995). Genera such as *Notaden, Limnodynastes, Bufo and Scaphiopus,* by contrast, are all successful desert inhabitants, but they do not produce cocoons.

Changes in concentrations of plasma and urinary electrolytes were measured in buried American desert toads (*Scaphiopus couchii*) over a seven month period by McClanahan (1967, 1972) and the osmotic pressure of both fluids increased dramatically, as seen in Figure 6.6. Analysis of these fluids revealed that, although concentrations of sodium and chloride increased progressively, the major osmolyte contributing to the increase was urea, to which the frog is very tolerant (see Figure 6.7) (McClanahan, 1964). Such a rise might be

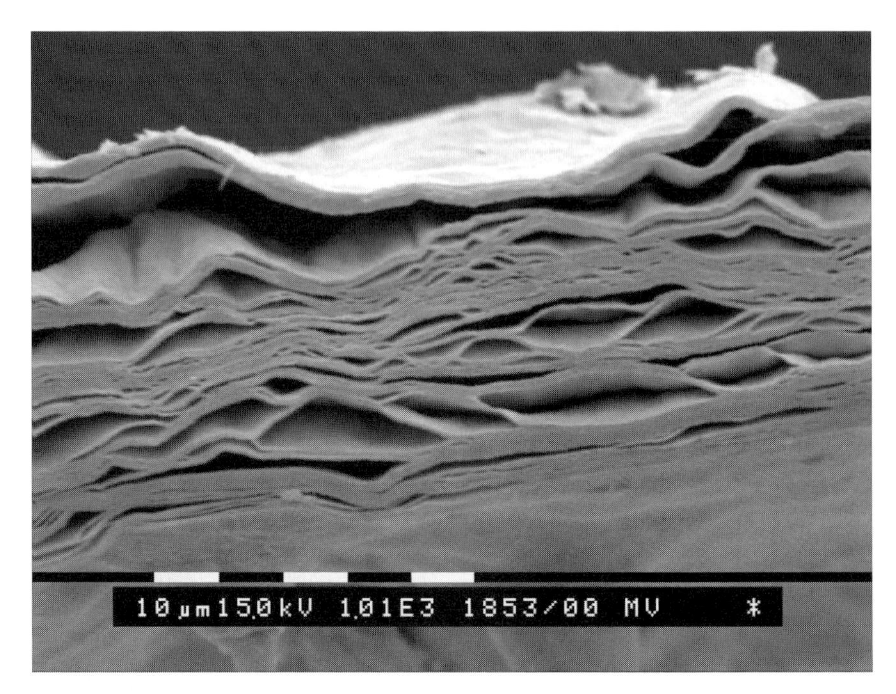

Figure 6.4. Scanning electron micrograph (SEM) of the cocoon from a desert burrowing frog, showing that the cocoon is made up of accumulated layers of shed skin. (Photo courtesy of Phil Withers.)

expected in a fasting animal that would slowly break down muscle protein to form glucose (gluconeogenesis), excreting urea in the process, but Jones (1980) has shown that actual rates of urea synthesis increase during dormancy. With the animal in a cocooned state, however, this urea could not be eliminated from the body and would thus gradually increase in all the body fluids. The interesting point about *Scaphiopus couchii*, however, is that it does not form a cocoon and this raises the very real possibility that the increased osmolality of its body fluids might be used, for a period of time, to absorb water from the surrounding soil. The urea, rather than being simply a metabolic by-product of protein degradation, could function as 'osmotic packing', as in marine elasmobranchs, thus enhancing the water economy of the frog (Withers and Guppy, 1996; Barker-Jorgensen, 1997; Withers, 1998). McClanahan (1972) presented data (see Table 6.1) showing that there is, for a period of time, a favourable osmotic gradient between the animal and the surrounding soil but that, as the soil dries out, its water potential increases to the point where, if the skin of the animal were to remain permeable, it would lose water to the soil (Hillyard, 1975, 1976). Water absorption from the soil has

Table 6.1. *Progressive changes in osmolality of the body fluids and soil moisture tension in buried* Scaphiopus couchii *in Arizona*

Season	Osmotic pressure of body fluids (mOsm kg^{-1})	Soil moisture tension (mOsm kg^{-1})
Autumn	286	150
Spring	295	197
Summer[a]	420	446

[a] Osmotic gradient negative for the frog.
From McClanahan (1972).

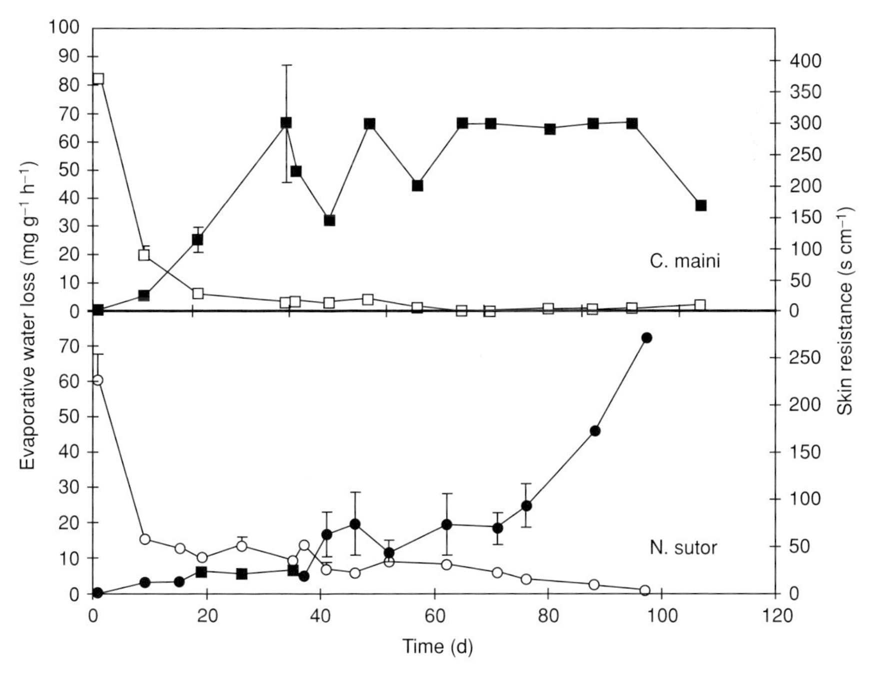

Figure 6.5. Changes in the rate of evaporative water loss (open symbols) and skin resistance (filled symbols) of two desert burrowing frogs, *Cyclorana maini* and *Neobatrachus sutor*, as a function of the time in days spent buried. (From Withers (1998), with permission.)

been demonstrated in the Australian frog *Heleioporus eyrei* (Packer, 1963) and Lee (1968), in an early paper, found that calling males lost considerable body mass overnight, which they subsequently reclaimed from the soil. Whether the pituitary hormone arginine vasotocin (AVT), which is known to increase

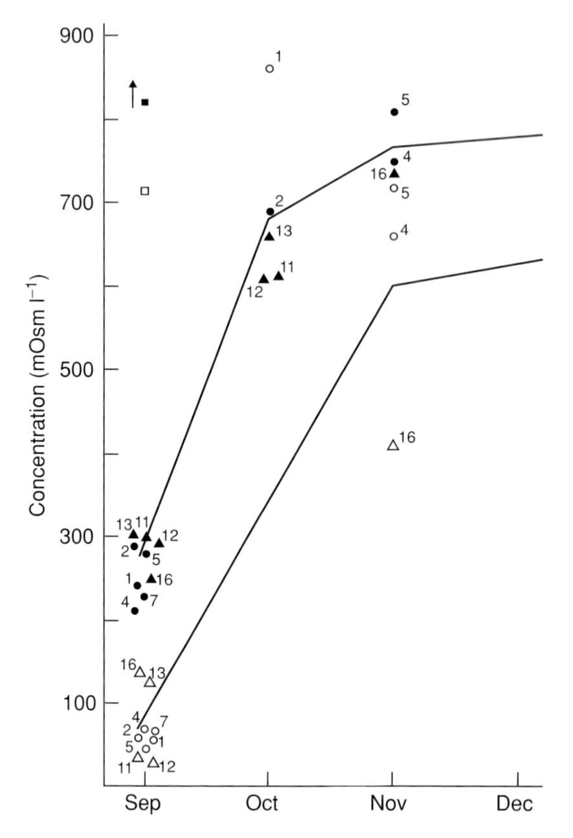

Figure 6.6. Progressive increase in the osmotic pressure of the plasma (filled symbols) and urine (open symbols) of buried spadefoot toads (*Scaphiopus couchii*) in soil with a moisture content of 5.0 ml per 100 g dry soil. (From McClanahan (1972) with permission.)

skin permeability in frogs (Bentley, 1971), is involved in this process is still very much debated (Bakker and Bradshaw, 1977; Jørgensen, 1992) and clear experimental evidence is lacking. What is more surprising is that, apart from the paper by Lee (1968), there appear to be no reports of anyone ever finding dehydrated frogs in the field. Certainly the first thing that buried desert frogs do on being disinterred is to urinate profusely, suggesting that they are far from being short of water. The reports of early European explorers in Australia also attest to the life-saving properties of the water stored in the bladder of the water-holding frog, *Cyclorana platycephala* (Spencer, 1896; Carnegie, 1898; Spencer and Gillen, 1912; Waite, 1929).

Withers (1995) and Withers and Thompson (2000) have studied cocoon formation and metabolic depression in desert frogs of the genera *Neobatrachus*

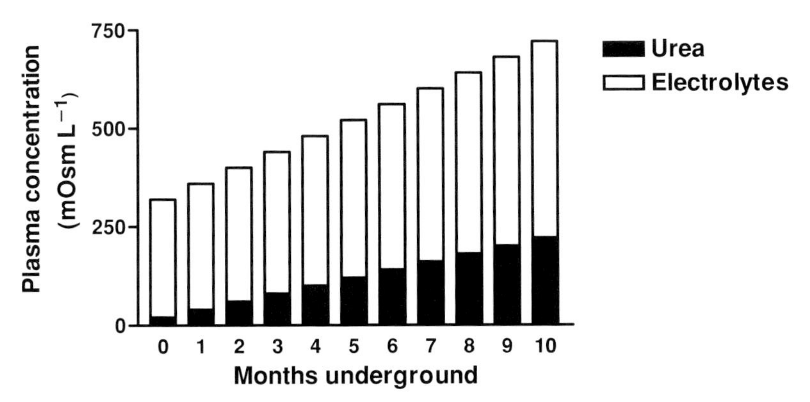

Figure 6.7. Progressive increase in plasma concentrations of electrolytes and urea in buried spadefoot toads (*Scaphiopus couchii*) in Arizona. Note that urea retention is primarily responsible for the increase in osmolality of the body fluids. (Adapted from McClanahan (1972).)

and *Cyclorana* and found that rates of oxygen consumption in cocooned frogs fall to about 20% of those in active animals. What is not clear, however, is the extent to which all tissues of the body participate in this depression, the suspicion being that the muscle tissue contributes most to the response (Guppy and Withers, 1999). Withers and Thompson (2000) have also been able to estimate the time that a frog has been cocooned from the thickness of its cocoon and the number of layers of sloughed skin of which it is composed. Rates of cocoon formation were found to range from 0.6 layers per day in *Cyclorana cultripes* to 0.7 in *Cyclorana australis*.

One of the most interesting and perplexing problems concerning desert frogs, which may remain underground for periods of up to nine years between rains (van Beurden, 1982), is how they are able apparently to anticipate when such rain will fall, and be in a physiological state ready to breed immediately. Following cyclonic rains in the arid parts of Australia, for example, frogs emerge in their millions and males immediately commence chorusing in the small pools and ponds of water that form. Mating with females occurs within a matter of hours (Heatwole, 1984) and clearly the gonads of both sexes are already in a state of readiness before the frogs' emergence. A recent study by Harvey *et al.* (1997) on *Scaphiopus couchii* has found evidence of preparedness for reproduction in pre-breeding animals, with elevated plasma concentrations of the sex steroids oestradiol and testosterone in females, and late stages of spermatogenesis already in males. Gametogenesis then continued during the period of breeding and in the weeks following coupling. This finding suggests that part of the very successful strategy of these desert-dwelling frogs lies in

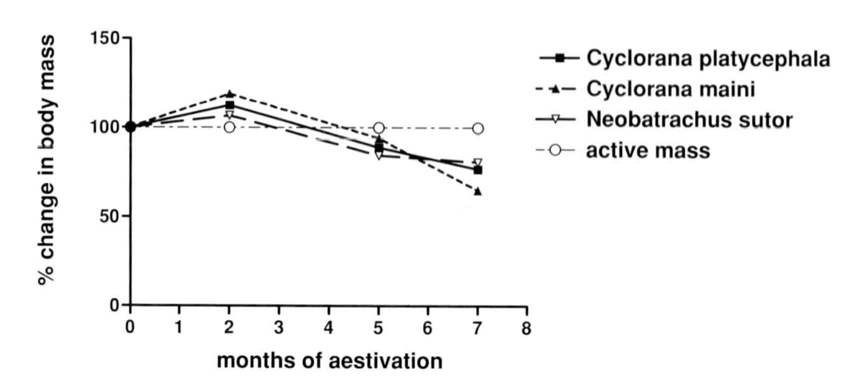

Figure 6.8. Progressive change in body mass of buried male Australian desert frogs over a seven month period in the laboratory. Note that all species increased their body mass over active mass in the first two months underground, suggesting that they absorb moisture from the soil in the initial stages of aestivation. (From Bayomy *et al.*, 2002).

commencing preparation for the next period of reproduction, during and at the completion of the current effort. These toads were collected only after they had emerged from their burrows and information is still lacking on the extent of the preparedness of burrowed frogs and toads a week or two before the rains fall. Another recent paper by Emerson and Hess (1996), on the significance of androgens in opportunistically breeding frogs, suggests that male sex steroids such as testosterone have a permissive rather than an activating role in the expression of mating behaviour and need not be maintained continually at high concentrations in the blood. Work by Hoff and Hillyard (1993) has also shown that water uptake via the skin in some frogs is affected by changes in barometric pressure, suggesting a means by which buried frogs may be able to anticipate approaching rain periods.

A recent study by S. D. Bradshaw *et al.* (in preparation) has investigated changes in body mass, plasma testosterone concentrations and gonadal histology in a number of Australian desert frogs over a 20 month period while buried underground. Frogs of four species (*Cyclorana platycephala, C. maini, C. australis* and *Neobatrachus sutor*) were collected from breeding choruses in the Kimberley region of Western Australia following heavy rain and then transported to Perth, where they were induced to burrow and cocoon in soil from the collection site. A sample of frogs was disinterred at regular intervals and killed, to determine temporal changes in the parameters under study. Initial changes in body mass of the burrowed frogs are shown in Figure 6.8. Interestingly, the frogs initially gain mass over the first two months, suggesting that they absorb water from the soil. Changes in the gonadosomal index (GSI) of

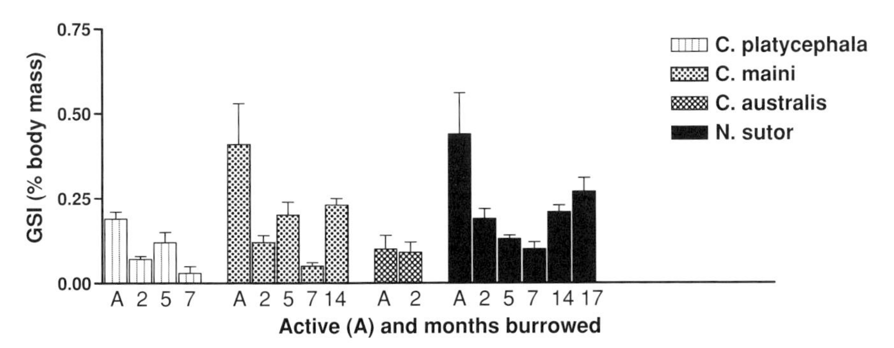

Figure 6.9. Variation in mean gonadosomal index (GSI) of aestivating Australian male desert frogs over a 17 month period in the laboratory. GSI is expressed as the mass of both testes as a percentage of the body mass of the frog. Data expressed as mean ± S.E. (From S.D. Bradshaw *et al.*, in prep.)

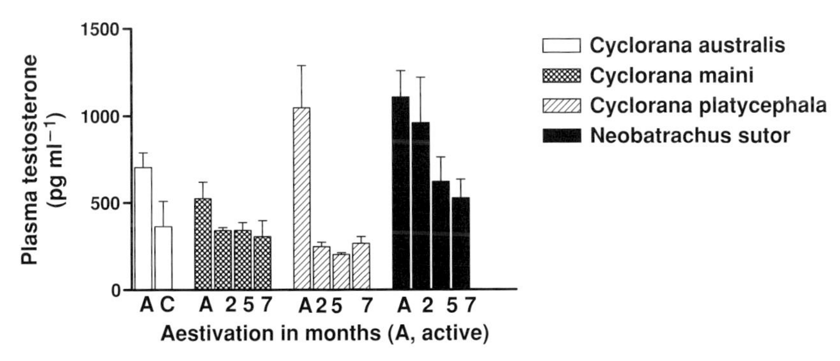

Figure 6.10. Progressive changes in plasma testosterone concentrations in aestivating Australian desert frogs in the laboratory. (From S.D. Bradshaw *et al.*, in prep.).

males, which is the ratio of the mass of the two testes to the actual body mass expressed as a percentage, are shown in Figure 6.9 and show a progressive decline in testis mass over a seven month period in all four species, followed by a significant increase in *Cyclorana maini* and in *Neobatrachus sutor* after 14 months, with a further increase in the remaining *N. sutor* after 17 months. These initial changes were also mirrored by concentrations of plasma testosterone (see Figure 6.10) and the diameter of the seminiferous tubules in the testes, which both decrease; further study should reveal whether plasma testosterone concentrations increase after 14 months with the recrudescence of the gonads. These data are the first to show that, despite the quiescent state of these buried and cocooned frogs, testicular recrudescence recommences spontaneously during æstivation and in the absence of any external environmental cues, as these frogs were maintained throughout at a constant temperature. We speculate that testes would continue to develop; mature sperm would be produced and stored, thus readying the frog for the advent of rain at some time in the near future. Changes in barometric pressure (Hoff and Hillyard, 1993), temperature, or soil moisture may be the ultimate cue inducing the frog to emerge from its cocoon and dig its way to the surface.

Lizards

Of all the vertebrates that inhabit deserts, reptiles are the most common and are usually considered to represent the acme of adaptation to this difficult environment. Numerous studies on lizards, snakes and tortoises from deserts in the African Sahara, North America, Australia and Central Asia have enhanced our understanding of how these animals contend with the ever-present problems of high diurnal temperatures, lack of free water and limited material resources for growth and reproduction that characterise the desert environment. Lizards, especially, have been the focus of many of these studies, because of their abundance, ease of capture and suitability as model organisms for ecological and ecophysiological studies (Huey *et al.*, 1983; Vitt and Pianka, 1994).

Agamid lizards of the genus *Ctenophorus* (formerly *Amphibolurus*) have been studied for three decades in Australian deserts, with much information on their ecology, population dynamics, reproduction and ecophysiology reviewed in Bradshaw (1981, 1986, 1997a) and Greer (1990). Despite their relatively large body size (30–50 g), many of these oviparous (egg-laying) lizards have proven to be 'annual' species, emerging from the egg late in summer, growing rapidly, breeding in spring and then dying in their first summer, as first shown by Storr (1967). An early theoretical paper by Cole (1954) proposed that the population advantages to be gained by such semelparity should make this a common life-history pattern, in contrast to the more usual reproductive pattern of iteroparity (multiple breeding). Gadgil and Bossert (1970) pointed to a defect in Cole's argument, in not taking juvenile mortality into account, but the discovery of a large number of semelparous species in Australia none the less heightened interest in this phenomenon and focused research on the physiological basis of this annual die-off in late summer.

The western netted dragon, *Ctenophorus nuchalis* (formerly *Amphibolurus inermis*) (see Figure 6.11), is abundant in sandy regions throughout the central deserts of Australia, north of latitude 29°S, and a population has been studied intensively over many years at Shark Bay, some 900 km north of Perth in Western Australia. This is still within a winter-rainfall area and juvenile *C. nuchalis* grow rapidly throughout the mild winter period and commence breeding in the spring (August–September) at age 7–8 months. Females lay

Figure 6.11. An 'elevated' male Western netted dragon, *Ctenophorus nuchalis*, photographed on a fencepost late in the afternoon at Shark Bay on the west coast of Western Australia when the air temperature was 39 °C.

up to five eggs in a clutch and some individuals may lay as many as three clutches before they die, contributing to a high reproductive potential of 2.02 for the species (Bradshaw, 1981, 1986).

A long-term ecophysiological study of *C. nuchalis* at Shark Bay focused on measurements of field metabolic rate (FMR), using the DLW method, water turnover, food intake (using sodium-22) and circulating concentrations of adrenal corticosteroids, in an effort to see whether the animals' demise was the result of chronic stress leading to 'pituitary–adrenal exhaustion'. Plasma corticosteroid and glucose concentrations were measured within minutes of capture of each lizard, and then some hours later after the animals had been confined in calico bags prior to routine processing. These two measurements gave some idea of the capacity of the lizards collected in different seasons (i.e. spring and late summer) to respond to a stressor, in this case that of confinement. At a later stage of the study, adrenal glands were also incubated *in vitro* in the field and the adrenal response to exogenous injections of adrenocorticotrophic hormone (ACTH) of hypophysectomised and dexamethasone-injected individuals was investigated (Bradshaw, 1978a,b). The attempt here was to see whether the summer die-off following breeding of both males and females of this species was due to some breakdown in the normal biosynthetic pathways in the adrenal glands, or to some defect in the negative feedback system by which the pituitary gland controls the secretion of steroid hormones such as corticosterone from the adrenal glands. What did emerge from this part of the study is that the response of the lizards to either the stress of confinement, or ACTH injections, is quite different in late summer and autumn from the response in spring. As seen in Figure 6.12, the response of the adrenal gland is quite 'normal' in spring, with roughly 60% of the population sampled responding with an elevation of the blood corticosterone concentration of 2–4 $\mu g\ dl^{-1}$. In summer, on the other hand, the response is more square than Gaussian, with roughly 30% of the individuals showing no, or hardly any, adrenal response to the ACTH, whereas other individuals respond massively to the same dose. This would seem to imply a progressive deterioration in the ability of the individual lizards to respond appropriately to the adrenal challenge by ACTH, reflecting the physiological heterogeneity of the population at that particular time.

Table 6.2 summarises seasonal changes in condition index (Bradshaw and De'ath, 1991), body water content, plasma glucose and corticosteroid concentrations and the state of the *milieu intérieur* in *C. nuchalis.* These data show clearly that the electrolyte constancy of the *milieu intérieur* is maintained in animals sampled in late summer and autumn, despite a significant loss of body condition and body water. Plasma glucose concentrations also fall significantly and, in late summer, plasma corticosteroid concentrations were elevated in animals bled minutes after their capture (4.52 ± 0.48 *versus* $0.4 \pm 0.1\ \mu g\ dl^{-1}$). Corresponding data for rates of water turnover are shown in Figure 6.13 and it is clear that late summer is associated with a significant increase in water efflux that is not matched by a corresponding increase in influx. The net result is dehydration (i.e. reduction in total body water content, TBW) due to a chronic negative water balance.

The adrenal glands of these lizards were not enlarged in late summer and autumn, as one might expect if they were experiencing chronic stress, but rates of secretion of corticosterone *in vitro* were increased significantly, suggesting

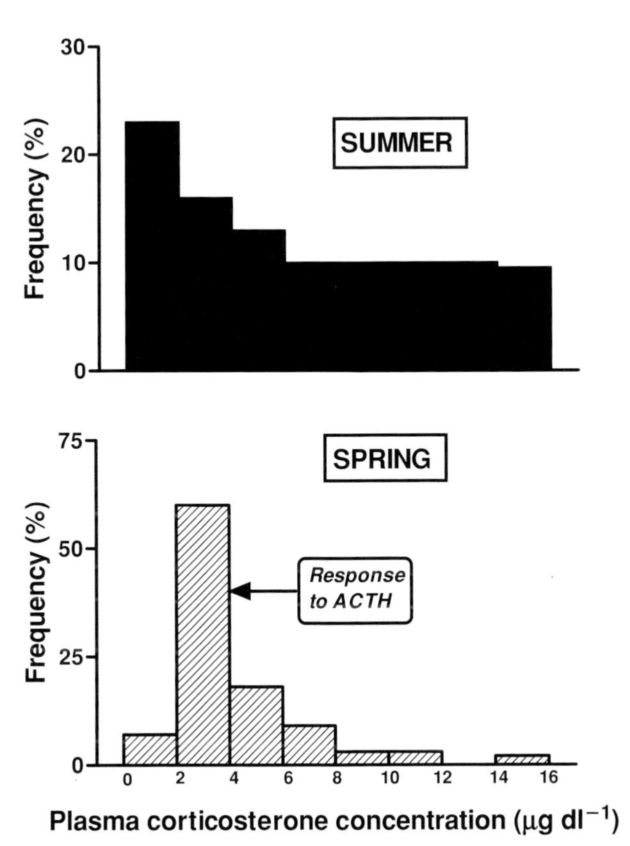

Figure 6.12. Adrenal response to ACTH in desert agamids. Frequency distribution of plasma corticosteroid concentrations in *Ctenophorus nuchalis* lizards injected with ACTH after dexamethasone blockade in spring and in late summer at Shark Bay in Western Australia. Note the 'non-normal' response of the animals in summer. (Modified from Bradshaw (1986).)

an appropriate response to a stressor, e.g. a lack of sufficient supplies of food and water. Changes in FMR and water turnover over the critical summer period are shown in Figure 6.14 and reveal a progressive decline in FMR between November and February, with a further fall in March. Rates of CO_2 production average 0.5 ml $g^{-1}h^{-1}$ in November and are very similar to those of similarly sized iguanid lizards (Nagy, 1982a, 1988b; Nagy *et al.*, 1999). The FMR falls progressively to just over 0.1 mlCO_2 g^{-1} h^{-1} by autumn, however, and it is clear that the animals have virtually ceased all activity. Rates of water influx and efflux are balanced in November, but become negative in February and remain so in March, with overall turnover falling progressively.

Table 6.2. *Seasonal changes in body condition, water content, plasma constituents and circulating corticosteroids in* Ctenophorus nuchalis *at Shark Bay, Western Australia*

Season	Condition index (%)	TBW (ml per 100 g)	Plasma concentrations (mmol l^{-1})			Osmolality (mOsm kg^{-1})	Plasma glucose (mg dl^{-1})	Plasma corticosteroids (μg dl^{-1})	
			Na^+	K^+	Cl^-			Pre-stress	Post-stress
Spring	100.5 ± 1.6	82.1 ± 0.9	170.4 ± 1.4	5.1 ± 0.1	128.6 ± 2.6	328.0 ± 8.2	320.1 ± 9.2	0.4 ± 0.1	8.74 ± 0.84
Summer	85.6 ± 2.1	74.7 ± 1.3	171.7 ± 1.5	5.0 ± 0.1	128.2 ± 2.5	294.2 ± 2.5	241.0 ± 2.5	4.52 ± 0.48	9.72 ± 0.6
Significance	$p < 0.001$	$p < 0.001$	NS	NS	NS	NS	$P < 0.001$	$p < 0.001$	NS

Modified from Bradshaw (1986).

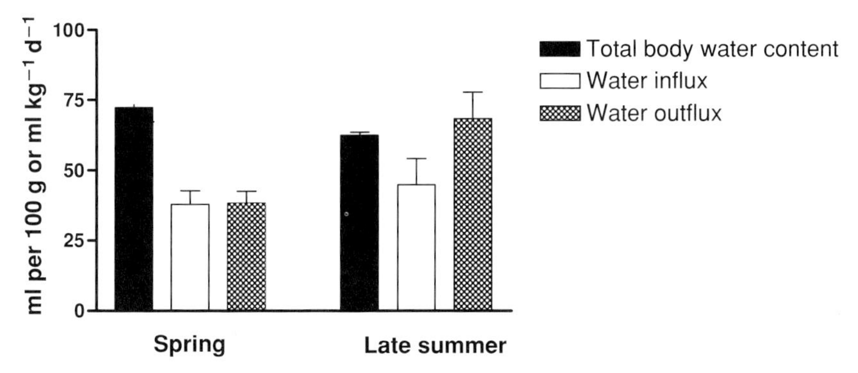

Figure 6.13. A comparison of the total body water content (TBW in ml per 100 g) and rates of water influx and efflux (in ml kg^{-1} d^{-1}) in the lizard *Ctenophorus nuchalis* in spring and in late summer at Shark Bay in Western Australia. (Adapted from Bradshaw (1986).)

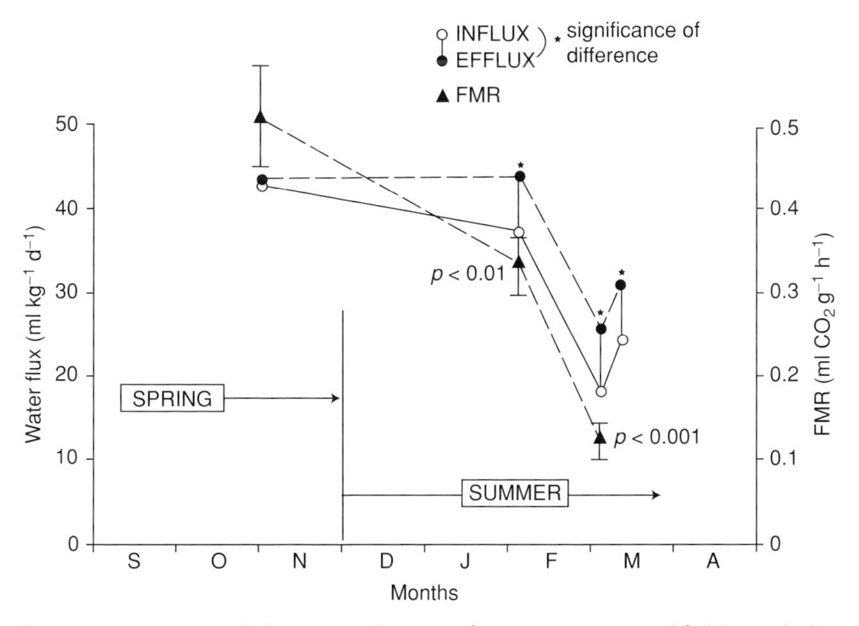

Figure 6.14. Seasonal changes in the rate of water turnover and field metabolic rate of *Ctenophorus nuchalis* at Shark Bay in Western Australia. Note that water efflux significantly exceeds water influx in February and March and that the FMR progressively falls throughout the summer period. (From Bradshaw (1986) with permission.)

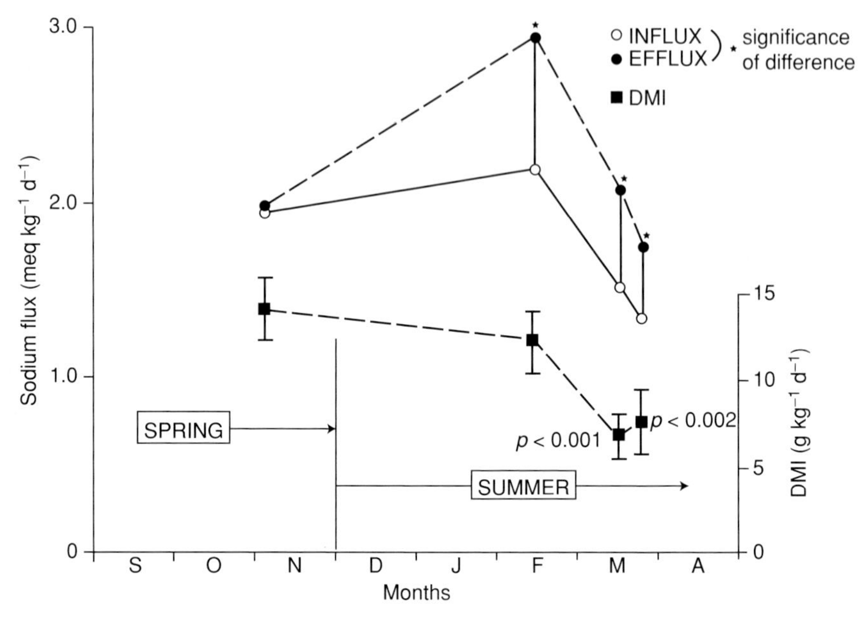

Figure 6.15. Seasonal variations in the rate of sodium turnover (influx and efflux measured with sodium-22) and estimated dry matter intake (DMI) of *Ctenophorus nuchalis* at Shark Bay in Western Australia. Note that sodium efflux exceeds influx in mid- to late summer and that DMI progressively falls over that period. (From Bradshaw (1986) with permission.)

Figure 6.15 shows estimated changes in dry-matter intake (DMI), calculated from turnover rates of both tritium and sodium-22; these follow a pattern similar to that of the FMR and water turnover data. DMI falls from roughly 14 g kg^{-1} d^{-1} in spring to a little over 5 g kg^{-1} d^{-1} in late summer and autumn, and it is clear that the lizards are starving to death. The reasons for this progressive decline in condition and loss of body water are not entirely clear. Rates of corticosterone production in summer are negatively correlated with body condition, suggesting that those individuals in poorest condition are responding appropriately to the lack of protein intake by mobilising body protein reserves. Excessively high environmental temperatures have been suggested by Nagy and Bradshaw (1995) as the proximate cause of the demise of *C. nuchalis* in late summer, with animals being forced to spend long periods of the day in high-temperature avoidance behaviour patterns that preclude feeding. Thus, even though insect food is still abundant, the adult lizards appear unable to avail themselves of this because of the very truncated opportunities for feeding. Paradoxically, however, juveniles that emerge from eggs at this time of the year manage to feed and grow rapidly and have FMRs equal to that of adults in spring (Nagy and Bradshaw, 1995). Ken Nagy, in a very interesting paper, has pointed to the fact that we know little of the energetic costs of growth in neonate reptiles and that this is an area needing more study (Nagy, 2000).

Dunlap and Wingfield (1995) described a somewhat similar situation to that of *Ctenophorus nuchalis* at Shark Bay in the North American lizard *Sceloporus occidentalis*, which they studied in a number of wide-ranging habitats in Oregon, California and Nevada. They found that the adrenocortical response to stress was greatest in populations living at the margins of the species' range, during the hottest and driest seasons, and in individuals with the largest decrements in physiological condition. The lizards similarly showed substantial mass loss, elevated concentrations of osmolytes, and reduced activity in the late summer, a period of low water and food availability.

Bradshaw *et al.* (1991) further explored the fascinating life history of *Ctenophorus nuchalis* by studying a second population, living much further north at Port Hedland in the summer rainfall area of the Pilbara region of Western Australia, and compared it with another sympatric species of the same genus, *Ctenophorus caudicinctus*. This second species only breeds in autumn, following cyclonic rains. A series of field trips over a period of eight years revealed that *C. nuchalis* could be found breeding in spring in some years, but in autumn in other years, at a time when all adults would have been dead had they been living in Shark Bay!

A considerable effort was needed to resolve this paradox. The paper of Bradshaw *et al.* (1991) compares the physiological condition of the two species (body mass and water content, turnover rates of water and sodium, and field metabolic rates) during years when there was appreciable winter rainfall, with those years when only cyclonic rains fell in late summer and autumn. Table 6.3 gives physiological data on *C. nuchalis* in years when they bred in spring at Port Hedland, compared with a year with only 5 mm of winter rainfall when they deferred breeding until late summer. A slight decrease in condition index is the only significant difference, and concentrations of testosterone and gonad size were identical, whatever the time of breeding. The dramatic loss of body condition and progressive dehydration seen in *C. nuchalis* at Shark Bay over the summer period is thus not mirrored further north at Port Hedland in years when the winter rainfall is insufficient to stimulate breeding.

Resource availability thus drives the reproduction of the two closely related species in differing ways. *Ctenophorus nuchalis* is essentially a spring (vernal) breeder and will breed in spring provided that it is preceded by sufficient winter rains to stimulate insect diversity. If good winter rains are then followed

Table 6.3. *Physiological condition of* Ctenophorus nuchalis *on Mallina Station in the arid Pilbara region of Western Australia when breeding in spring or in late summer*

Data given as means ± S.E. (n)

			Turnover				
			Water		Sodium		
Reproductive status	Condition index	Total body water (%)	Influx (ml 100 g^{-1} d^{-1})	Outflux (ml 100 $g^{-1}d^{-1}$)	Influx (mmol kg^{-1} d^{-1})	Outflux (mmol kg^{-1} d^{-1})	FMR ml g^{-1} h^{-1}
Spring breeding	1.19 ± 0.04 (13)	82.2 ± 2.2 (10)	5.74 ± 0.49 (3)	7.38 ± 0.94 (3)	2.13 ± 0.56 (3)	3.15 ± 0.75 (3)	0.251 ± 0.13 (3)
Autumn breeding	1.02 ± 0.03 (30)	82.6 ± 0.9 (22)	8.58 ± 1.13 (8)	9.32 ± 1.29 (8)	2.15 ± 0.56 (8)	2.50 ± 0.85 (8)	0.331 ± 0.03 (5)
Statistical significance	$p = 0.02$	NS	NS	NS	NS	NS	NS

Modified from Bradshaw *et al.* (1992).

by an early cyclone at Port Hedland, this species will continue breeding; in 1988–89 they bred continuously for a period of six months. If the winter rainfall is inadequate, however, the lizards will 'defer' breeding until the summer rains arrive and thus breed at the same time as *C. caudicinctus*; this occurred in three of the eight years of the study. *Ctenophorus caudicinctus* is, on the other hand, an obligate summer (æstival) breeder and we saw no instance of its varying this pattern, which was invariably associated with cyclonic rains. Both species die, however, after breeding and one can only conclude that it is the process of reproduction itself that renders the adults no longer able to maintain homeostatic processes that are essential for life. Although we have searched for clues to this post-reproductive breakdown in regulatory processes (for example, by studying the functioning of the pituitary–adrenal axis (Bradshaw, 1997a)), it still remains unresolved. A study of steroid-binding characteristics in breeding individuals, similar to that described for the small dasyurid marsupials in Chapter 5, could prove very profitable. An interesting recent study by Jennings *et al.* (2000) has found that plasma from different colour morphs of the American lizard *Urosaurus ornatus* displays different corticosterone binding characteristics, which affect the way in which the animals respond to stress.

Tortoises

The large (up to 5 kg) terrestrial tortoise *Gopherus (Xerobates) agassizii* was once widespread in the deserts of southwestern North America but is now restricted to a small number of reserves in southern Nevada, California, Arizona and northwestern Mexico (van Devender *et al.*, 1976; Berry, 1984). Recent population declines have been attributed to increased levels of human disturbance alienating tortoise habitat and altering floristic patterns, plus spreading disease (Brown *et al.*, 1999), but debate currently rages over the species' conservation status, with some claims that the surveys attesting to its rareness are flawed (Brown, 2000). Early attempts were made to model the desert tortoise's habitat by Schamberger and Turner (1986) in an effort to better predict population trends, but daily and seasonal activity patterns, particularly of juveniles, need to be better understood before reliable population estimates can be made (Morafka, 1994; Rautenstrauch *et al.*, 1998; Wilson *et al.*, 1999a,b). The study by Nagy *et al.* (1997) on survival of juvenile desert tortoises found that they remained relatively inert and hibernated in burrows during winter but lost mass rapidly in rainless summers and were at high risk of dying. Christopher *et al.* (1999) have published an exhaustive set of values describing the variations in haematologic and biochemical values to be found in field-caught animals, which should prove invaluable in interpreting future studies, especially if the tortoises are likely to encounter stressful situations on a more frequent basis.

Osmoregulation and energetics of 11 immature individuals in a 9 ha enclosure in Nevada were studied by Nagy and Medica (1986) over the period March 1976 to April 1978, and a population of smaller individuals was studied by Minnich (1977) and Minnich and Ziegler (1976) near Barstow in California. Tortoises hibernate over the winter period and emerge in spring to feed on succulent annuals. These provide excess water and potassium ions that are stored in the urinary bladder, but energy intake, measured with doubly labelled water, was found to be less than required to meet energy expenditure via respiration (Nagy and Medica, 1986). Thus, although the tortoises gained in body mass, this was essentially due to water alone and body solids actually declined in this study over spring. As the water content declined in the foodplants in late spring, the tortoises eventually achieved a positive energy balance while eating grasses, but they were then in negative water balance and their body mass accordingly fell.

Plasma osmolality increased from 293 to 355 mOsm kg^{-1}, with plasma sodium increasing from 136 to 174 mmol l^{-1} by July, and the osmolality of the urine increased to become isosmotic with the plasma by mid-summer. Plasma potassium rose to extraordinary concentrations in these tortoises in mid-July with a mean of 15.5 ± 1.8 mmol l^{-1}, a value that would be lethal for any mammal and one of the highest values recorded for any vertebrate.

As summer progressed the tortoises reduced their activity, spending much of their time in æstivation, but they emerged to drink when there were infrequent thunderstorms and constructed small depressions in which the rain collected (Medica *et al.*, 1980). The effect of drinking was dramatic; both plasma and urine osmolalities fell to early spring values, as seen in Figure 6.16. Urine with high osmotic pressure was first voided by the animals and the bladder then filled with dilute urine, essentially storing water by this mechanism which is later used to dilute dietary salts.

On the basis of their data, Nagy and Medica (1986) speculated that desert tortoises relinquish maintenance of internal homeostasis on a daily basis during most of the year and tolerate large imbalances in their water, energy and salt budgets. By our working definition, these would clearly appear to be stressed animals, although to date concentrations of regulatory hormones such as AVT and corticosterone have yet to be measured. Concentrations of the thyroid hormone, thyroxine (T_4) have been monitored by Kohel *et al.* (2001) on a seasonal basis at the Desert Tortoise Conservation Center in Las Vegas, USA, and found to correlate with feeding rates. Interestingly, T_3 was undetectable and the low T_4 concentrations suggest that *Gopherus agassizii* is one of those

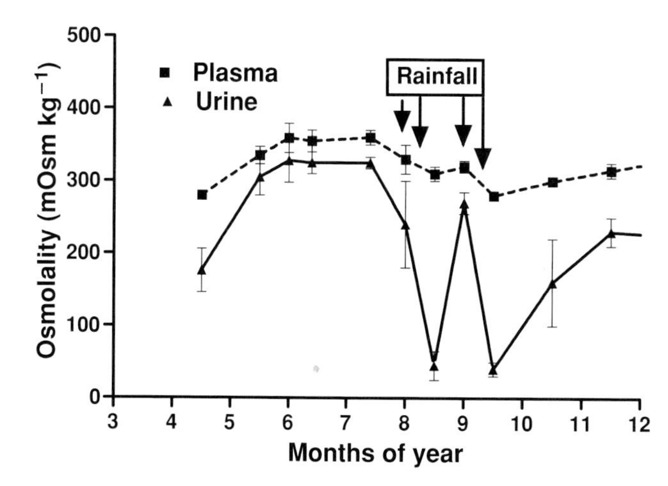

Figure 6.16. Seasonal changes in plasma and urine osmolality of the desert tortoise, *Gopherus agassizii,* in a 9 ha outdoor enclosure in Nevada, showing the marked effect of rainfall and drinking by the animals on these parameters. (Adapted from Nagy and Medica (1986).)

chelonians that does not possess a specific thyroid hormone binding protein (Licht *et al.*, 1991).

Peterson (1995) first coined the term 'anhomeostasis' to describe this tendency of the desert tortoises to osmoregulate opportunistically and tolerate significant deviations of the *milieu intérieur* during late spring and summer. A somewhat similar pattern of regular perturbations of the *milieu intérieur* due to the storage of sodium ions in the extracellular fluid was described many years ago in the agamid lizard *Ctenophorus (Amphibolurus) ornatus* by Bradshaw and Shoemaker (1967) and fuelled the debate over whether such changes are adaptive, or simply result from the inability of the animal to maintain homeostasis (Bradshaw, 1997a). Turner *et al.* (1986) first studied egg production in the desert tortoise, noting that, amazingly – even in drought years – the females were able to reproduce. Brian Henen carried out an exhaustive study of seasonal and annual energy budgets of female *Gopherus agassizii* over a two-year period in a number of sites in the Mojave Desert in southern California (Henen, 1997). He found that desert tortoises rarely achieved a seasonal energy balance and rarely maintained a constant body condition. He also found that FMR varied as much as 10-fold between seasons and 6–10-fold for the same season in different years. Despite this, the tortoises managed to produce eggs in every year of the study and they did this in years with low productivity of winter annual plants by relaxing their control of energy and water homeostasis (see also the study on egg production by Wallis *et al.*, 1999). Typically, the tortoises increased body energy content (both lipid and body protein) before winter hibernation and used this reserve the following spring to produce eggs. Henen *et al.* (1998) studied the effect of climate on field metabolism and water relations of desert tortoises over a 3.5 year period in the Mojave Desert and confirmed the extreme variability of these parameters. Estimates of FMR varied 28-fold and the water economy index (WEI), which is the ratio of water influx to FMR, showed 237-fold differences between seasons, as seen in Figure 6.17.

His conclusion was that desert tortoises either have a highly variable *milieu intérieur* (termed 'heterostasis') or temporarily relinquish homeostasis from time to time ('anhomeostasis') and espouse a 'bet-hedging' strategy in order to survive. Behavioural and physiological plasticity is thus the key to this animal's survival. Another interpretation, however, would be that this is a highly stressed animal that is having extreme difficulty in surviving in its current habitat. How can one distinguish between a strategy (i.e. an evolutionarily stable strategy, or ESS, that has evolved in response to environmental dictates) and a simple unfortunate consequence that reflects, instead, the animal's last chance for survival?

The papers of van Devender *et al.* (1976) and van Devender and Moodie (1977) provide evidence that the desert tortoise was once much more widespread and occupied more mesic habitats in the Pleistocene in North America, and the present populations may thus only be relics that are barely surviving under the more rigorous conditions that prevail today. Our tendency is always to interpret biological situations in adaptationist terms (i.e. any difference that we observe must have arisen through natural selection) but, as Gould and Lewontin (1979) have argued, the excesses of what they called the 'adaptationist programme' have often led to 'Panglossian' conclusions[5], particularly in the case of desert reptiles (Bradshaw, 1988).

Desert tortoises lack a functional nasal salt gland (Dantzler and Schmidt-Nielsen, 1966) and eating their normal diet is obviously osmotically stressful because of the high concentrations of ingested osmolytes (Minnich, 1977). Nagy and Medica (1986) argue that it is the inability of the tortoises to cope with dietary potassium that leads to the reduction and eventual cessation of feeding as drought progresses and they are forced to tolerate extraordinarily high blood potassium concentrations. As mentioned above, this pattern of enforced osmolyte retention in the body fluids, which is ultimately alleviated by rain, is very reminiscent of the case of sodium ions in the agamid lizard

[5] After Dr. Pangloss in Voltaire's *Candide*, who thought that all ends were the best of all possible ends and thus thought that the 'function' of noses was, of course, to support glasses!

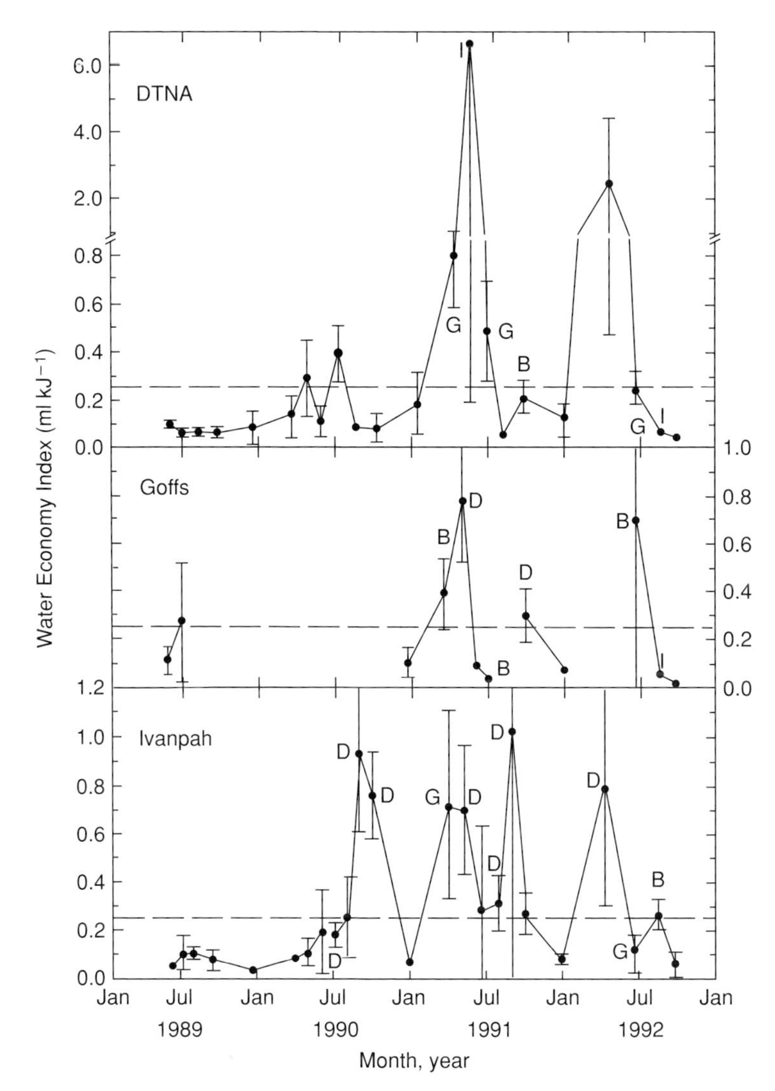

Figure 6.17. Seasonal and spatial variations in the water economy index (WEI) of desert tortoises (*Gopherus agassizii*) at three California sites (Desert Tortoise Research Natural Area (DTNA), Goffs and Ivanpath) in the Mojave Desert over a three-year period. WEI values greater than 0.25 (dashed line) indicate probable drinking of free-standing water. (From Henen *et al.* (1998) with permission.)

Ctenophorus ornatus; this pattern may be more common in reptiles than is appreciated. The universal acceptance of Claude Bernard's concept of the *milieu intérieur* and Walter Cannon's genial paradigm of homeostasis may have clouded our critical faculties when we are confronted with animals that fail to conform to these dictates. It is natural that we would attempt to view them as somehow still attempting to respect these 'laws' , even if it means their temporary abandonment (e.g. 'anhomeostasis', 'heterostasis' and 'allostasis'). Another interpretation, perhaps less charitable, is that these are animals going down an inevitable path to extinction and, by accident as it were, we are able to glimpse the terminal phases of their struggle. Brian Henen, Ken Nagy and Xavier Bonnet have recently commenced a research programme on the relatively abundant desert steppe tortoise of Uzbekistan (*Testudo horsfieldii*) (Bonnet *et al.*, 2001) and future data to compare with *Gopherus agassizii* will be most valuable. Only further research will resolve this problem, hopefully before nature itself closes the story on the North American desert tortoise!

Rodents

Any discussion of the ecophysiology of desert rodents needs to pay tribute to the pioneering work on these intriguing mammals by Knut and Bodil Schmidt-Nielsen in the 1950s and 1960s. They did much to establish the paradigm of a group of uniquely adapted animals capable of surviving in the most inhospitable of arid environments. The kangaroo rats, *Dipodomys spectabilis* and *D. merriami*, became the acme of desert adaptation: small, burrowing, nocturnal granivores with a nasal countercurrent heat exchanger and the most efficient kidneys known for the conservation of water (Schmidt-Nielsen and Schmidt-Nielsen, 1953; Schmidt-Nielsen, 1964).

Like all successful paradigms, however, the excellence of the model obviates the need for further thought on the subject and it is often only when contradictory or counterintuitive data emerge that anyone bothers to question the basic premises of that paradigm. A good example is provided by the story of osmoregulation in teleost fish. Two researchers, the Dane August Krogh, (Bodil Schmidt-Nielsen's father) and the American Homer Smith, were responsible for unravelling the basic osmoregulatory mechanisms that enable fish to survive in fresh and sea water (see Krogh, 1939; Smith, 1932). In fresh water, where they tend to gain water by osmosis, they eliminate this water in the form of copious dilute urine and balance lost ions by active uptake through the gills. In sea water, which is more concentrated than their body fluids, they lose water by osmosis and are forced to drink sea water to maintain their water balance, as they cannot form an hyperosmotic urine. The extra salts that they absorb in this way are then actively excreted, again via the gills.

This model has held sway since the 1930s and is repeated in some form or another in every first-year and high-school textbook of biology. It explains, in a very neat way, all the basic facts known about fish when living in either fresh or saline media and highlights the role of the gills as the primary effector organs involved in osmoregulation. What happens, however, in the case of a fish that moves regularly between fresh and sea water, such as trout or salmon (so called 'euryhaline' teleosts) that cross this osmotic barrier in their reproductive migrations? Following the paradigm, there should be no problem: the gills would pump excess Na^+ and Cl^- out of the fish when in sea water and then switch to pump the ions inwards as the fish commenced its upstream river migration to its traditional spawning grounds. But this would require some form of reversible ion pump that is capable of pumping ions in either direction or, failing that, two pumps, each pumping either in or out across the gill membrane. In fact, biochemical studies have only ever identified a single ion pump (Na^+/K^+-activated ATPase) located on the outer, mucosal surface of the gill membrane. This pump is not bidirectional and only transports sodium ions outwards from the fish (outflux) (Rankin and Davenport, 1981). How is it possible, then, for the fish to apparently pump sodium ions in either direction when needed?

It was only with the advent of studies using radioactive isotopes of sodium (^{22}Na and ^{24}Na) that a solution to this paradox was found. René Motais and Jean Maetz, two French biologists working at Villefranche-sur-Mer, found that turnover rates of sodium were astonishingly high when measured in sea water; as much as 30% of the total sodium pool of the fish was exchanged with the surrounding water per hour (see Motais, 1967; Maetz, 1971). It became very apparent that the small net fluxes of sodium measured in fresh and sea water were thus the resultant of very high simultaneous influxes and outfluxes that could only be observed if ^{23}Na ions were tagged with radioactive isotopes.

Homer Smith's and August Krogh's model is thus a static description or representation of what is, in reality, a highly dynamic phenomenon. The fish in sea water achieves a net loss of sodium ions by ensuring that outflux is just greater than influx. Similarly, in fresh water, the fish depresses the outflux until it is just lower than the influx and thus, effectively, takes up sodium ions from the medium. By controlling only one flux, outflux, the fish is able to either gain or lose ions by altering the balance between inflow and outflow. This outflux is controlled by the pituitary hormone prolactin in fish, and this is the means by which euryhaline teleosts are able to adapt to and survive in fresh water (Hazon and Balemont, 1998).

This example illustrates the power of paradigmatic models to enthral their readers. It is only when one pushes the models to their extremes that their limitations may become apparent. Similarly, in the case of heteromyid rodents, the inherent assumption in studies of their extraordinary capacity to elaborate highly concentrated urine is that this adaptation must have arisen through natural selection in the desert environment where water is in such short supply. One imagines, therefore, that desert rodents must, in the normal course of their existence, face severe water deprivation in summer, which their kidneys enable them to overcome. One would also predict that concentrations of antidiuretic hormone (ADH), the hormone that regulates the concentrating mechanism in the mammalian kidney, would be elevated in the blood of these animals during summer, and particularly during periods of drought.

An early paper reporting rates of water turnover of desert rodents was that of de Rouffignac and Morel (1973), who used tritiated water to measure rates of water influx in two desert rodents, *Gerbillus gerbillus* and *Meriones shawii*, and compared them with mice and white rats in the laboratory. As expected, the two desert rodents had lower rates of water turnover and they reduced these further when deprived of drinking water. The first studies to be undertaken of rates of water turnover in free-living desert rodents in the field were those of Mullen (1970, 1971) on *Perognathus formosus, Dipodomys merriami* and *D. microps.* Mullen used the doubly labelled water technique with deuterium as the hydrogen isotope. Surprisingly, he found that water turnover rates were much higher than would have been predicted from previous laboratory studies, such as that of Richmond *et al.* (1960) using tritiated water. The half-life ($t_{1/2}$) of injected tritium in *D. deserti* was 15.2 d and in the pocket mouse, *Perognathus formosus*, 15.6 d in the laboratory, but Mullen (1970) found that this was reduced to 3.8–6.5 d in the latter species in its natural habitat (i.e. the water turnover rate was much faster). Mullen (1970) concluded from this that '. . . laboratory determinations of these phenomena in desert rodents more often than not are exercises in the determination of the rodent's ability to adapt to extreme environments'. He then went on to speculate that heteromyid rodents in general are probably never stressed to the same extent in their natural environment (where they are nocturnal and have the protection of a humid burrow during the day) as in the various laboratory studies on which the paradigm is based.

Bradshaw *et al.* (1976) used tritiated water to measure turnover rates of two desert rodents, *Meriones shawii* and *M. libycus*, in the Tunisian Sahara in mid-summer with daytime shade temperatures reaching 48 °C. Measured rates of water turnover under these conditions were extraordinarily high for animals that might be thought to be short of water, averaging close to 24% of the total body water pool exchanged per day in *M. shawii* and 13% in *M. libycus.* Burrow temperatures during the day were low and the humidity exceeded 95% at all times. The water content of seeds stored in the meriones' burrows was not measured but, as Nagy (1987a) points out, these seeds are often hygroscopic and may reach water contents as high as 30%. Morton and MacMillen (1982) have also pointed to the importance of seeds as potent sources of metabolic water in desert rodents.

Lachiver *et al.* (1978) extended this field study on the two *Meriones* species in southern Tunisia, and Ben Chaouacha-Chekir *et al.* (1983) added a third species, the fat sand rat, *Psamommys obesus*, which they studied in the coastal region of Mahrès and in the desert, south of the village of Tatahouine. Rates of water turnover were even higher in *Psammomys* than in *Meriones*, ranging from 41.6 to 47.4% of the total body water pool per day, with *Meriones shawii* and *M. libycus* again averaging 20.5% and 12.9% per day, respectively.

Petter *et al.* (1984) reviewed these studies on Tunisian rodents and interpreted the turnover data from *Psammomys obesus* in the context of the unusual diet of this species. The fat sand rat is not a granivore and feeds almost exclusively on halophytic (salt-rich) plants belonging to the family Chenopodiaceae (e.g. *Arthrophytum scoparium, A. schmittianum, Salsola foetida, Atriplex halimus* and *Traganum nudatum*), which store large amounts of salt in their leaves. Petter *et al.* (1984) calculated that the *Psammomys* eat approximately 80% of their body mass in plant matter per day, consuming in the process approximately 1.2 g of sodium chloride and 2 g of oxalic acid. This very high salt load is eliminated through the production of a copious (1 ml h^{-1}) and very concentrated urine with a sodium concentration of almost 2000 mmol l^{-1} and an overall osmotic pressure of 6500 mOsm kg^{-1}.

Very similar results were obtained by Degen *et al.* (1990, 1991) working with fat sand rats in Israel. Degen *et al.* (1988) and Kam and Degen (1988) studied in detail the surprising ability of fat sand rats to subsist on a diet of saltbush (*Atriplex halimus*), which has a very high salt content and low nutritive value. They found that fat sand rats routinely scraped off the outer layers of the leaves of the plant with their teeth, thereby removing 14–19% of the ash content of the leaves and substantially reducing their electrolyte intake. Although *Atriplex halimus* represents a poor-quality food source, fat sand rats are able to survive on this diet for a number of reasons: it is available all year round and thus provides a more stable diet than seeds; sand rats have no competitors for this salt-rich food; the burrows of fat sand rats are at the base of saltbush plants and they thus expend minimal energy in foraging for their food. *Psammomys obesus* is in many ways a most unusual animal with an unique diet that would be unpalatable to most other species, and it first came to notice owing to its tendency to develop diabetes mellitus when maintained on standard rodent laboratory diets (Hackel *et al.*, 1965; Haines *et al.*, 1965).

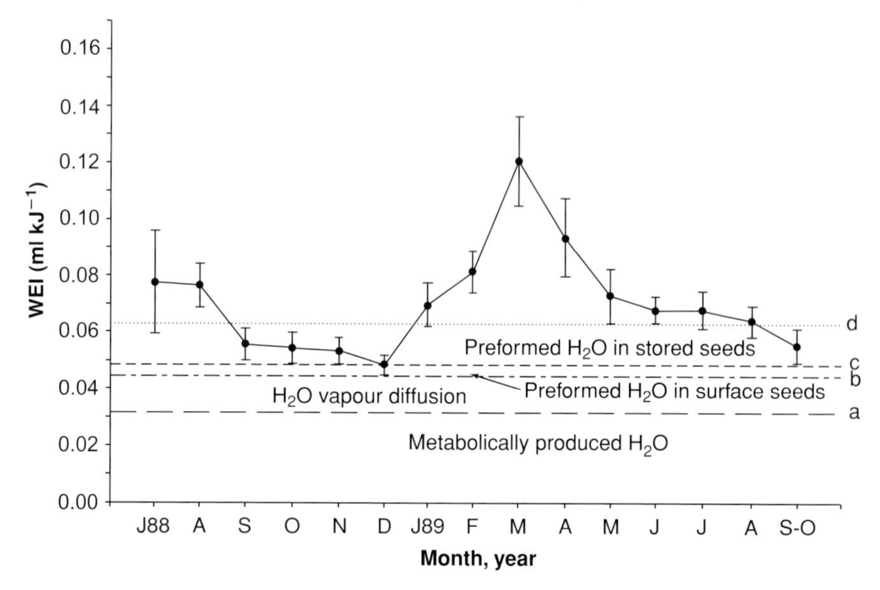

Figure 6.18. Variations in the water economy index (WEI) of Merriam's kangaroo rats (*Dipodomys merriami*) over a 15 month period in the Mojave Desert. The successive horizontal lines represent (a) metabolic water production from a diet of dry seeds, (b) the added effect of water vapour exchange expected given the high humidity in burrows, (c) extra effect due to preformed water in dry seeds on the surface, (d) added effect due to preformed water in seeds stored in humid burrows. (From Nagy and Gruchacz (1994) with permission.)

Much of Allan Degen's exceptional work on the ecophysiology of rodents inhabiting the Israeli deserts is summarised in his recent monograph (Degen, 1997) to which reference should be made by those interested in greater detail.

Nagy and Gruchacz (1994) carried out an extensive field study of Merriam's kangaroo rat, *Dipodomys merriami*, in the Mojave Desert and found that it subsisted essentially on seeds, favouring the creosote bush (*Larrea*), although some green vegetation was consumed during the breeding season. The water economy index (WEI) (see p. 124) was above the theoretical minimum for a seed-eating rodent of 0.047 ml kJ^{-1}, and during spring they were gaining more than twice as much water as expected for a diet of air-dried seeds, as seen in Figure 6.18. Seed storage in burrows, with hygroscopic absorption of water, was thought to be the factor responsible for this extra water input (Nagy, 1994a). Similar results were obtained by Degen *et al.* (1997) studying rodents in the Negev Desert highlands in Israel, where *Gerbillus henleyi* was found to be turning over 29% of its body water pool per day in winter and 16% in summer. Comparable figures were obtained by Degen *et al.* (1992)

working with *Gerbillus allenbyi* and *G. pyramidum*, in which rates of water influx fell in summer, but still remained between 14 and 16.5% of the body water pool per day. Degen *et al.* (1986) had also shown when working with three 30–40 g desert rodents in the Judean desert (*Acomys russatus*, *A. cahirinus* and *Sekeetamys calurus*) that their rates of water influx in summer ranged from 18.7 to 22.2% of the total body water pool per day (see also Degen, 1994). Downs and Perrin (1990) measured field water turnover rates with tritium in three *Gerbillurus* species in the Namib Desert and, again, these ranged from 12 to 35% of the body water pool per day. These figures only need to be compared with turnover rates for larger desert marsupials, such as the spectacled hare wallaby, with daily turnover rates of only 3.5% of the body water pool (the lowest of any mammal) (Bakker and Bradshaw, 1989) and 5.5% for the Barrow Island euro (Bradshaw, 1997b), as seen in Table 5.2 and Nagy and Bradshaw (2000), for one to conclude that desert rodents in their natural habitat show little sign of water deprivation.

Another North African desert rodent that has been studied to some extent is the gundi, *Ctenodactylus vali*, which is a species that does not burrow, inhabiting instead rocky scree slopes in Tunisia and Morocco. It is most unusual in being diurnal rather than nocturnal (Grenot, 1973). Thus, at first sight, this would appear to be an animal exposed to the full onslaught of the arid situation, and one that would need to exercise a high level of water economy. Paradoxically, however, urine samples collected from animals in the field were actually hypo-osmotic (S. D. Bradshaw & M. Séguignes, unpublished), that is, less concentrated than the body fluids, a condition that usually only occurs in mammals under conditions of maximal water loading and diuresis. In the laboratory, de Rouffignac *et al.* (1981) found that captive animals showed only a slight ability to concentrate their urine, with a maximum value of 1600 mOsm kg^{-1}being recorded, which is lower than that of the white rat, which can reach 2900 mOsm kg^{-1}. A study of the anatomy of the kidney of this fascinating species reveals that, although the pyramid and papilla are well developed in the gundi, and the medulla has long-looped nephroi, the structure of the outer medulla is much simpler than that of species, such as *Psammomys obesus*, which have a high concentrating capacity.

Figure 6.19 compares the medullary microstructure of the gundi with that of *Meriones shawii*. The structure of the outer medulla (OM) of the gundi is much simpler than that of the merione, lacking any separation into outer and inner stripe (OS and IS), which is found in all other desert rodents. de Rouffignac *et al.* (1981) note that the characteristic fusion of vascular bundles in the IS to form secondary and giant bundles, seen in the merione and in *Psammomys obesus*, is absent in the kidney of the gundi and conclude that

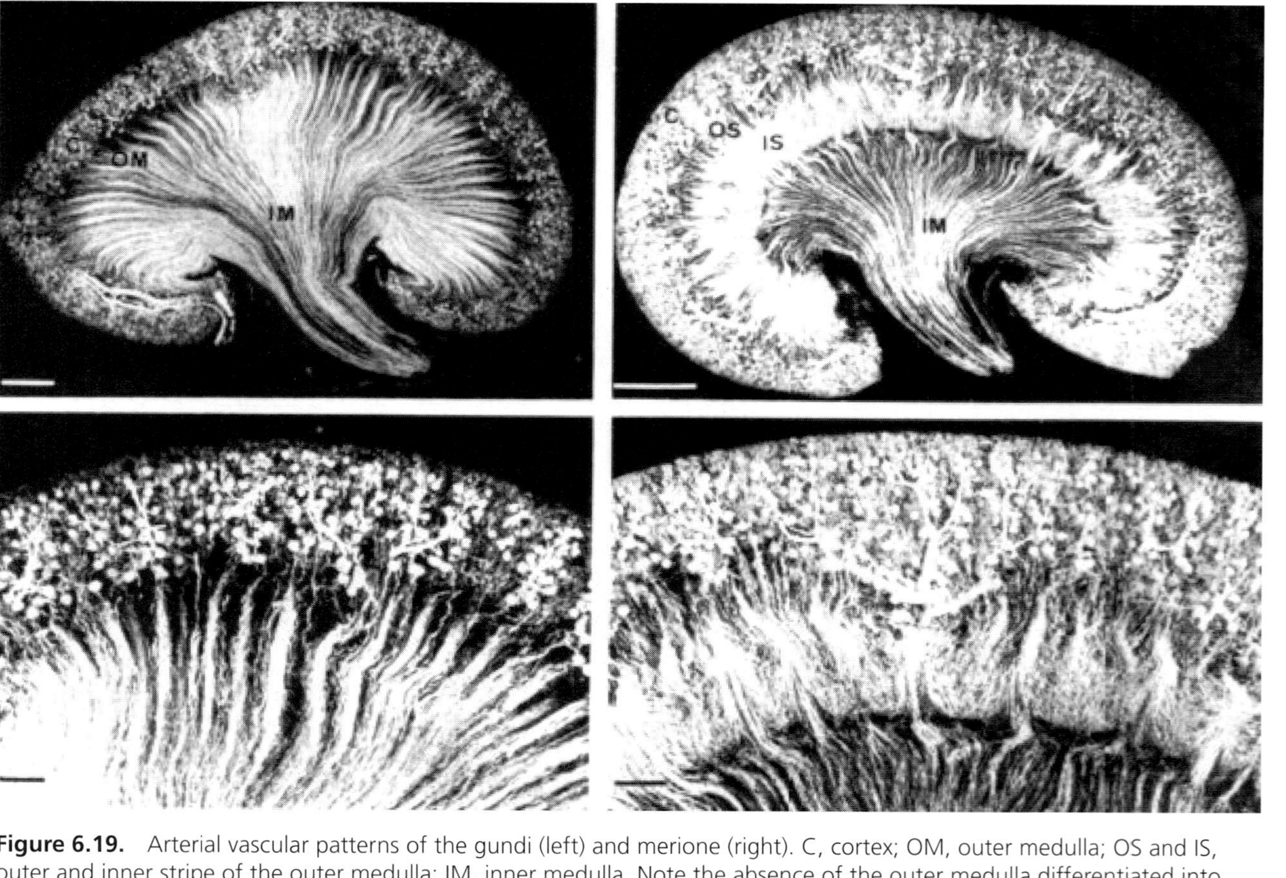

Figure 6.19. Arterial vascular patterns of the gundi (left) and merione (right). C, cortex; OM, outer medulla; OS and IS, outer and inner stripe of the outer medulla; IM, inner medulla. Note the absence of the outer medulla differentiated into

this must be the explanation for their limited concentrating capacity. This, in some ways, is a cautionary tale as it demonstrates that a mammal with a well-developed medullary segment and very long loops of Henle may still not be an effective concentrator; the correlation between relative medullary area and maximum urinary concentration is extremely poor in the gundi (Beuchat, 1990).

The other expectation from laboratory studies, that desert rodents would have high circulating concentrations of antidiuretic hormone (ADH) in the plasma in mid-summer, has yet to be substantiated. Some workers assume, somewhat uncritically, that this is indeed the case (see, for example, Sicard and Fuminier, 1996; Degen, 1997; Lacas *et al.*, 2000) but the few published data are worth careful examination. Plasma concentrations of arginine vasopressin (AVP) average 3–6 pg ml^{-1} in hydrated laboratory Wistar rats (Windle *et al.*, 1993) and approximately 10 pg ml^{-1} in hydrated *Meriones shawi* (Sellami *et al.*, 2002). Baddouri *et al.* (1981, 1984) reported extraordinarily high concentrations of AVP of 479 ± 59 pg ml^{-1} in the jerboas *Jaculus orientalis* and *J. deserti*, maintained on a dry diet in the laboratory, and this only fell to 130 ± 30 pg ml^{-1} when water-loaded by gavage (see also Baddouri *et al.*, 1987). These very high figures have yet to be confirmed. Stallone and Braun (1988), by contrast, using a specific radioimmunoassay for AVP, reported much lower circulating AVP concentrations in the kangaroo rat *Dipodomys spectabilis*, which averaged 6.0 ± 0.7 pg ml^{-1} when fed a normal dry grain diet and increased progressively with dehydration to reach a maximum of 68.8 ± 4.4 pg ml^{-1} after 192 h of dehydration, during which the rodents lost approximately 20% of their initial body mass.

Sellami *et al.* (2002) compared changes in circulating AVP concentrations in *Meriones shawi* and white rats with prolonged water deprivation and found that the levels were similar in both species, rising to a mean of 18–20 pg ml^{-1} after 30 d. Stallone and Braun (1988) compared the rate of increase of measured AVP concentrations of a number of species of mammals in the literature in relation to the concomitant increase in plasma osmotic pressure. This is a measure of the 'sensitivity' of the AVP response that was first suggested by Robertson *et al.* (1973) and the comparison of pg ml^{-1} of AVP per mOsm kg^{-1} has the kangaroo rat at 2.99, well above the laboratory rat at 1.41 and the dog at 0.24 (Gray and Simon, 1983). Stallone and Braun (1988) concluded from this that *Dipodomys* possesses a much more sensitive AVP response to changes in plasma osmotic pressure than these other non-desert mammals – the only flaw in this argument being that the domestic cat registers a figure of 4.04, and its kidney has nothing like the concentrating abilities of that of the kangaroo rat!

Hewitt (1981) found that the Australian desert rodent *Notomys alexis* was extraordinary in that, during a period of total water deprivation, there was no increase in either plasma osmolality or haematocrit, and plasma sodium concentrations actually decreased by as much as 30 mmol l^{-1}. This is in contrast to what is seen in gerbils and kangaroo rats, where both the haematocrit and plasma osmolality increase significantly (Wright and Harding, 1980). Weaver, *et al.* (1994) tried to assess the role, if any, of the kidney enzyme renin and AVP in enabling *Notomys alexis* to cope so effectively with total water deprivation over a period of some 30 d, and concluded that neither was involved in controlling this adaptation. The interesting thing about these results is that the desert rodents did not become dehydrated, despite their 30 d without access to water, and their plasma osmotic pressure remained unchanged. This means, of course, that since the stimulus for the secretion of AVP in mammals is an increase in the osmotic pressure of the plasma, there would be no signal to the pituitary gland to release the hormone into the blood and we would not expect plasma AVP concentrations to increase. So far as I am aware, no-one has yet measured AVP concentrations in field-caught desert rodents, although we now have data from desert wallabies (see later), and the question is thus still unanswered.

The effects of thyroxine on metabolic rate and water balance have recently been investigated in *Dipodomys merriami* by Banta and Holcombe (2002). Although implants of T_4 had the expected effect of increasing metabolic rate, they had no impact on overall rates of water loss. Water loss in this species was also found to be profoundly influenced by acclimation to either wet or dry conditions during development in the laboratory (Tracy and Walsberb, 2001), hence the need for measurements of physiological parameters under actual field, rather than laboratory, conditions.

If subsequent work on desert rodents confirms what we know to date – that their rates of water turnover in mid-summer are high – and also what we suspect, that circulating concentrations of AVP are not unusually elevated, we need to account for the fact that, with the exception of the gundi, all species excrete highly concentrated urine in the field. The case of the fat sand rat, *Psammomys obesus*, is relatively straightforward. By feeding exclusively on halophytic plants, this species has an enormous daily intake of sodium chloride, which it must excrete in very concentrated urine if it is not to lose large amounts of body water and dehydrate. This species' kidney would thus appear to have evolved its spectacular concentrating capacity in order to cope with the need to excrete salt, rather than because of any lack of available water in the animal's environment.

What of the other species, however, such as the North African meriones and the North American kangaroo rats, which are essentially granivores? A close examination of the chemical composition of the medullary osmotic gradient used by these species to concentrate their urine shows that urea is the major osmolyte, followed by sodium chloride (Schmidt-Nielsen *et al.*, 1961; Schmidt-Nielsen and Robinson, 1970). In *Psammomys*, on the other hand, the gradient is composed almost entirely of sodium chloride and urea contributes little to the osmotic gradient (de Rouffignac and Morel, 1969; Imbert and de Rouffignac, 1976; Jamison *et al.*, 1979).

These differences accord with the differing diets of these rodents: the granivores consume a diet that is rich in carbohydrates and protein but low in salt, whereas the diet of *Psammomys* is very rich in salt but with a much lower nitrogen (protein) content (Kam and Degen, 1988). Proteins need to be deaminated, and nitrogenous wastes are eliminated in the form of urea through the kidneys. Water is 'osmotically obligated' to this excretory process but, the higher the concentrating capacity of the kidney, the less water is expended. One may postulate, therefore, that seed-eating desert rodents, with a particularly high intake of protein, have a need to eliminate large quantities of urea and their kidneys have evolved structurally to meet this need. The extraordinary concentrating systems that we see may thus have developed as a means of excreting high urea loads *without excessive loss of water*, rather than primarily as water-conserving devices.

This hypothesis would certainly accord with the fact that, in their natural environment, the desert rodents that have been studied do not appear in any way to be deprived of adequate supplies of water. This is because their small body size results in the production of large amounts of metabolic water, and their habit of storing seeds in humid burrows also results in an extra water gain. Coupled with their nocturnal habit and nasal countercurrent heat exchangers, their losses of water are quite small and readily met from their diet. In the absence of the ability to excrete urea at very high concentrations, however, their rates of urinary water loss would be substantially higher and survival in such habitats would be impossible. Natural selection may thus have favoured the development of complex renal countercurrent osmotic multipliers in all of the seed-eating rodents with high protein intakes and the case of *Psammomys* may be one of parallel evolution, or convergence, where the problem is essentially the same, i.e. the excretion of large amounts of an osmolyte, but where the osmolyte is sodium chloride rather than urea.

It is also interesting to speculate that other habitual seed-eaters living in the desert, such as many of the Australian parrots, would appear to have solved this same problem by excreting nitrogenous waste products in the form of insoluble uric acid, and thus have no need of an exceptional concentrating kidney.

Jackrabbits

The jackrabbits are large hares occurring widely in a range of habitats in North America, including the desert regions of California, Arizona and Nevada. An early paper by Schmidt-Nielsen *et al.* (1965) on two desert-dwelling species, *Lepus californicus* and *L. alleni*, brought these strange animals to the notice of the scientific community. Jackrabbits are not small like desert rodents; they weigh 2–5 kg. They do not dig deep burrows and they are apparently diurnal, being visible during much of the day. In addition, their kidneys possess only a moderate capacity to concentrate the urine. Their ability to maintain thermal and water balance in the challenging habitats of the North American deserts during summer was thus of considerable interest and led to a number of important ecophysiological studies in the late 1970s (Costa *et al.*, 1976; Nagy *et al.*, 1976; Shoemaker *et al.*, 1976).

The early work of Schmidt-Nielsen *et al.* (1965) and Dawson and Schmidt-Nielsen (1966) focused on the ability of the jackrabbit to thermoregulate in its challenging environment and revealed its rather limited capacity to endure environmental temperatures much above 40 °C. Rates of evaporative water loss increase exponentially above ambient temperatures of 35 °C but evaporative cooling does not exceed metabolic heat production, even at ambient temperatures of 45 °C. These authors advanced the novel idea that the very large and prominent ears of the jackrabbits may function as radiant heat exchangers and assist the animal to lose heat during the hottest part of the day. They speculated that jackrabbits that are often seen sitting in small shaded depressions (called 'forms') beneath bushes during the middle of the day, could orient their large ears towards the cold north sky and then radiate significant amounts of heat to that sink. Radiation exchange between two bodies occurs in accordance with the Stefan–Boltzmann equation:

$$R = 4.92 \times 10^{-8} e T^4 \text{ kcal m}^{-2}\text{h}^{-1},$$

where R = radiation flux, e = emissivity and T = absolute temperature.

The absolute temperature of the north sky in southern Arizona at 4 pm was measured at 16 °C and thus a jackrabbit with a body temperature of 40 °C would automatically lose body heat to the sky if shaded from direct sunlight. The calculations were approximate, but they estimated that as much as one third of the animal's BMR could be dissipated in this way from the large ears functioning as radiant emitters (Schmidt-Nielsen, 1964). What is most interesting is the fact that the desert-dwelling species of the genus

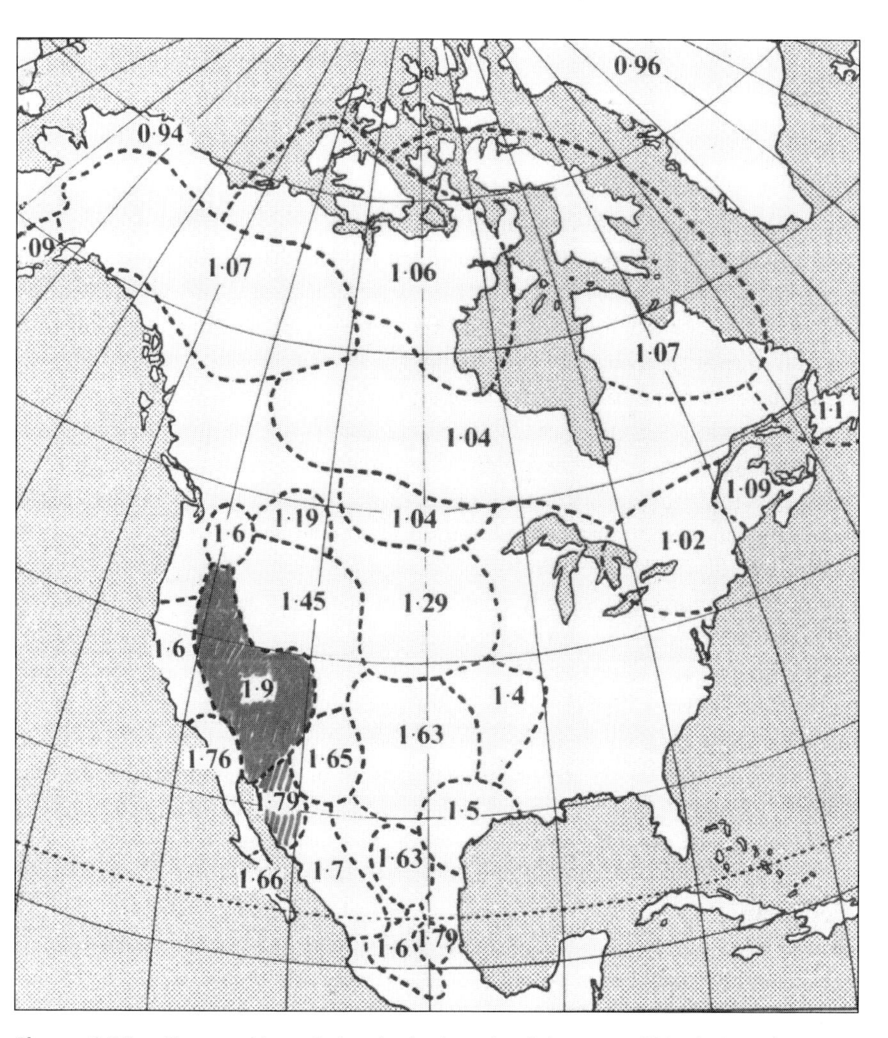

Figure 6.20. Geographic variation in the length of the ears of North American hares (genus *Lepus*) in relation to overall skull length. Note that the ears of jackrabbits inhabiting the Great Basin desert region are relatively the largest on the continent. (Adapted from Hesse (1928).)

Lepus in North America do have relatively much larger ears than species that are restricted to the cooler parts of the continent, as seen in Figure 6.20. Although this contention has never been tested experimentally, the jackrabbit remains one of the few animals studied to date in which radiant heat exchange is thought to play an important role in its daily thermoregulation, the camel being another.

No-one has yet carried out an ecophysiological study on free-ranging jackrabbits, as they are difficult to catch and fragile to handle in captivity. Nagy *et al.* (1976) and Shoemaker *et al.* (1976), however, carried out a detailed seasonal study of thermoregulation and water and energy utilisation with a small number of jackrabbits (*Lepus californicus*) maintained in 0.4 ha enclosures containing natural desert vegetation. Doubly labelled water was used to assess both rates of water turnover and field metabolic rates (FMR) throughout a full year. The jackrabbits maintained mass and were in positive water balance when eating green vegetation in spring but their condition deteriorated progressively as the vegetation dried out. Despite the fact that jackrabbits are capable of producing unusually dry faeces, with a water content of only 38% (Nagy *et al.*, 1976), their daily water requirements are high at 120 ml kg^{-1} d^{-1} and their plant diet must contain no less than 66% by mass of water if they are to remain in balance. Diet utilisation studies showed that the animals were unable to maintain nitrogen balance on the winter diet available to them, which consisted of dried leaves, bark and woody stems. Figure 6.21 shows graphically the nutritional plight of jackrabbits, which deteriorates markedly in summer and autumn, to reach its nadir in winter.

Few jackrabbits survive the winter, and mass mortalities are commonly reported at this time (Hayden, 1966). Some survive, however, in moist habitats (oases) and the high reproductive potential of the jackrabbit ensures that these reinvade the drier desert areas with the coming of the spring rains. To quote Nagy (1988a)

> jackrabbits do not persist in deserts by relying primarily on physiological, morphological or behavioral adaptations which facilitate maintenance of water and energy balance, but instead by means of opportunistic exploitation of oases, coupled with a great reproductive output when conditions are favourable.

Island wallabies

Desert islands often figure prominently in people's imaginations, and the popular literature, as places of calm and refuge from the strife and tension of modern life, where one can 'get away from it all' and live a simpler existence. Such islands do exist, but it is a moot point whether the animals that are forced to inhabit them enjoy such an idyllic lifestyle. Those islands that lie off the northwest coast of Australia, for example, are extremely arid and vegetated by a series of plants and grasses of very poor nutritional value. Mammals now found on these islands were isolated there many thousands of years ago by the eustatic rise in sea level that occurred at the close of the Pleistocene era (Main, 1961) and have thus been subjected to selective pressures that may

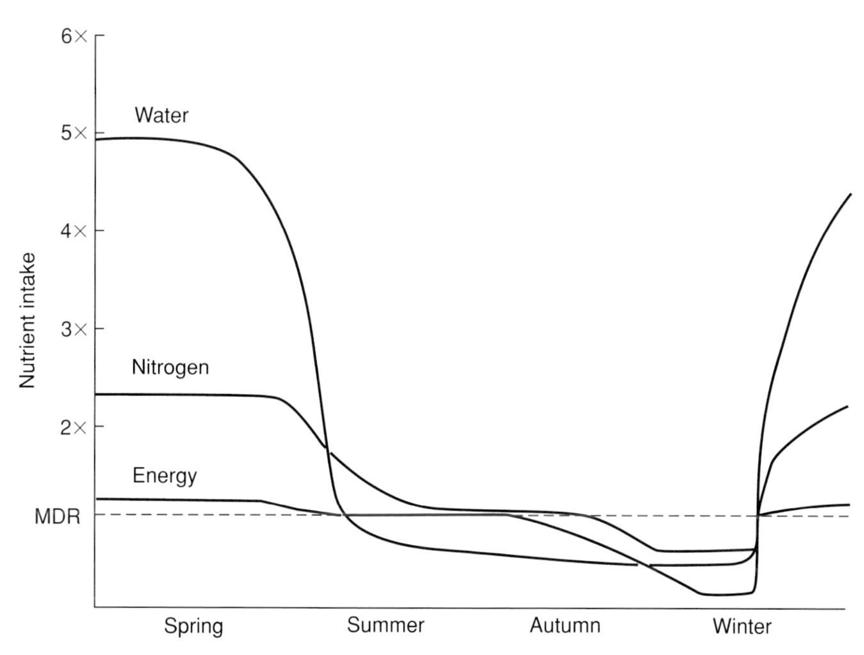

Figure 6.21. Schematic showing the relation between daily nutrient intake and minimum daily requirements (dotted line) for jackrabbits eating natural diets in the field. As the summer drought progresses, the diet firstly ceases to provide adequate amounts of water; then, in autumn, digestible energy and nitrogen fall below minimum requirements before the arrival of the winter rains. (From Nagy (1987) with permission.)

differ to some extent from those operating on the mainland continent (Main and Bakker, 1981). An absence of competitors and predators is a feature of some of these islands (e.g. Rottnest, Garden and Carnac Islands off the coast of Perth) but this is balanced in most cases by low resource availability.

One of the most difficult problems faced by island animals is the maintenance of adequate genetic variability, which is often imperilled by two processes, founder effect and genetic drift, both of which may lead to inbreeding depression (Amos *et al.*, 2001). The genetic diversity normal to an out-breeding population of a species may be impoverished if only a very small number of individuals is isolated by chance from the main population and then forced to subsist alone: this is the 'founder effect'. Equally, small populations forcibly have high levels of inbreeding and alleles may either be lost by chance, or become fixed through random events rather than by natural selection, leading to 'genetic drift' and a loss in fitness and the overall adaptedness of the population. Loss of fitness through inbreeding (inbreeding depression) has been documented in many species and is particularly likely to occur in small, insular populations of animals (Amos *et al.*, 2001; Seymor *et al.*, 2001). A good case in point is the small population of the black-footed rock wallaby, *Petrogale lateralis*, found on Barrow Island off the Pilbara coast of Western Australia, which has been found to have the lowest level of genetic variation of any mammal yet studied in the world (Eldridge *et al.*, 1999). Inbreeding depression and stressors experienced during the period of growth and development of an animal can also result in morphological asymmetries in the adult. As mentioned in Chapter 3, the study of 'fluctuating asymmetry' is currently a popular topic, especially in threatened species (Palmer and Stobexk, 1986; Parsons, 1990; Sarre and Dearn, 1991; Gilligan *et al.*, 2000).

A fascinating example of either founder effect or genetic drift appears to have occurred in the dwarf form of the euro kangaroo, *Macropus robustus isabellinus*, which is restricted to Barrow Island in Western Australia (see Figure 6.22). Studies carried out by myself and colleagues in the 1990s as part of a major ecophysiological study of the vertebrate fauna of this island revealed that the euros were seriously anaemic, with a blood haematocrit of approximately 30% instead of 40–45%, which is the norm for mammals. At first we thought this might be the result of heavy tick infestations – ticks being blood-feeding parasites – but we could find no correlation between the number and size of ticks on an animal and its blood haematocrit and haemoglobin content. Blood parasites were also eliminated as a possible factor, as was potential pollution from the oil field operating on the island. Table 6.4 compares haematological parameters of island compared with mainland populations of the euro and shows significant reductions in red cell (erythrocyte) numbers and haemoglobin content, which must seriously compromise the respiratory capacities of the Barrow Island individuals. Billiards *et al.* (1999) reported significant alterations in the activity of a number of enzymes in the erythrocytes of the Barrow Island euro, which would appear to result from genetic mutation. At this stage, the best interpretation that we have of this intriguing phenomenon is that, either by founder effect or by genetic drift, a potentially deleterious mutation has occurred that affects the production and haemoglobin content of erythrocytes in the Barrow Island individuals. On the mainland, such a mutation – were it to occur – would be rapidly eliminated, as the individuals carrying such a trait would be seriously disadvantaged in terms of overall vigour and respiratory performance compared with their conspecifics. On the island, however, in the absence of any large predators such as foxes and dingoes, such a drop in respiratory performance is perhaps of little consequence and the euros can carry on leading their normal, rather sedentary, existence. The anaemia affects juvenile euros as well as adults on the island (J. M. King, pers. comm.)

Figure 6.22. The Barrow Island euro, *Macropus robustus isabellinus*, a dwarf form of the mainland euro, photographed in the shade of a giant termite mound. (Photo courtesy of WAPET Pty Ltd.)

Table 6.4. *Haematology of Barrow Island and mainland kangaroos, showing anaemia of the insular population*

Asterisks indicate differences statistically significant at $p < 0.05$.

Blood parameter	Barrow Island *M. robustus isabellinus* ($n = 48$)	Mainland *M. robustus erubescens* ($n = 7$)
Hb (g l⁻)	*97.02 ± 2.07	118.57 ± 4.20
PCV ($l\ l^{-1}$)	*0.31 ± 0.01	0.36 ± 0.01
RBC ($\times 10^{12}\ l^{-1}$)	*4.22 ± 0.08	4.82 ± 0.17
MCV (fl)	73.53 ± 0.76	75.67 ± 0.48
MCH (pg)	23.12 ± 0.23	24.61 ± 0.16
MCHC ($g\ l^{-1}$)	*314.46 ± 1.87	325.33 ± 2.67
WBC ($\times 10^{9}\ l^{-1}$)	5.66 ± 0.22	4.87 ± 0.64
TSProt ($g\ l^{-1}$)	55.74 ± 0.57	58.29 ± 1.30

From J. M. King and S. D. Bradshaw (unpublished).

and raises the important question of whether island populations of threatened species can always be considered as a pristine resource for reintroduction programmes following extinction processes (Short *et al.*, 1992; Caughley, 1994).

Studying the ecophysiology of insular populations thus offers a number of interesting problems. What are the particular constraints posed in the island habitat that differ from those in the species' main area of distribution? How has the isolated population managed to contend with these and has the isolation been long enough for genetic adaptation and divergence to occur? Has the normal genetic diversity of the insular population been compromised by the processes mentioned above, and what effect has this had on fitness?

A recent study by Bradshaw *et al.* (2001) compared the water and electrolyte homeostasis and kidney function of two species of desert wallaby (see Figure 6.23), the spectacled hare wallaby (*Lagorchestes conspicillatus*), living on Barrow Island, and Rothschild's rock wallaby (*Petrogale rothschildi*) on Enderby Island off the arid Pilbara coast in Western Australia. Rock wallabies were captured at night in Bromilow traps (see Chapter 3), baited with apple, whereas hare wallabies were captured with handnets. They were then taken to a field laboratory on either island for immediate processing. A blood sample was taken as soon after capture as possible for the measurement of circulating hormone concentrations, the antidiuretic hormone, lysine vasopressin (LVP), being the hormone of primary interest in this study. Measurements of kidney function followed the procedures that have already been described in detail

Figure 6.23. Photographs of the spectacled hare wallaby, *Lagorchestes conspicillatus* (above), and the black-footed rock wallaby, *Petrogale lateralis* (below), on Barrow Island, which lies 80 km off the arid Pilbara coast of Western Australia.

in Chapter 3, and voided urine was collected from the wallabies over a 12 h period in metabolism cages before they were released.

Data were collected on three field trips to Barrow Island and four to Enderby Island over the period 1986–1992 and it was fortunate that one of these, in November 1990, coincided with the driest year ever recorded in the area with a total of only 122 mm of rain falling on Barrow Island. Changes in body condition, along with rates of turnover of water and sodium and electrolyte measures of homeostasis in both hare wallabies and rock wallabies, are summarised in Tables 6.5 and 6.6, with the data grouped into wet and dry seasons.

Hare wallabies lost body mass in the dry season (November–December) but their total body water content did not change significantly. Rates of water turnover in the dry season were, at 27.5 ml $kg^{-0.82}$ d^{-1}, the lowest recorded for any marsupial and equal the lowest measurement from any desert mammal in the world, and they are also less than the lowest rates ever measured in the desert kangaroo rat *Dipodomys merriami* (Nagy and Gruchacz, 1994). In the wet season (March–April) rates of influx increased enormously and the sodium influx also fell as the plants became hydrated and thus had lower concentrations of sodium and other electrolytes. Most significantly, however, there was no change in the concentration of plasma electrolytes (sodium and potassium) between the seasons, nor any change in the osmotic pressure of the plasma. This means that the hare wallabies were able to withstand the driest period yet recorded on Barrow Island without any significant perturbation of their *milieu intérieur*.

The situation is quite different, however, with the rock wallabies that shelter in caves and rockpiles on Enderby Island. Their total body water content fell significantly in the dry season owing to dehydration, and their water influx, at 64.8 ml $kg^{-0.82}$ d^{-1}, was over double that of the hare wallabies. Plasma electrolyte concentrations were also not regulated well between the seasons and the higher osmotic pressure of the plasma of rock wallabies collected in the dry season confirms that they were experiencing dehydration.

A detailed study of kidney function, measuring both rates of blood flow and glomerular filtration (GFR) was carried out with both species in both wet and dry seasons and the results are summarised in Tables 6.7 and 6.8. Hare wallabies reduced blood flow to the kidney (C_{PAH}) and GFR (C_{IN}) in the dry season and manifested a very effective antidiuresis, with rates of urine production falling from 50.2 to 6.9 ml kg^{-1} d^{-1}. As would be expected, the fractional reabsorption of filtrate (FR_{H_2O}) increased and was exceptionally high in the dry season, with 99.6% of the fluid filtered by the glomeruli being reabsorbed in the nephroi of the kidney.

Table 6.5. *Water and electrolyte turnover and homeostasis of spectacled hare wallabies on Barrow Island*

Data are means ± S.E.

Season	Body mass (kg)	Total body water (%)	Water influx (ml $kg^{-0.82}$ d^{-1})	Water efflux (ml $kg^{-0.82}$ d^{-1})	Sodium influx (mM kg^{-1} d^{-1})	Sodium efflux (mM kg^{-1} d^{-1})	$[Na^+]_p$ (mM l^{-1})	$[K^+]_p$ (mM l^{-1})	$[OP]_p$ (mOsm kg^{-1})
Dry	2.31 ± 0.06	76.4 ± 1.6	27.5 ± 2.0	36.3 ± 2.5	6.4 ± 0.7	7.3 ± 0.8	151.9 ± 1.1	5.6 ± 0.2	287.3 ± 2.1
Wet	2.76 ± 0.04	79.7 ± 0.7	139.1 ± 5.9	138.8 ± 5.6	3.1 ± 0.3	3.0 ± 0.5	146.7 ± 0.7	5.6 ± 0.1	288.2 ± 3.3
p	< 0.01	NS	< 0.001	< 0.001	< 0.001	< 0.001	NS	NS	NS

From Bradshaw *et al.* (2001).

Table 6.6. *Water and electrolyte turnover and homeostasis of rock wallabies on Enderby Island*

Data are means ± S.E.

Season	Body mass (kg)	Total body water (%)	Water influx (ml $kg^{-0.82}$ d^{-1})	Water efflux (ml $kg^{-0.82}$ d^{-1})	Sodium influx (mM kg^{-1} d^{-1})	Sodium efflux (mM kg^{-1} d^{-1})	$[Na^+]_p$ (mM l^{-1})	$[K^+]_p$ (mM l^{-1})	$[OP]_p$ (mOsm kg^{-1})
Dry	3.17 ± 0.13	73.8 ± 1.52	64.8 ± 5.1	74.1 ± 4.6	3.63 ± 0.43	4.12 ± 0.36	139.4 ± 4.0	4.1 ± 0.15	286.9 ± 2.7
Wet	2.53 ± 0.11	81.3 ± 1.46	160.1 ± 15.6	157.9 ± 14.6	4.91 ± 1.35	4.79 ± 1.30	146.9 ± 1.13	5.17 ± 0.20	283.9 ± 2.57
p	< 0.001	< 0.001	< 0.001	< 0.001	< 0.001	< 0.001	< 0.001	< 0.001	< 0.001

Modified from Bradshaw *et al.* (2001).

Table 6.7. *Renal parameters in spectacled hare wallabies on Barrow Island in dry and wet seasons*

C_{PAH}, clearance of para-aminohippuric acid (ml kg^{-1} min^{-1}); C_{IN}, clearance of inulin (ml kg^{-1} min^{-1}); C_{osm}, osmolar clearance (ml kg^{-1} d^{-1}); C_{H_2O}, free-water clearance (ml kg^{-1} d^{-1}); V, rate of urine production (ml kg^{-1} d^{-1}); FR_{H_2O}, fractional reabsorption of filtrate (%); U/P_{osm}, urine to plasma quotient of osmolytes; ECFV, extracellular fluid volume (%); LVP, plasma concentration of lysine vasopressin (pg ml^{-1}). Data are means ± S. E.; $n = 6$.

Season	C_{PAH}	C_{IN}	C_{osm}	C_{H_2O}	V	FR_{H_2O}	U/P_{osm}	ECFV (%)	LVP
Dry	6.0 ±0.4	1.5 ±0.1	57.5 ±3.2	−50.5 ±2.8	6.9 ±2.2	99.6 ±0.1	8.4 ±0.3	18.3 ±0.7	32.7 ±5.6
Wet	9.0 ±1.3	2.7 ±0.2	91.6 ±6.9	−39.6 ±3.9	50.2 ±4.5	97.8 ±0.3	2.0 ±0.1	18.2 ±1.4	10.7 ±1.8
p	0.04	0.001	0.001	0.059	0.001	0.001	0.001	NS	0.001

Modified from Bradshaw *et al.* (2001).

Table 6.8. *Renal parameters in rock wallabies on Enderby Island in dry and wet seasons*

For explanation of abbreviations, see Table 6.7. Data are means ± S. E.; $n = 7$.

Season	C_{PAH}	C_{IN}	C_{osm}	C_{H_2O}	V	FR_{H_2O}	U/P_{osm}	ECFV (%)	LVP
Dry	5.8 ±0.1	0.8 ±0.1	34.8 ±6.7	−28.1 ±5.4	6.7 ±1.4	99.4 ±0.1	5.4 ±0.7	12.2 ±0.5	3.5 ± 1.5
Wet	7.9 ±0.5	1.3 ±0.1	59.2 ±6.2	−46.3 ±4.5	14.3 ±2.1	99.2 ±0.1	5.4 ±1.1	18.8 ±1.1	6.0 ± 0.9
p	0.008	0.005	0.02	0.02	0.01	NS	NS	0.001	NS

Modified from Bradshaw *et al.* (2001).

This increased reabsorption of water was accompanied by a fall in the rate of excretion of osmolytes, and the osmolar clearance (C_{osm}) was lower in the dry than in the wet season. The free-water clearance (C_{H_2O}) was more negative in the dry than in the wet season, indicating enhanced water reabsorption, which is reflected in the very significant increase in the U/P_{osm} quotient from 2.0 to 8.4. There was no change in the volume of the extracellular fluid (ECFV), which is consistent with the lack of any evidence of dehydration in the hare wallabies. Most importantly, concentrations of the antidiuretic hormone (ADH) in this macropod marsupial – lysine vasopressin – increased from a low of 10.7 pg ml^{-1} in the wet season to a mean of 32.7 pg ml^{-1} in the dry season. These, along with the concentrations of LVP in the quokka wallaby, *Setonix brachyurus*, mentioned in Chapter 2 (see Jones *et al.*, 1990), are the first measurements to be made of concentrations of ADH in free-ranging mammals living in their natural habitat and, in the hare wallaby, make a compelling case for this hormone acting to economise water during periods of drought.

The situation, however, is very different with the rock wallaby, as seen in Table 6.8. Blood flow and GFR were reduced in the dry season, but the GFR falls to a much lower rate than in the hare wallaby. The antidiuresis achieved by the rock wallabies is quite comparable with that of the hare wallabies (6.7 compared to 6.9 ml kg^{-1} d^{-1}, respectively) and the fractional reabsorption of filtrate (FR_{H_2O}) at 99.4% is virtually the same. What is extraordinary about these data, however, is that the FR_{H_2O} does not change between wet and dry seasons, and neither does the U/P_{osm}, which reflects the concentration of the urine being produced. The U/P_{osm} averaged 5.4 in both seasons, which means that, when water is scarce, the rock wallabies respond not by voiding a more concentrated urine, but simply by voiding less of the same.

This, so far as I am aware, is an unparalleled case among mammals, but it becomes even more intriguing when we look at the plasma LVP concentrations. These also do not change significantly between wet and dry seasons, so the antidiuresis would appear to be achieved without the aid of an antidiuretic hormone. There is a marked reduction in the extracellular fluid volume (ECFV) in the dry which confirms the dehydration already remarked upon from the data in Table 6.6. Osmolar clearance falls in the dry season in the rock wallabies but the free-water clearance (C_{H_2O}) is less negative in the dry, confirming that their kidneys are indeed conserving less water at tubular levels than in the wet season.

These data are new but they present some startling revelations that appear to contradict everything that we have learnt and come to expect about kidney function in mammals. They suggest, very clearly, that the rock wallabies lack a hormone-controlled system for water reclamation in their kidneys. Figure 6.24, which plots plasma LVP concentrations in the two species against the U/P_{osm} quotient of the urine, makes this point very clearly. Astonishingly,

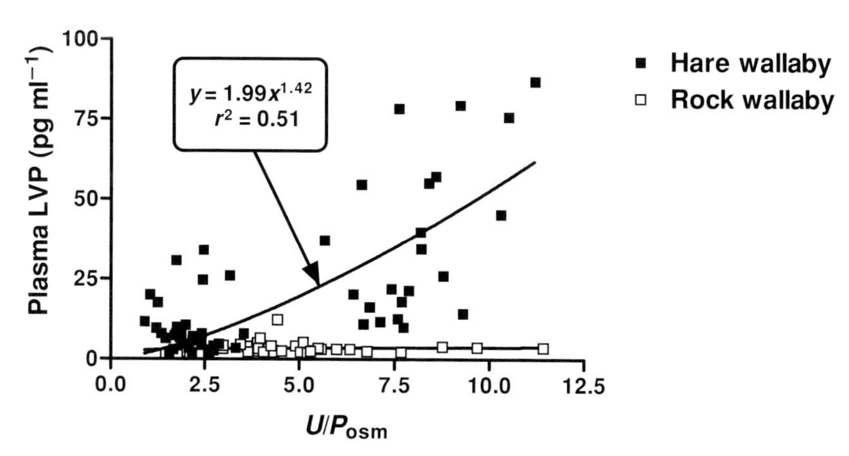

Figure 6.24. Variation in plasma concentrations of the macropodid antidiuretic hormone, lysine vasopressin (LVP), as a function of U/P_{osm} concentration of the urine. Note that increasing concentrations of rock wallaby urine are achieved without increase in LVP. (Modified from Bradshaw *et al.* (2001).)

the concentration of LVP in the rock wallabies shows absolutely no change with increasing concentration of the urine, in complete contrast to the expected correlation that is seen in the hare wallaby.

Such a surprising result needs careful consideration and one needs to ask whether, perhaps, we are measuring the wrong hormone in LVP? We owe to Roger Acher and his colleagues in Paris the discovery that lysine vasopressin (LVP), rather than arginine vasopressin (AVP), is the main neurohypophysial peptide secreted by all the macropodid marsupials studied to date (some six species) (Acher and Chauvet, 1997); both rock and hare wallabies are macropods. Prior to Acher's work, LVP was thought only to occur in the Suina (pigs, peccaries, warthogs, etc.) and the Peru strain of mice. This group working in Paris also discovered a completely novel neuropeptide in kangaroo pituitaries, which they named phenypressin (Phe^2–Arg^2 vasopressin) (Chauvet *et al.*, 1980). Could it be that this hormone rather than LVP is acting as an antidiuretic hormone in the rock wallaby?

Although this seems a very plausible explanation for our unusual results, it can be ruled out, as the antibody used in the heterologous LVP assay also cross reacts with phenypressin (see Figure 3.15). Thus our 'LVP concentrations' would also include phenypressin if there were any being secreted at that time. Mesotocin, instead of oxytocin, is also present in the pituitary gland of macropodid marsupials (Chauvet *et al.*, 1983) but its actions parallel those of oxytocin on the mammary gland and reproductive tract of the female and this hormone has not been found to have any effect on kidney function.

As well as in the quokka, Wilkes and Jannsens (1986) have also demonstrated that LVP functions as an antidiuretic hormone in the developing tammar wallaby, *Macropus eugenii*, and there thus seems no reason to doubt our interpretation that the rock wallaby lacks a hormonally-mediated response to water deprivation. We can, however, make some sense of this unusual finding by examining the microhabitats of the two species.

As would be expected from its name, the nocturnal rock wallaby inhabits areas where there are caves and rockpiles, and it passes the day deep inside these refugia, which remain cool and humid. Figure 6.25 shows temperature and humidity records during one week in April 1986 taken from within a rockpile on Enderby Island that was used by rock wallabies as a refuge during daylight hours. As may be seen, the air temperature remains virtually constant within a degree of 30 °C and the relative humidity averages over 85%. Rock wallabies are thus supremely protected from the rigours of the harsh arid climate by the nature of their rock habitat and, arguably, have no need of sophisticated physiological adaptations for the conservation of water.

The microhabitat occupied during the day by the hare wallaby on Barrow Island is very different from that of the rock wallaby and consists of large spinifex bushes (*Triodia angusta* and *T. pungens*), shown in Figure 6.26. The hare wallabies shelter in the centre of these bushes during the day but, as seen in Figure 6.27, the temperature within these bushes is only marginally lower than the air temperature, which reaches 46 °C on some days. This figure also contrasts the low and constant temperatures recorded over a two week period in an underground warren where another macropod, the burrowing bettong (*Bettongia lesueur*), shelters during the day on Barrow Island.

These major differences between the habitual microclimate of the two species of wallaby holds the key to interpreting the radical difference between their physiological capacities and kidney physiology. The hare wallabies respond to the challenge of water deprivation in the same way as eutherian mammals: by mobilising their antidiuretic hormone (LVP), which reduces water loss through increasing the concentration of the urine being voided. The rock wallabies respond simply by reducing blood flow to the kidneys, which has the effect of cutting down the GFR and ultimately the volume of urine that is finally voided. This urine, however, is no more concentrated than that produced when the animal has access to unlimited supplies of water.

The rock wallabies are enormously advantaged compared with the hare wallabies in terms of their overall water economy by the humid caves and

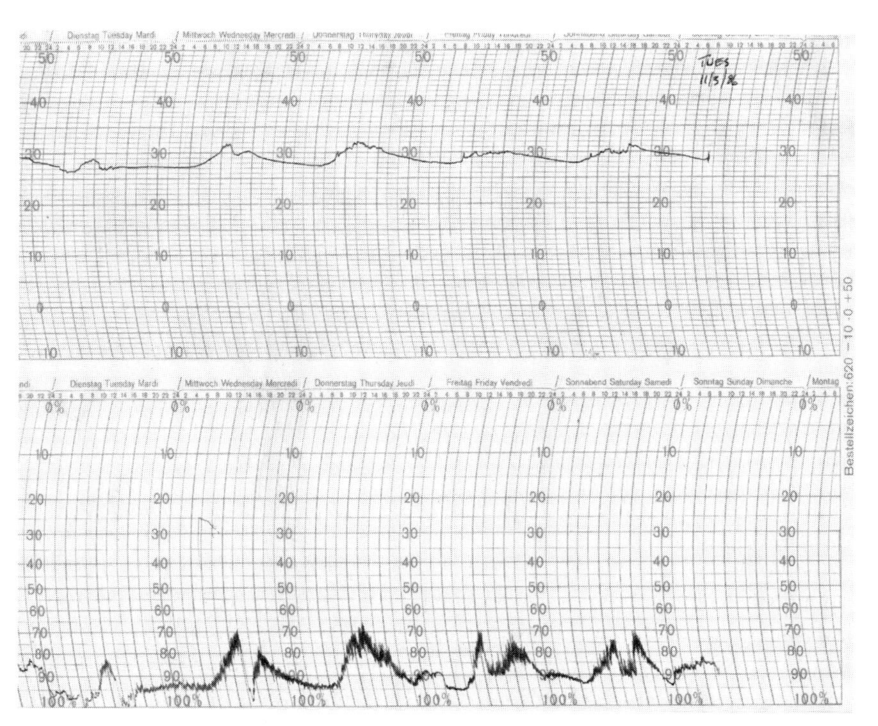

Figure 6.25. Hygrothermograph trace of temperature (above) and relative humidity (%) (below) changes over a one week period in a rockpile cave in which rock wallabies sheltered during the day on Enderby Island in the Pilbara region of Western Australia. Note that the air temperature never exceeded 32 °C and the RH was always greater than 70%.

Figure 6.26. Large clumps of native spinifex grass of the genus *Triodia*, which provide daily refuge sites for the spectacled hare wallaby, *Lagorchestes conspicillatus*, on Barrow Island in Western Australia.

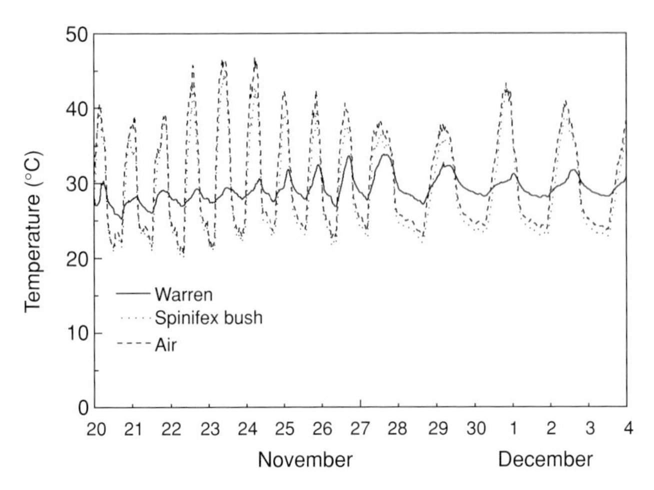

Figure 6.27. Temporal variations in air temperature and that measured within a large spinifex bush used as a daily refuge by the spectacled hare wallaby on Barrow Island, compared with the temperatures recorded with a data logger in a subterranean burrow of the burrowing bettong (*Bettongia lesueur*). The superior protective thermal régime of the burrow is obvious.

rockpiles within which they spend the daylight hours. In order to survive, hare wallabies, by contrast – which have no cool or humid shelter during the heat of the day – must be able to reduce to an absolute minimum their water losses, and it is no surprise that their rate of water turnover during the dry season equals the lowest yet measured in any desert mammal.

Comparing the ability of the two species to maintain overall water and electrolyte homeostasis reveals that, despite its protected environment, the rock wallaby experiences wider perturbations of its *milieu intérieur* between wet and dry seasons than does the hare wallaby. Seasonal variations in condition index, total body water content and plasma electrolyte concentrations, shown in Table 6.9, are 2–3 times greater in the rock wallaby and emphasise the extraordinary ability of the hare wallaby to maintain homeostasis in its natural environment.

This ecophysiological study has thus produced two fascinating discoveries: firstly, that all mammals do not necessarily share the same hormonally controlled system of kidney function; secondly, that behavioural solutions to survival in desert regions prove just as effective as physiological ones. Both species of wallaby are exquisitely adapted to their arid *milieu,* but one has achieved this through sophisticated physiological adaptations, the other by behaviourally exploiting a rare microclimate. The burrowing bettong is similar in that it builds an extensive underground burrow system to avoid the

Table 6.9. *Body composition homeostasis in desert wallabies: maximum percentage variation in parameters recorded over all field trips*

Species	Body mass	Condition index	Total body water content	Plasma Na^+	Plasma K^+	Plasma osmolality
Hare wallaby	21.2	14.7	4.3	5.9	7.7	2.9
Rock wallaby	31.2	28.2	15.7	9.0	36.9	5.6

From Bradshaw *et al.* (2001).

heat of the day and I predict that its kidney physiology, when studied, will prove to be similar to that of the rock wallaby.

A final word on the likely long-term future for these two species, both of which were once widespread on mainland Australia and now only survive in any numbers in island refugia. Both depend on the persistence of their habitat for their long-term survival. However, as recently pointed out by McDonald *et al.* (2002), the current wildlife legislation in Western Australia protects the habitat of a threatened plant species, but not that of an animal! It is to be hoped that legislators will take note of this anomaly and act to ensure that the places where animals live are given the adequate protection that scientific studies have shown are essential if our current biodiversity is to be preserved.

Birds

Desert birds are non-fossorial, diurnal animals that, potentially, encounter the full brunt of the desert climate and thus might be expected *a priori* to display unique adaptations to their particular habitat. Recent allometric studies on the rate of food intake of free-living animals, such as that of Nagy (2001), also highlight the fact that birds require, on average, 45% more fresh food each day than does a typical mammal and they are thus among the most 'expensive' of vertebrates to maintain. The study of desert birds, however, poses an unique problem that is not encountered with any other vertebrate group: the fact that very few species are actually permanent residents in desert regions and thus may never be exposed to selective pressure there (Keast, 1959; Serventy, 1971; Weins, 1991). Their extreme mobility means that birds can readily enter and leave arid zones in response to changes in resource availability, and populations may thus fluctuate widely on a seasonal basis. As Davies (1984) points out, 'nomadism' is a common characteristic of many desert birds. Dawson (1984) summarised most of what was known of the physiology of desert birds some two decades ago, and Maclean (1996) has recently published a monograph on the ecophysiology of desert birds, reinforcing the accepted paradigm that desert birds are not uniquely adapted to such habitats. Rather than review the field as he has done, I will concentrate on a number of case studies, which highlight some of the ways in which birds contend with the desert environment.

The chukar (*Alectoris chukar*) and the sand partridge (*Ammoperdix heyi*) are two species that are permanent residents in the Negev Desert in Israel, and they have been the subject of study by Allan Degen and his colleagues over a number of years. Chukars have a wide southern Palearctic distribution and only inhabit deserts at the margins of their range, whereas sand partridges are wholly restricted to xeric habitats in southwestern Asia and northeastern Africa. In Israel, they occur principally in the Negev, Arava and Judean Deserts.

Degen *et al.* (1982) measured rates of water flux using tritiated water in chukars under laboratory conditions and found that they had relatively low water fluxes when compared with mesic birds. Procedural difficulties over the correct measurement of total body water content using dilution of tritiated water were resolved by Crum *et al.* (1985), and Alkon *et al.* (1984) used tritiated water to estimate rates of water turnover of free-living chukars and sand partridges in the Negev Desert in 1980–81. They found that water influx was lowest in the dry autumn before the winter rains, averaging 15% of the total body water pool per day and then increasing to 45.3% in late winter. Interestingly, this was only the second time that tritiated water had been used to measure the water turnover of any free-ranging bird. Degen *et al.* (1985) carried out a comparative study of rates of water turnover of both free-ranging chukars and sand partridges in the Negev Desert in 1981 and found mean water influxes of 44.1 ml d^{-1} for chukars and 20.8 ml d^{-1} for sand partridges. Chukars are almost twice the size of sand partridges, however, and on a mass-specific basis the water influx was higher in sand partridges than in chukars (122.5 compared with 100.6 ml kg^{-1} d^{-1}). This raises the question of how best to compare rate functions between species that differ significantly in body size – a question that has been explored in great detail by Peters (1983) in his fascinating book.

The most extensive study of the scaling of water flux rate with body size in both invertebrate and vertebrate animals is that of Nagy and Peterson (1988). They derive an allometric relationship using data from 27 species of birds that has a slope of 0.694. This slope does not differ significantly from most of the slopes derived by other workers, which vary between 0.66 and 0.73, but does differ from slopes of 0.596 and 0.75 derived by Degen *et al.* (1985), Walter and Hughes (1978) and Hughes *et al.* (1987). Using Nagy and Peterson's (1988) slope of 0.694, we can calculate that chukars have a water influx of 75.5 ml $kg^{-0.694}$ d^{-1}, which compares with a figure of 69.3 ml $kg^{-0.694}$ d^{-1} for sand partridge. These means are probably not significantly different and suggest that the apparently more arid-adapted sand partridge does not have spectacularly lower rates of water turnover than the chukar, when measured together in the same habitat. A similar result is found if one uses the allometric equation derived by Williams *et al.* (1993) where the body mass in grams is raised to the power 0.682: figures of 0.61 and 0.67 ml $g^{-0.682}$ d^{-1} are obtained for the sand partridge and chukar, respectively. Degen *et al.* (1985) in fact concluded that both species were drinking free water when only dry forage was available and Thomas *et al.* (1987) found that neither chukars nor sand partridges were ever deprived of access to free water in the Negev Desert.

The lowest rates of water turnover recorded for any birds in the literature are those for the burrowing owl (*Speotyto cunicularia*) and the Petz conure (*Aratinga canicularis*), both desert-living species studied by Chapman and McFarland (1971). These rates were established in captivity, however, and given that the data were based on only a single individual in each case, they need to be accepted and cited with considerable caution. Nagy and Peterson (1988) have shown that the rates of water turnover of birds in the field are about 50% higher than those of birds held in captivity, and field-based measurements are thus what is needed in assessing the ecophysiological performance of different species.

The spinifex pigeon, *Geophaps plumifera*, is a bird that inhabits rocky and extremely hot areas of the arid Pilbara region of Western Australia. Its rates of metabolism and evaporative water loss were studied by Withers and Williams (1990), who found that its BMR was very low and only 68% of allometric predictions. This bird was found to have an extremely high operant temperature when studied in the field, with body temperatures averaging 43.6 °C between 1200 and 1400 hours, and individuals can be readily observed in the middle of the day sitting in full sunlight on granite rocks with surface temperatures in excess of 60 °C (Williams *et al.*, 1995). Despite this, the birds need to drink every other day and are never found far from permanent sources of water (Serventy and Whittell, 1967; Fisher *et al.*, 1972).

The water requirements, thermoregulation, and field metabolism of free-ranging spinifex pigeons were studied in detail by Williams *et al.* (1995), who predicted that, in view of their very low BMR, their FMR would be similarly reduced. This in fact proved to be the case, and the FMR averaged 73.5 kJ d^{-1}, which is 38.7% lower than allometric expectations. Williams *et al.* (1993)

developed an alternate allometric equation for desert birds alone and this predicts an FMR of 120.1 kJ d^{-1}, with the FMR of the 87 g spinifex pigeon still being only 63.3% of this value. Water influx of the spinifex pigeons averaged 18.4 ml d^{-1} but only 4% of this came from their diet of seeds. Metabolic water production was insufficient to make up this deficit, so 83.5% of their daily water influx thus comes from drinking.

Goldstein and Nagy (1985) studied rates of resource utilisation in Gambell's desert quail (*Callipepla gambellii*), using doubly labelled water, and similarly found that the FMR was some 40% lower than that predicted for a bird of similar body mass. They attributed this reduction in energy expenditure to much reduced resting rates of metabolism. When the FMR of Gambell's desert quail and the spinifex pigeon are adjusted to $mass^{0.674}$ according to the allometric relation derived by Williams *et al.* (1993) they both have similar rates of energy expenditure in the wild during summer of 3.2 and 3.7 kJ d^{-1}, respectively. Nagy *et al.* (1999) offer a more recent equation for desert birds with FMR in kJ $d^{-1} = 6.35$ (body mass in g)$^{0.671}$ and the resultant figures for the two species are almost identical at 3.24 and 3.67 kJ d^{-1}. Ambrose and Bradshaw (1988a,b) working with the white-browed scrubwren, *Sericornis frontalis*, found that arid-habitat sub-species of the genus also displayed both reduced BMRs and lower rates of water turnover in summer.

Williams *et al.* (1995) argue that the very low FMR of birds such as the spinifex pigeon, when compared with birds from mesic habitats, may be a consequence of their conservative lifestyle with periods of inactivity during the middle of the day, but the primary contributing factor is probably their low BMR. Reduced BMRs and, as a consequence, lower FMRs could be a common feature in desert birds, but Williams *et al.* (1991) did not find this to be the case with a group of Australian parrots and both BMRs and FMRs of four desert species from Western Australia were indistinguishable from those of other non-passerine birds (see Aschoff and Pohl, 1970; Bennett and Harvey, 1987). Rates of evaporative water loss (EWL) in the laboratory and rates of water turnover in the field of the desert-living species were, on the other hand, significantly lower than those of parrots from more mesic environments and below allometric predictions derived from 35 non-passerine species by Crawford and Lasiewski (1968). Estimates of the water economy index (WEI) of Nagy and Peterson (1988) for these parrots were the lowest that have yet been reported for desert-adapted birds.

Table 6.10 compares published rates of water influx in free-living desert birds that have been measured by isotopic turnover, ranked by body mass; they are all low when contrasted with mesic species, ranging from 0.55 to 1.33 ml $g^{-0.682}$ d^{-1}. Because the BMR of these species are not all low, it is

Table 6.10. *A comparison of rates of water influx in free-ranging desert-adapted birds measured by isotopic turnover*

Species	Body mass (g)	Location	Water influx (ml $g^{-0.682}d^{-1}$)	Source
Budgerigar (*Melopsittacus undulatus*)	27.9	W. Australia: arid Pilbara region	1.13	1
Dune lark (*Mirafra erythrochlamys*)	28.5	Namib Desert	0.55	2
Spinifex pigeon (*Geophaps plumifera*)	87	W. Australia: arid NW coast	0.88	3
Gambell's quail (*Callipepla gambelii*)	143.4	Colorado desert	0.63	4
Port Lincoln parrot (*Barnardius zonarius*)	145.0	W. Australia: semi-arid wheatbelt	0.84	1
Sand partridge (*Ammoperdix heyi*)	176.0	Negev Desert, Israel	0.61	5
Pink galah (*Cacatua roseicapilla*)	307.0	W. Australia: semi-arid wheatbelt	0.89	1
Chukar (*Alectoris chukar*)	460.8	Negev Desert	0.67	5
Ostrich (*Struthio camelus*)	88 250	Namib Desert	1.33	6

Sources: 1, Williams *et al.* (1991); 2, Williams (2001); 3, Williams *et al.* (1995); 4, Goldstein and Nagy (1985); 5, Degen *et al.* (1982); 6, Williams *et al.* (1993). Table modified from Williams *et al.* (1995).

clear that some desert birds have evolved mechanisms for reducing EWL and water turnover that do not simply rely on an overall reduction in the rate of metabolism. At this stage, it is not clear what these mechanisms may be and whether they are physiological or behavioural, or a combination of both

(Tieleman and Williams, 1999a). An interesting case is the small Australian diamond dove, *Geopelia cuneata*, which has recently been studied in the laboratory by Schleucher *et al.* (1991). Its BMR is normal for a columbid bird but it allows its body temperature to rise at ambient temperatures above 36 °C, thus storing heat that it would otherwise need to dissipate by evaporation of water. It displays gular fluttering at rates of up to 960 min^{-1} at high ambient temperatures but, overall, total evaporative water loss is 19–33% lower than expected and it thus displays impressive water conservation mechanisms. Schleucher (1993) also radiotracked diamond doves in the field and found that they can remain active during much of the hottest period of the day, when most other birds shelter. The manifest advantages of hyperthermia in birds in conserving body water that would otherwise be lost have also recently been reviewed by Tieleman and Williams (1999b).

Another very interesting species in this context is the small zebra finch, *Taeniophygia castanotis* (now *T. guttata*), which has been the subject of many physiological studies over the years (Weins, 1991). Although zebra finches will drink in the field (Fisher *et al.*, 1972), they show a remarkable ability to survive with little drinking water in the laboratory – as little as 1 ml per week – and they are also able to tolerate highly saline water (Maclean, 1996). In an early visit to Western Australia, Oksche *et al.* (1963) found that zebra finches collected near the wheat-belt town of York in Western Australia were able to maintain mass when given only 0.8 mol l^{-1} sodium chloride to drink, but this could not be confirmed by Lee and Schmidt-Nielsen (1971), working with captive birds.

Other workers, also studying aviary-bred birds, were unable to confirm this early observation; maximum salt concentrations tolerated were usually no higher than 0.3–0.5 mol l^{-1} (Cade *et al.*, 1965). Skadhauge and Bradshaw (1974) had an opportunity to examine this problem first hand and found that Zebra finches, when collected wild from the field in Western Australia, were indeed able to tolerate salt solutions up to 0.8 mol l^{-1}. Skadhauge and Bradshaw (1974) were also puzzled by the fact that the maximum urinary concentration of the finches of 1000 mOsm kg^{-1} was less than that of the saline water that they were drinking. Tritiated water was used to measure the rate of metabolic water production (MWP) of the zebra finches and this was found to be a very high 31.3% of their daily water influx. This additional influx of water had the effect of diluting the saline fluid being drunk to a point where all the salt could be excreted by the kidneys. Small body size is thus not necessarily a disadvantage for a desert bird, provided that it has access to adequate supplies of carbohydrate, in the form of grains, as its high rate of metabolism automatically results in a high MWP.

This point has also been reinforced by Williams' (1996) study in which he corrects the slope of the first allometric equation relating evaporative water loss to body mass in birds, originally proposed by Crawford and Lasiewski (1968), and shows that small birds do benefit more than large birds in terms of production of metabolic water, thus also correcting an idea first proposed by George Bartholomew in 1972, and based on Crawford and Lasiewski's (1968) regression. Metabolic water production has been shown to be an important component of the water balance of dune larks (*Mirafra erythrochlamys*) living in the hyper-arid Namib desert by Williams (2001) and accounted for 47% of their daily water influx.

The other very important point to emerge from the work of Skadhauge and Bradshaw (1974) with the zebra finch is that birds in captivity are usually highly inbred and clearly lose some of the physiological capacities (salt tolerance and renal concentrating capacity) that are retained in wild birds and are essential for their survival.

Studies by Marder and Gavrieli-Levin (1987) with pigeons have also revealed that desert species may have surprising, previously unsuspected, abilities to lower their body temperature through cutaneous evaporation. Birds have long been thought to lose water only via the mechanisms of panting and gular fluttering at high environmental temperatures, but Marder *et al.* (2002) have found that rates of cutaneous water loss vary considerably, and in relation to habitat aridity. Rates of cutaneous water evaporation (CWE) were found to vary from 2.1 $mgH_2O\ cm^{-2}\ h^{-1}$ for rainforest species, to 5.6 $mg\ cm^{-2}\ h^{-1}$ in temperate-zone birds and peaking at 13.5 $mg\ cm^{-2}\ h^{-1}$ in arid-zone pigeons (diamond dove, rock pigeon (*Petrophassa albipennis*) and spinifex pigeon). Marder and Gavrieli-Levin (1987) speculated that this novel mechanism of cutaneous evaporative cooling (CEC) is a recently evolved mechanism in birds, compared with panting. Marder *et al.* (2002) further speculate that CEC evolved as a cooling mechanism in altricial hatchlings of pigeons in a temperate-zone climate and has achieved its highest effectiveness in both young and adult desert pigeons.

There has also been a number of interesting speculations regarding the origin of the low rates of energy exchange seen most characteristically, but not always, in desert birds (Tieleman and Williams, 1999). The idea that birds are able to live in deserts by virtue of general avian characteristics, rather than because of specific physiological adaptation, as has been shown in reptiles (Bradshaw, 1986), was first questioned by George Williams (1966). This point was specifically investigated by Joe Williams (1996) who found that laboratory-measured rates of total evaporative water loss were lower in desert species and this difference persisted when phylogenetically independent contrasts (see Harvey and Pagel, 1991) were computed. Tieleman and Williams (2000) then carried out a large multi-species comparison, also using phylogenetically independent contrasts, and showed that birds from deserts had significantly lower rates of metabolism and water flux than species from more mesic areas. Williams and Tieleman (2002) have recently performed more detailed analyses for basal metabolic rate, field metabolic rate, and field water flux, and found that these traits are generally reduced in desert birds. The question thus arises: are these genuine adaptations to the desert environment, or are they merely the result of the phylogenetic history of these lineages of birds? They could also result from phenotypic plasticity in view of the results of a study by Williams and Tieleman (2000) in which the BMR of hoopoe larks (*Alaemon alaudipes*) was significantly altered by acclimation to high and low temperatures in the laboratory.

Williams *et al.* (1995) attempted to answer this question by reviewing the published data on BMR for the various species of the family Columbidae, which includes the spinifex pigeon, and analysing these by using the method of phylogenetically independent contrasts (Felsenstein, 1985). Values for the BMR were mass adjusted by dividing by (body mass)$^{0.893}$, where the exponent is that of the equation that relates BMR to body mass among pigeons. Values for the mass-adjusted BMR of ancestral nodes were estimated by means of the squared-change parsimony reconstruction method of Maddison (1991).

They found that, although occupants of the hot, dry interior of Australia such as the spinifex pigeon and the crested or topknot pigeon (*Ocyphaps lophotes*) have the lowest mass-adjusted BMR so far reported for pigeons, some species that live in relatively mesic areas, such as *Columba palumbus*, also have a relatively low mass-adjusted BMR. Williams *et al.* (1995) were thus unable to conclude whether a reduced BMR in Australian pigeons was the consequence of ecological adaptation or phylogenetic constraint. A similar situation is found when considering rates of water flux in the field. The water flux rates of desert birds are, on average, 41% lower than those of birds of mesic regions but this difference disappears when phylogenetically independent contrasts are calculated (Williams and Tieleman, 2001) and the question thus cannot be resolved at the present time.

A final cautionary tale is perhaps appropriate to conclude this section on the varied nature of the adaptive strategies that enable birds to survive in desert environments. The spinifex bird or Carter Desert bird, *Eremiornis carteri*, is a small (10 g) bird that is found living permanently in the Australian arid zone, where it nests and retreats to large clumps of spinifex grasses of the genera *Triodia* and *Plectrachne*. It is an insectivore that is not found near sources of free water and thus does not drink, and would seem *a priori* the perfect example of a desert-adapted bird.

As part of an ecophysiological study of the vertebrate fauna of Barrow Island, Ambrose *et al.* (1996) measured the FMR and water influx rate of free-ranging spinifex birds during the dry season, as well as their BMR and EWL in the laboratory. Surprisingly, rates of water flux in the spinifex bird, instead of being very low as anticipated, were amazingly high, averaging 77.6% of the total body water pool per day in the dry season and increasing to 91.6% in late autumn. The problem was to identify the source of water, as Barrow Island has no free water and radiotracking of spinifex birds confirmed that they are quite sedentary, rarely moving more than 200 m. The solution to this enigma was found on collecting spinifex birds early in the morning, before they had emerged from their evening refugia. Typically, they spend the night period in the centre of a large spinifex bush; birds collected from such sites before dawn were extremely wet. Dew formation is quite prevalent on Barrow Island and it would appear that dew was condensing on the cooler outer feathers of the birds during the night. In the morning, the first thing that the birds were observed doing was to preen themselves and ingest all of the free water that collected in and on their feathers. We thus, fortuitously, uncovered an unusual water source, which evaporated within a few hours as ambient temperatures rose on the island, but one that was being very effectively exploited by the spinifex birds. Once again, the search for a 'genuine' desert-adapted bird was thwarted and it would appear that spectacular physiological adaptations are more the exception than the rule in these vertebrates.

Camels

No chapter on deserts would be complete without a few words about the 'ship of the desert' – the camel – which has served mankind faithfully as a beast of burden for probably thousands of years, and continues to do so. Firstly, though, a word of caution about the 'function' of the camel's hump, or humps as the case may be. The story that the hump is a place where camels store water for their long journeys across the desert is apocryphal, but it has even survived after evidence to show that the hump is filled not with water, but with fatty tissue (see Figure 6.28). The idea was then proposed that since fat, on being oxidised, produces more water than its own mass (it is true that 1 g of fat, when oxidised completely, results in the production of 1.071 g of water), then the hump is still a water store, albeit a biochemically disguised one!

Some stories die hard and it is worth doing a very simple calculation to prove that this one should have been buried long ago. Imagine that we have a 500 kg camel with a BMR of 10 000 kcal d^{-1} (41.87 MJ d^{-1}) and we wish

Figure 6.28. Old habits die hard! The truth about the camel's hump.

to calculate the amount of either fat or carbohydrate needed to produce that amount of heat. We need to calculate how much water will be produced from each source to see whether stored fat can function as a water store. Table 6.11 shows the amount of oxygen needed to metabolise each of the three fuels, and the amount of water that will be lost by pulmonary evaporation in taking that oxygen into the lungs and eliminating carbon dioxide. The *balance* shows very clearly that, in both cases, the animal loses more water than it gains in oxidising each of these fuels and that, in fact, it loses more in the case of fat oxidation than with carbohydrate. Fat is thus not a disguised water store: it is simply an energy store in the camel, as it is in every vertebrate.

Having dispensed with that, let us look at some of the ways in which camels are able to survive in desert regions, keeping in mind, however, that almost all of the published data come from camels held in pens, rather than free-ranging.

Table 6.11. *Water balance of a 500 kg camel when metabolising fat and carbohydrate to maintain a daily BMR of 41.87 MJ*

Fuel	Water produced (ml g^{-1} fuel)	Amount of fuel required for BMR (kg)	Water formed (l)	Oxygen used (l)	Water lost[a] (l)	Balance (l)
Carbohydrate	0.556	2.39	1.33	1980	1.70	−0.37
Fat	1.071	1.06	1.13	2130	1.80	−0.67

[a] Pulmonary water loss estimated in dry air.

Again, we are fortunate in having at our disposal a recent monograph dedicated to the subject (Wilson, 1989) and I will focus just on highlights. Early studies by Knut and Bodil Schmidt-Nielsen (Schmidt-Nielsen *et al.*, 1957) revealed that the body temperature of the camel varies appreciably throughout the day, as seen in Figure 6.29, and may fluctuate by as much as 6.2 °C when dehydrated. This 'heterothermy' of the camel was initially interpreted as evidence for poor thermoregulatory ability but it became apparent that, by allowing its body temperature to fluctuate between 34.5 °C and 40.5 °C, the camels were effectively conserving large volumes of water that would otherwise be needed to maintain the body temperature constant. A temperature increase of 6 °C in a 500 kg camel enables 1.26×10^7 J of heat to be stored in the body, equivalent to the conservation of approximately 6 l of water if sweating had to be used by the animal to dissipate the same amount of heat. The camel easily radiates the stored heat to the cold desert sky at night and this is our second example, along with the jackrabbit, of an animal that relies heavily on radiation to maintain its heat balance.

Camels are also renowned for their tolerance of dehydration. The two Schmidt-Nielsens, in a very early study, showed that camels can sustain a loss of up to 30% of their total body water content over a period of 9 d (Schmidt-Nielsen *et al.*, 1956). This water loss is also not distributed evenly throughout the body but comes preferentially from the gut (50%) and the intracellular fluid volume (30%), with only 20% from the extracellular space. This has the effect of 'protecting' the circulating fluid volume (plasma and blood volume) from haemoconcentration and an increase in viscosity, which is the usual cause of death in dehydrating mammals. 'Plasma sparing' has also been described in the spiny mouse, *Acomys cahirinus*, by Horowitz and Borut

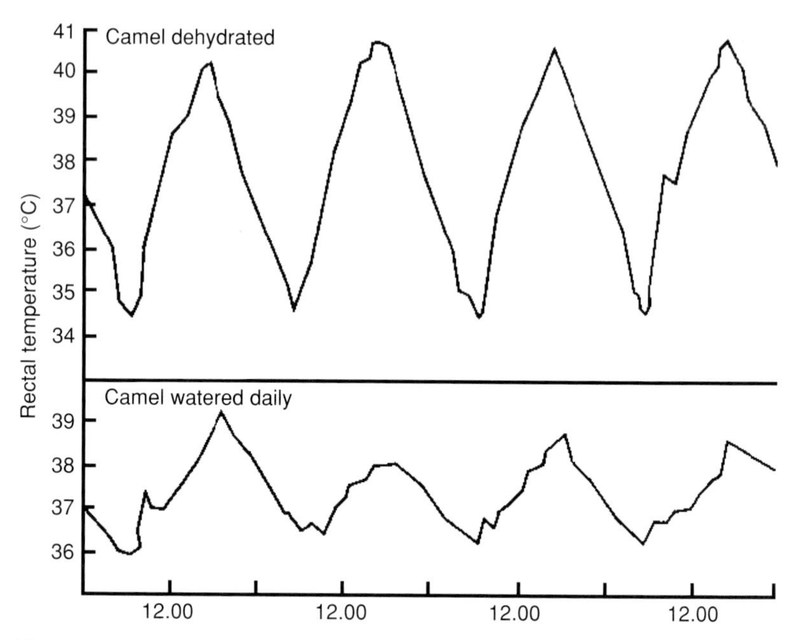

Figure 6.29. Temporal variation in the body temperature of camels as a function of their state of hydration, showing that dehydrated animals allow their body temperature to fluctuate to a much greater extent than 'watered' individuals. (Adapted from Louw and Seely (1982) after Schmidt-Nielsen *et al.* (1957).)

(1970) and Borut *et al.* (1972) as well as in the agamid lizard *Ctenophorus ornatus* (Bradshaw, 1986). It may well be a common feature in desert-living animals that are exposed to chronic dehydration and it has even been proposed for the jackrabbit (Reese and Haines, 1978). Detailed studies (Horowitz and Borut, 1973, 1975, 1994) suggest that low capillary permeability and high colloid osmotic pressure, due to an increase in albumin synthesis, are at the basis of this phenomenon.

Just as impressive as their tolerance of dehydration is the spectacular ability of camels to rehydrate rapidly after dehydration. Following a 20% loss of body mass, camels can restore their original body mass in less than 10 min by drinking. Schmidt-Nielsen *et al.* (1956) record one female drinking 66.5 l (33.1% of body mass) in one session and a male drinking 186 l in two sessions! This raises interesting physiological questions of how the animals are able to avoid massive dilution of plasma electrolyte concentrations; Yagil and Etzion (1979) show that plasma aldosterone concentrations are greatly elevated during rehydration and this may help to conserve sodium ions that might otherwise be lost via the kidneys with this massive haemodilution.

This paper also shows that plasma AVP concentrations increase from 1.17 to 5.16 pg ml^{-1} during dehydration and then fall to 0.21 pg ml^{-1} one hour after rehydration.

As mentioned above, however, none of the above studies has been carried out with free-ranging camels and their true ecophysiology remains a field yet to be explored. Macfarlane *et al.* (1962, 1963) carried out early studies in the Australian desert, using tritiated water, which showed that the rate of water turnover of camels was considerably lower than that of sheep and cattle and averaged 188 ml $kg^{-0.82}$ d^{-1}. Macfarlane *et al.* (1967) also reported that AVP has the unusual effect of increasing electrolyte excretion in camels and the normal antidiuresis seen in hydrated animals with AVP injection was transformed into what was apparently an osmotic diuresis in camels that had low rates of urine flow (Siebert and Macfarlane, 1971).

As may be seen, much of this research was carried out a number of decades ago and there would appear to be a real opportunity now, with the development of more sophisticated techniques – such as the doubly labelled water method for estimating material and energy balance of free-ranging animals – to focus once again on camels. Although burdened with the reputation of having a foul temperament as an experimental animal, the camel none the less has the distinct advantage of large body size and could readily carry instrument packages that could transmit vital body parameters as well as GPS co-ordinates on a real-time basis. It is to be hoped that some intrepid ecophysiologists will take up this challenge in the future.

7

Torpor and hibernation in cold climates

Endothermic thermoregulation

The ability of endotherms to reduce significantly their body temperature during periods of intense cold or low food availability would seem to be an eminent adaptation for vertebrates that have high basal rates of metabolism. Unlike ectotherms, endotherms elevate their body temperature above that of their surroundings by the constant oxidation of foodstuffs and the heat produced from this process (i.e. resulting from the thermodynamic inefficiency of the various biochemical pathways, otherwise known as 'futile' proton cycling) is not wasted, but used to raise the body temperature of the animal (see Brand *et al.*, 1991; Rolfe and Brand, 1997). This is in contrast to ectothermic vertebrates such as reptiles and amphibians, which use external sources of heat, such as the sun, to elevate and maintain a constant body temperature while active.

As a consequence of this difference in sources of heat production, ectotherms are more productive than endotherms, with much higher efficiencies of biomass conversion. Pough (1983) calculated that 16 species of small birds and mammals converted only 1.4% of the energy they assimilated into biomass, compared with an average conversion efficiency of 46.3% for eight species of amphibian and reptile. Thus, despite their low rates of energy flow, ectotherms are of great importance in terrestrial ecosystems because of their high conversion efficiencies and the energy they make available to other organisms. In a forest ecosystem in New Hampshire, for example, Burton and Likens (1975) estimated that the annual energy production per hectare of the one population of red-backed salamanders (*Plethodon cinereus*) was five times greater than that of the entire avian community of the forest.

Endothermy is thus an expensive mode of living, as a very large proportion of the energy consumed by the animal is ultimately dissipated in the form of heat.

The evolution of endothermy in birds and mammals is perhaps one of the most significant and far-reaching changes to have occurred in the reptile–mammal lineage and involved not only a change in the *level* of metabolism, as is well appreciated, but also in its *quality* or biochemical source (Hulbert, 1987; Hulbert and Else, 1989; Ruben, 1995). Endotherms fuel their activity aerobically, but ectotherms must rely on anaerobic metabolic pathways to generate energy for high or sustained levels of activity, a discovery attributed to the late Walt Moberly (Moberly, 1968). Bennett and Licht (1972) showed that over 60% of the total energy utilised by the common iguana, *Iguana iguana*, during 5 min of activity was derived from anaerobiosis and 80–90% of the total energy expended during the first 30 s of activity in lizards is associated with lactate formation (Bennett, 1982).

Ectothermic reptiles also differ most notably from birds and mammals in the actual body temperature associated with activity. As we have seen in Chapter 5, many reptiles regulate their body temperature during the day with great precision but they do not all maintain the same, or even closely related, temperature. Preferred or 'eccritic' temperatures of reptiles (i.e. the temperature maintained behaviourally in a photothermal gradient) vary from as low as 24 °C in the gecko *Underwoodisaurus milii* to 39 °C in the desert iguana, *Dipsosaurus dorsalis* (Bradshaw, 1986). One needs to compare this situation with that of mammals, which all maintain a body temperature between 36 and 38 °C, or birds, where the range is from 40 to 42 °C, in order to appreciate that endotherms have specialised and selected a narrower range of operating temperatures and hence a more limited set of biochemical pathways to sustain their activity (Hulbert and Else, 1981; Ruben, 1995).

Table 7.1 lists preferred body temperatures (PBT for a number of species of lizard and compares these with the temperature at which, *in vitro*, the myosin ATPase from their muscles has its peak activity and the temperature at which muscle tension is maximal. As may be seen, the correspondence between the three values for each species is close and it is apparent that lizards must possess a series of isozymes that ensure maximal physiological efficiency is achieved at, or close to, the activity body temperature. Birds and mammals, on the other hand, have opted for narrower ranges of operating temperatures and enzymes that operate most effectively over those narrow ranges.

Dropping the body temperature is a daily occurrence in ectotherms but is potentially hazardous for endotherms. Else and Hulbert (1987) and Hulbert and Else (1990, 1999) argue that the cells of endotherms are much

Table 7.1. *Preferred body temperatures (PBT) of a number of lizard species compared with the thermal optimum of their myosin ATPase and temperature at which maximal muscle tension is achieved*

Species	PBT (°C)	T_{max} for myosin ATPase (°C)	T_{max} for maximal muscle tension (°C)
Dipsosaurus dorsalis	38.8	40	39
Uma notata	37.5	39	36
Sceloporus undulatus	36.3	35	34
Eumeces obsolatus	34.5	34	34
Gerrhonotus multicarinatus	30	32	31.8
Underwoodisaurus milii	24	25	26

From Licht (1964).

'leakier' than those of ectotherms and more energy is required to maintain transmembrane ionic gradients through the action of the membrane-bound enzyme pump, Na^+/K^+-activated ATPase. Lowering the body temperature is thus more likely to result in the degradation of ion gradients in endotherms than in ectotherms, as the latter possess tight junctional complexes in their cell membranes that maintain transmembrane ion gradients, even in the face of hypoxia and anoxia (Guppy *et al.*, 1987). Hulbert and Else (1989, 1999) have also found that the degree of polyunsaturation of membrane phospholipids is correlated with cellular metabolic activity, being characteristically higher in endotherms than ectotherms. Thyroid hormones have been implicated in this process, specifically by increasing the polyunsaturation of omega-6 fatty acyl chains in cell membranes (Hulbert and Else, 2000), and Hulbert (2000) argues in a recent major review that this is the means by which these hormones increase the metabolic rate. He goes further to argue that the various iodothyronines are actually normal constituents of biological membranes in vertebrates and assist in increasing membrane rigidity, with their effects on metabolism being due primarily to membrane acyl changes. These are fairly radical suggestions but they find support in a recent paper by Eales (1997), who suggests that the iodothyronines are more akin to vitamins than to hormones and coined the word 'vitamone' to describe such a rôle. The superfamily to which the thyroid hormone nuclear receptors belong also includes those for vitamin D and retinoic acid (Jensen, 1991), and the thyroid hormones may prove not to be 'regulatory' hormones in the usual sense, but act permissively in the tissues, as do some steroid hormones.

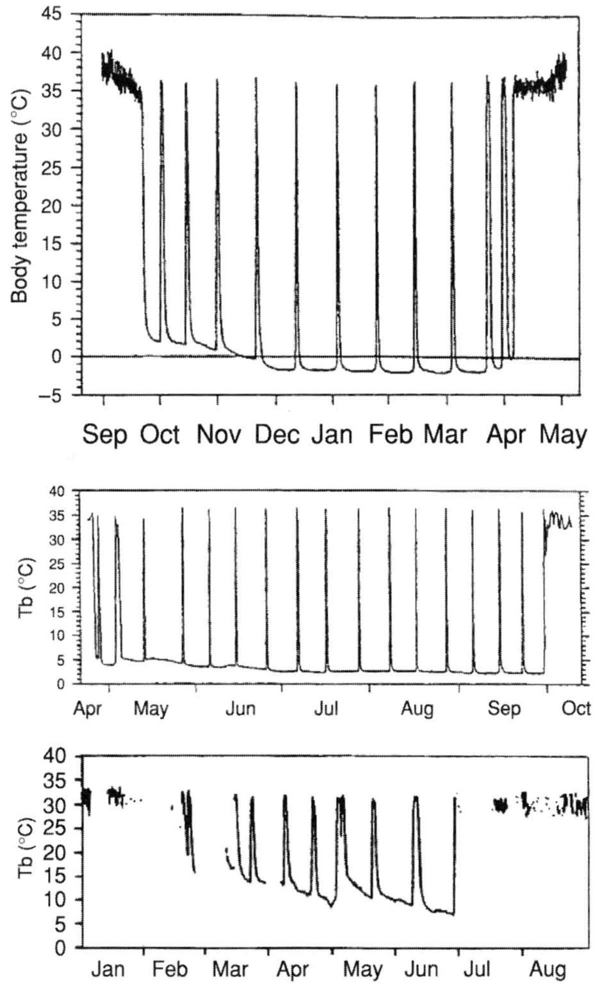

Figure 7.1. Seasonal records of field body temperature from three different mammalian hibernators: the Arctic ground squirrel (*Spermophilus parryi*, top), the mountain pygmy possum (*Burramys parvus*, middle), and the short-beaked echidna (*Tachyglossus aculeatus*, bottom). (From Grigg and Beard (2000) with permission.)

Torpor

The study of the processes that allow birds and mammals to lower their body temperature during torpor and hibernation, sometimes by as much as 30 °C below normal (see Figure 7.1), and without disrupting basic cellular functions, is thus of considerable interest to both physiologists and ecophysiologists. Wang (1989) reviewed what was known of this field a decade ago and recent updates and insights may be had in the last two volumes of the International Hibernation Symposia (Geiser *et al.*, 1996; Heldmaier and Klingenspoor, 2000).

Torpor involves a resetting of the 'thermostat' for body temperature regulation but little is still known of the neural mechanisms involved, despite substantial progress in the study of mammalian torpor (Wang, 1989). Whether ectotherms display torpor, in the sense that they will voluntarily select and maintain a lower body temperature than that normally associated with activity, is debatable. Rismiller and Heldmaier (1988, 1991) have shown that the green lizard, *Lacerta viridis*, will select a temperature as low as 12 °C in a thermal gradient in winter and also choose 20 °C at night compared with 30 °C during the day in autumn. If this is indeed a form of torpor, it differs from that of endotherms, however, which are always capable of spontaneous arousal and warming by the use of endogenous heat production, even in the absence of brown fat (Holloway and Geiser, 2001). Torpor thus involves a specific downregulation of body temperature from which spontaneous recovery is always possible (Malan, 1996), but it is typically short-lived in comparison with hibernation, and normally lasts less than 24 h (Wilz and Heldmaier, 2000).

The rate of heat production falls below the BMR during torpor and the accepted explanation for this for many years has been that this results from a Q_{10} effect of the lower body temperature reducing tissue metabolism (Heldmaier and Ruf, 1992; Malan, 1993). Q_{10} effects, however, are often above 2.0 to 3.0 and metabolic depression, in addition to temperature effects, has been invoked as a mechanism in recent years (Geiser and Kenagy, 1988; Storey and Storey, 1990). Current thought is that the metabolic rate may be actively downregulated during torpor and independent of the body temperature effect (Heldmaier and Ruf, 1992; Song *et al.*, 1995).

Torpor is clearly an energy-conserving strategy, which is not necessarily confined to situations of cold exposure but may also be deployed during times of food shortage (Nicol and Anderson, 1996). Holloway and Geiser (1995) measured the amount of energy saved by going into daily torpor in the small dasyurid marsupial *Sminthopsis crassicaudata*, and found that bouts of 10 h reduced metabolic costs by 30–50%. Withers *et al.* (1990) studied torpor in the honey possum, *Tarsipes rostratus*, and found that a combination of food deprivation and an ambient temperature of 12 °C was sufficient to initiate torpor. Daily torpor in response to food deprivation has also been recently documented in elephant shrews (*Elephantulus rozeti* and *E. myrus*) from Morocco and southern Africa, respectively, by Lovegrove *et al.* (2001). In this case, the animals were maintained at a constant ambient temperature of 18 °C, thus demonstrating that cold *per se* is not an essential stimulus for torpor to occur. Schmid and Speakman (2000) also showed that the small lemurian primate *Microcebus murinus* enters torpor spontaneously over a wide range of temperatures in the dry season in Madagascar, but not in the wet season. The only marsupials known to enter deep and prolonged torpor are the mountain pygmy possum, *Burramys parvus*, the pygmy possum, *Cercatetus nanus* (Geiser and Ruf, 1995), and the feathertail glider, *Acrobates pygmaeus*, in which Jones and Geiser (1992) reported body temperatures falling to 0 °C for several days.

Hibernation

In the northern hemisphere, the dual challenge of low environmental temperatures and food shortage coincide on an annual basis in winter and many animals accumulate fat reserves in autumn which will enable them to pass the winter period in a secure hibernaculum (Jallageas and Assenmacher, 1986; Bertolino *et al.*, 2001). Body temperatures may fall as much as 35 °C below normal and rates of metabolism reach 2% of normal euthermic values (Wang, 1989). Bears, for example, hibernate for between three and seven months and neither urinate nor defaecate during that period; body protein stores are preserved by nitrogen recycling mechanisms (Nelson, 1980). Annual energy expenditure was measured in golden-mantled ground squirrels (*Spermophilus saturatus*) with DLW by Kenagy *et al.* (1989) and found to average only 13–17% of the total for the 7.5 months that they spent in winter hibernation.

There has been considerable debate in the literature concerning the physiological distinction between torpor and hibernation, other than their obvious temporal aspect (Geiser, 1994; Geiser and Ruf, 1995). Diet has been shown to affect the length of torpor bouts in hibernating chipmunks (*Eutamias amoenus*) and was significantly increased when the animals were fed a diet high in polyunsaturated fats (Geiser and Kenagy, 1987). A similar effect has been observed in the deer mouse, *Peromyscus maniculatus* (Geiser, 1991).

This seems to be a general effect brought about by changes in the fatty acid composition of the animal's tissues and mitochondrial membranes (Geiser, 1990) and is not restricted to rodents, as a polyunsaturated fat diet also lowered the minimum body temperature of a hibernating marsupial, the feathertail glider (*Acrobates pygmaeus*) and significantly lengthened torpor bouts (Geiser *et al.*, 1992). Interestingly, Geiser *et al.* (1992) have also shown that an unsaturated fat diet will lower the body temperature selected by the lizard *Tiliqua rugosa* in a photothermal gradient, adding credence to Hulbert's (2000) recent thesis on the fundamental importance of polyunsaturated fatty acids in membranes.

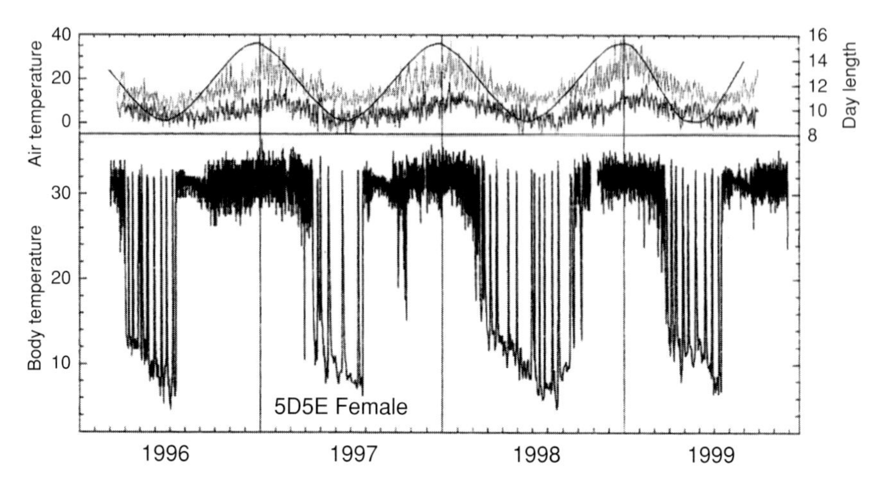

Figure 7.2. Body temperatures recorded in the field from a female echidna (*Tachyglossus aculeatus*) over a period of four years at Lovely Banks in Tasmania. Note that this monotreme enters into hibernation in late summer and autumn and then awakens to breed in mid-winter, when air and ground temperatures are at their lowest. (From Nicol and Anderson (2000) with permission.)

The ability to hibernate was originally thought to be a highly derived (apomorphic) condition that had evolved in a small group of cold-adapted mammals (Bartholomew, 1982). As more and more comparative research has been done, however, the taxonomic spread has increased to now include rodents, insectivores, bats, carnivores, primates, marsupials (Geiser and Broome, 1991; Geiser, 1994) and even monotremes (Grigg *et al.*, 1989) and a plesiomorphic status for hibernation seems much more plausible (Augee and Gooden, 1992).

Recent field research on the monotreme echidna, *Tachyglossus aculeatus*, in Tasmania and the Snowy Mountains of Australia has helped to overturn the long-standing paradigm that hibernation is a cold- and hunger-induced, energy-conserving adaptation. The echidna has been found to fall into torpor and enter hibernation in late summer and autumn, before temperatures fall, and it reawakens to breed in mid-winter, at the coldest time of the year (see Figure 7.2) (Nicol and Anderson, 2000). Grigg and Beard (2000), building on the suggestion of Malan (1996) that animals may utilise torpor or hibernation for energetic advantage – rather than energetic necessity – argue that hibernation in echidnas is an adaptation to anticipated food shortage. They propose that:

> . . . echidnas in cool-cold climates, having regained their fat reserves after the breeding season, can take advantage of falling ambient temperatures underground as

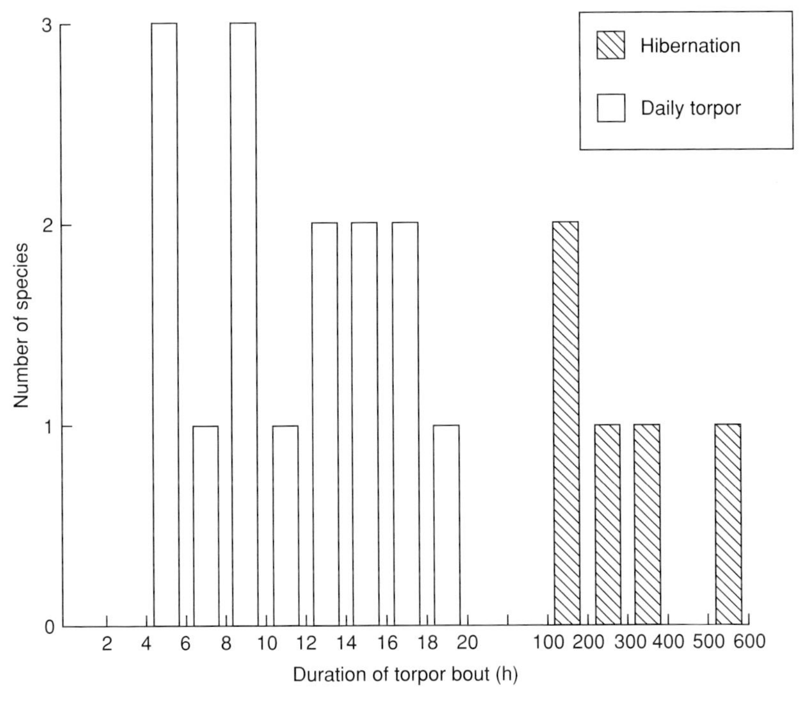

Figure 7.3. Frequency distribution of the duration of the longest torpor bout recorded in 20 species of marsupial and compared with the duration of hibernation bouts. (From Geiser (1994) with permission.)

> autumn advances by entering hibernation and, thus, conserving their energy rich state at minimum cost until the next breeding season.

They believe that this is the first time that it has been suggested that hibernation may occur in the absence of environmental stress, just to conserve stored energy. What is more, they cite a number of instances in the eutherian literature of species displaying a similar proclivity to torpidate or hibernate under mild climatic conditions, and the phenomenon is probably widespread. Somewhat surprisingly, Lovegrove (2000) has shown that the Afrotropical and Australasian regions have the highest number of species, genera and orders of mammals exhibiting daily and summer torpor, adding weight to the argument that torpor and hibernation are not primarily cold-induced responses.

Attempts to find a fundamental physiological or biochemical distinction between torpor and hibernation, other than duration (see Figure 7.3), have proven less than successful over time and the field has benefited greatly from the comparative analysis of torpor patterns by Geiser and Ruf (1995).

Lovegrove *et al.* (2001), working with elephant shrews, concluded that the deep torpor shown by these species was identical in all but duration to the changes seen in seasonal hibernation. Wilz and Heldmaier (2000) have gone even further and conclude that the metabolic depression and fall in body temperature seen in daily torpor, hibernation and æstivation are all forms of dormancy based on the same physiological mechanisms of thermal and metabolic regulation. The term 'hibernation' has also been invoked in the case of reptiles, such as the lizard *Lacerta vivipara,* that overwinters and supercools in winter hibernacula in France (Gregory, 1982; Storey and Storey, 1992; Grenot *et al.*, 1996) and we should not be surprised that the evolutionary origins of this very effective survival strategy will be found in the reptiles, where heterothermy is a daily reality (Malan, 1996). Rismiller and McKelvey (2000) have recently documented 'spontaneous arousal' in a varanid lizard throughout the autumn and winter months on Kangaroo Island, South Australia, although it is important to note that the lizard raises its body temperature above ambient by using radiant rather than endogenous sources of heat.

8

Marine birds and mammals

Vertebrates that live in and by the sea are faced with serious challenges, as the sea is essentially a desert in terms of the availability of fresh water needed for osmoregulation and the maintenance of water and electrolyte homeostasis. How birds and mammals contend with this is still, in many cases, a matter of speculation, especially in the case of mammals such as seals and cetaceans. Ecophysiological studies of marine birds and mammals are still very much in their infancy but recent advances in radiotelemetry and satellite monitoring are rapidly revolutionising what we know of them. Some recent data will be reviewed in this chapter.

Albatrosses

Perhaps the most exciting data to emerge in recent years concern the flight paths of the wandering albatross, *Diomedea exulans*, in the Southern Ocean between the continents of Africa, Australia and Antarctica. These spectacular birds (Figure 8.1) spend more than 95% of their life in the open ocean, ten years of this as immature birds, and half of their mature life during non-breeding years. The long-term survival of these birds is now threatened by the activity of Asian long-line fisheries operating in the Southern Ocean. Brothers (1991) estimated that in one year alone, 1988, some 44 000 albatrosses were killed by taking baits set behind Japanese long-line fishing boats, 9600 of these being wandering albatrosses. Weimerskirch *et al.* (1997) found that the pronounced decline in numbers of wandering albatrosses in the Indian Ocean over the past three decades appears to have reversed since 1986, with a slow increase in numbers which they attribute to decreased fishing effort and a concentration outside the central Indian Ocean by Japanese fisheries. The

Figure 8.1. A pair of wandering albatrosses, *Diomedea exulans*, which mate for life, greeting each other (above) and on the nest (below). (Photo by courtesy of Henri Weimerskirch.)

Amsterdam albatross, *Diomedea amsterdamensis*, on the other hand, remains close to extinction.

Jouventin and Weimerskirch (1990) reported the first results from satellite tracking of breeding wandering albatrosses captured and instrumented on the Crozet Islands, which lie at latitude 46°43′S and longitude 51°86′E.

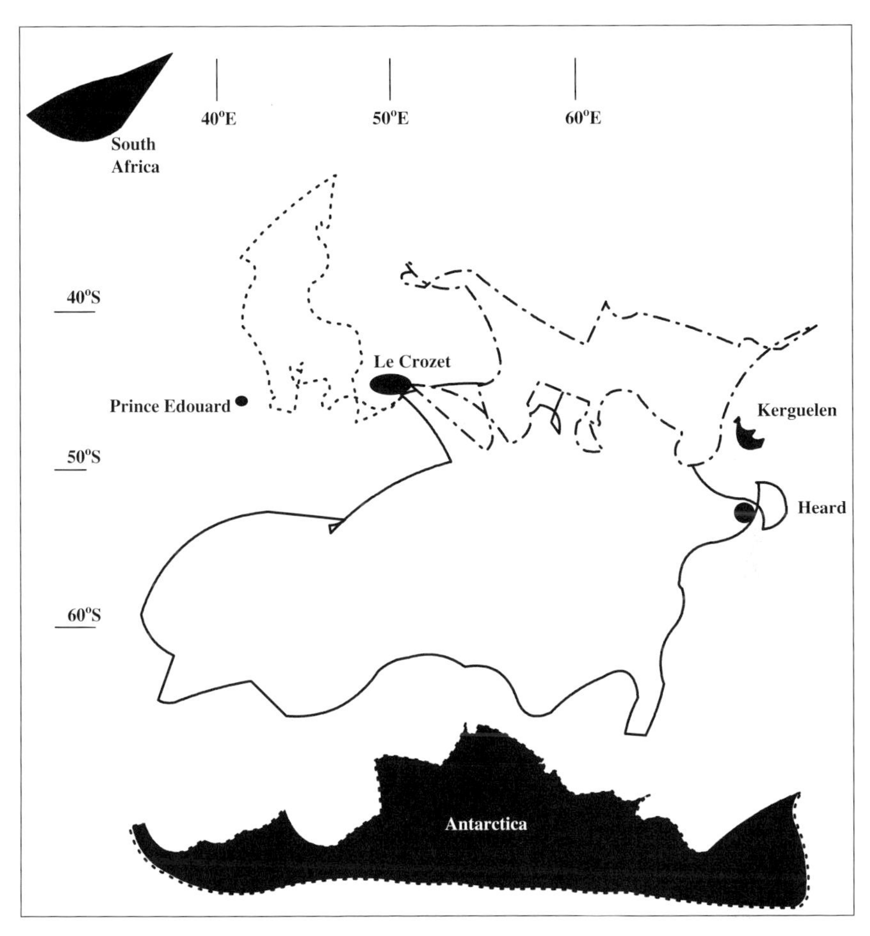

Figure 8.2. Paths of three wandering albatrosses (*Diomedea exulans*) in the region of the Crozet Islands as revealed by satellite tracking. (Adapted from Jouventin and Weimerskirch (1990).)

The tracks of three birds are shown in Figure 8.2. Distances of between 3600 and 15 000 km were recorded between visits to the developing chick, with birds covering up to 900 km in a single day at speeds ranging from 80 to 130 km h^{-1}. Fernandez *et al.* (2001) have recently reported similar extensive foraging paths for three species of *Phoebastria* albatrosses that frequent tropical and subtropical waters at lower latitudes, off the west coasts of North and South America.

The doubly labelled water technique has been used successfully to measure the energy expenditure of free-ranging wandering albatrosses that are feeding

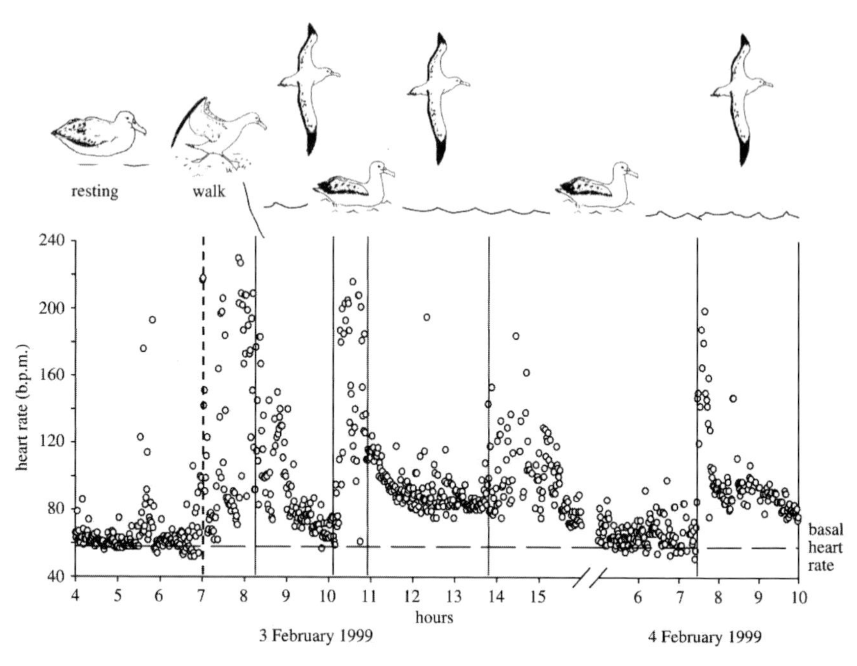

Figure 8.3. Changes in heart rate in wandering albatrosses in different behavioural situations, showing that heart rate falls to near basal levels when flying. (From Weimerskirch *et al.* (2000) with permission.)

growing chicks (Adams *et al.*, 1986; Arnould *et al.*, 1996); their FMRs when flying are extremely low. Daan *et al.* (1990) concluded that wandering albatrosses have one of the lowest ratios of foraging to resting energy expenditure of any bird; early studies on flight energetics models suggested that this was due to the economy of their soaring mode of flight (Pennycuik, 1982). Weimerskirch *et al.* (2000) have recently published the results of a very interesting study in which seven male wandering albatrosses were each fitted with three miniature electronic devices with a total mass of 80 g (0.75% of the body mass of the bird) incorporating a heart-rate transmitter, an activity recorder and a satellite transmitter. Heart rates were measured every 60 s over 5 d during each foraging trip at sea and the activity pattern (sitting on water or in flight) was measured with an activity recorder. Some of the results are shown in Figure 8.3. Heart rates were lowest when the birds were resting on land (65 bpm) and highest when the birds were walking on land or taking off from either land or sea (230 bpm). When in flight, the heart rate fell virtually to resting levels, indicating how economic in terms of energy expenditure their method of dynamic soaring flight can be.

This ecophysiological result, achieved with free-ranging animals, also shows how misleading laboratory-based estimates of energy expenditure may be and it is worth noting that Baudinette and Schmidt-Nielsen (1974) and Hedenström (1993) had estimated that gliding flight incurs a cost of about three times the standard metabolic rate in marine birds. Flint and Nagy (1984) used the DLW method to measure the flight energetics of free-living sooty terns (*Sterna fuscata*) on Tern Island, off the coast of South America, and found that the overall FMR was equal to 4.8 times the standard metabolic rate (SMR), much lower than previous estimates of flight metabolism that were based on theoretical predictions from laboratory studies (Tucker, 1973).

Weimerskirch *et al.* (2000) have also shown that, in order to have the highest probability of experiencing favourable winds, wandering albatrosses use predictable weather systems to engage in stereotypic flight patterns of large looping tracks. When heading in a northerly direction they fly in anticlockwise loops and change to clockwise loops when heading south. Satellite tracking has shown that the Crozet birds do not wander aimlessly but instead travel in flyways that provide favourable winds throughout their foraging trip. Birds at the extreme north and south parts of the range use the prevailing easterly winds to move westwards without a head wind. In the region of the Crozet Islands the birds switch to a figure of eight pattern to counter the strong west–east headwinds (the 'roaring forties') typical of that latitude. When foraging, two discrete search patterns are used by the albatrosses (Weimerskirch *et al.*, 1997): long curvilinear search routes over oceanic waters, with infrequent landings, and concentrated searching in neritic waters, with frequent landings and take-offs in areas where gyres bring favoured prey such as squid to the surface.

The work of Pierre Jouventin, Henri Weimerskirch and their colleagues at the Centre d'Études Biologiques de Chizé in western France has been pivotal in unravelling many of the mysteries of this amazing bird and it would appear that the name 'wandering albatross' is also something of a misnomer! Even in their year 'off' – what Henri calls their 'sabbatical' – when the pair has fledged their young, the birds do not wander aimlessly for a year before returning to breed on land again, as has generally been assumed (Murphy, 1936; Nicholls *et al.*, 1997). Weimerskirch and Wilson (2000) have shown from satellite tracking of four birds (two males and two females) that they all flew to a specific oceanic sector where they overwintered, returning to their breeding place in late December (see Figure 8.4). These sectors ranged from tropical and subtropical waters just south of Madagascar in the case of the females, to sub-Antarctic and Antarctic waters in the case of the males, the distances varying from 1500 to 8500 km from the breeding grounds at the Crozet Islands.

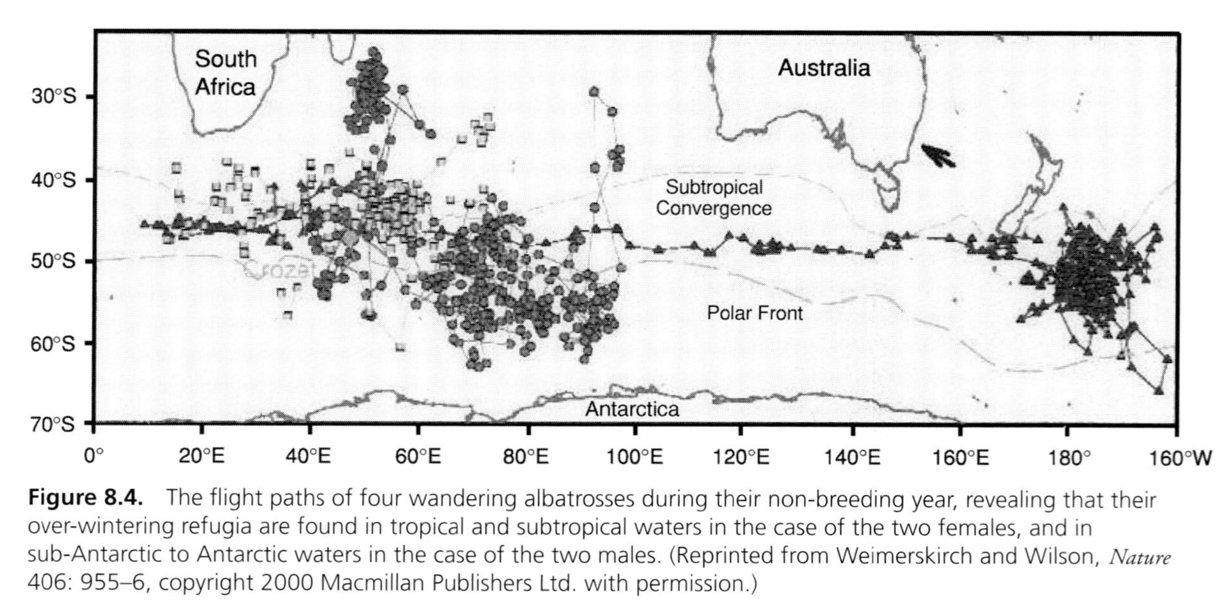

Figure 8.4. The flight paths of four wandering albatrosses during their non-breeding year, revealing that their over-wintering refugia are found in tropical and subtropical waters in the case of the two females, and in sub-Antarctic to Antarctic waters in the case of the two males. (Reprinted from Weimerskirch and Wilson, *Nature* 406: 955–6, copyright 2000 Macmillan Publishers Ltd. with permission.)

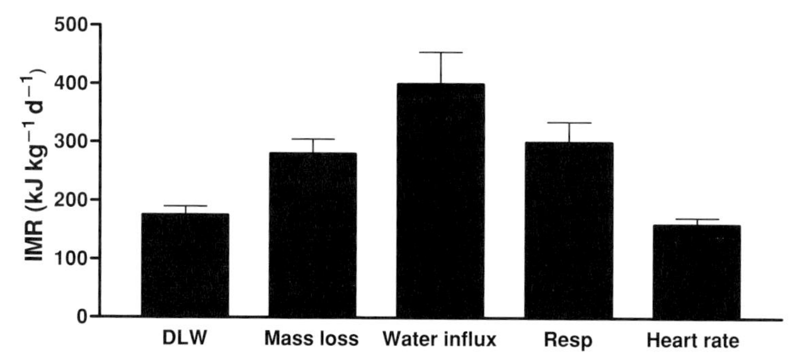

Figure 8.5. A comparison of estimates of incubation metabolic rate (IMR) of wandering albatrosses, using different methods. Note that estimates derived from measurements of heart rate are in close agreement with those measured by using doubly labelled water (DLW). (Modified from Shaffer *et al.* (2001).)

Feeding rates of foraging wandering albatrosses have been ingeniously documented by having birds swallow stomach temperature recorders that transmit a drop in temperature of 2–9 °C each time the bird feeds on very cold prey (see Grémillet and Plos, 1994). It was found that they fed, on average, once every 4.4 h and consumed 2.1 kg of food daily (Weimerskirch *et al.*, 1994).

Various methods have also been used to estimate the energy expenditure of female wandering albatrosses when incubating their one egg. These include the loss in body mass of the female (Croxall and Ricketts, 1983), the rate of water turnover, heart rate (Bevan *et al.*, 1995) and the doubly labelled water (DLW) technique (Pettit *et al.*, 1988). Shaffer *et al.* (2001) have recently compared the effectiveness of these various approaches and conclude that DLW and heart rate give the best estimates of the costs involved. Figure 8.5 compares incubation metabolic rates (IMR) estimated by the various methods and shows the low variance and good agreement between estimates made with DLW and from heart rate telemetry. Using the DLW data, Shaffer *et al.* (2001) estimate that a breeding pair of wandering albatrosses expends 124–234 MJ to incubate the egg for 78 d and that incubation metabolic rates (IMR) were no greater than the best estimates of BMR, suggesting that heat production from adult maintenance metabolism is sufficient to incubate the egg.

Petrels and penguins

Obst *et al.* (1987) used the DLW method to study Wilson's storm-petrels (*Oceanites oceanicus*), which are the smallest endotherms (40 g) living and breeding in the Antarctic and might, therefore, be expected to have high energy requirements for maintenance and growth. The BMR was found to be 37 kJ d^{-1}, which was 25% higher than predicted by avian allometric models. The FMR of incubating birds was 81 kJ d^{-1}, which is 2.2 times the BMR, and significantly higher than that of other birds, but the FMR of storm-petrels in free flight was comparable with that of other highly aerial birds, at 4.2 times BMR. Rates of energy utilisation were maximal for the adults late in the chick-rearing period and greatest during the brooding period. Obst *et al.* (1987) calculated that Wilson's storm-petrels need to consume between 120 and 154% of their body mass in krill per day to satisfy their energy requirements, and that of their developing young, during the breeding season.

The DLW technique has also been used most effectively to estimate rates of energy expenditure of Arctic birds and to quantify their importance in the marine food chain. It has also served to check the accuracy of earlier methods, which were based on time–energy budget (TEB) studies of individual species. Williams and Nagy (1984) and Weathers and Nagy (1980) had suggested that TEB estimates with birds could involve large errors. Gabrielsen *et al.* (1987) used the DLW method in studying black-legged kittiwakes (*Rissa tridactylus*) in the Barents Sea. They found that the FMR of non-foraging birds averaged 596 kJ d^{-1}, or 1.9 times BMR, whereas that of foraging birds was 992 kJ d^{-1}, equal to 3.2 times the BMR. When translated into food intake, these data indicated a consumption of 315 g of fresh matter per bird every other day; these authors estimate that this kittiwake colony, on Hopen Island (76°30′N, 25°03′E) in the Svalbard archipelago, would consume approximately 1.25 t of fresh fish daily.

Green and Brothers (1989) used a combination of the radioactive isotopes, tritium and sodium-22, to estimate rates of food intake in free-living fairy prions (*Pachyptila turtur*) and common diving petrels (*Pelecanoides urinatrix*), which have a circumpolar distribution, on Albatross Island, off the coast of Tasmania (40°23′S, 148°10′E). Influx rates of both sodium and water were high and similar in both species: approximately 800 ml kg^{-1} d^{-1} of water and 170 mmol kg^{-1} d^{-1} of sodium. They estimated that 21–35% of the sodium intake of the two species came from the ingestion of sea water and that rates of krill ingestion varied from 704 to 854 g kg^{-1} d^{-1}. Seabirds feeding on invertebrate species such as krill thus have an extremely high salt intake, which must place severe constraints on their osmoregulatory capacities. Birds, as amniotes, possess a metanephric kidney and a countercurrent osmotic multiplier system similar to that of mammals, but one that has much more limited concentrating capacities, achieving a maximum of no more than 2.5 times plasma osmolality (Skadhauge, 1981). Electrolyte excretion in marine birds, particularly of sodium chloride, is on the other hand vastly enhanced through the functioning of specific salt glands located above the eye sockets. These have been extensively researched in the past few decades (Peaker, 1971; Peaker and Linzell, 1975; Hughes, 1995).

Salt glands

The existence of these salt glands was first discovered by Fänge and Schmidt-Nielsen (1958) in the herring gull and it was these authors who also carried out the first experiments into the mechanism of control of secretion of these fascinating glands (Fänge *et al.*, 1958) (see also Holmes and Phillips, 1985). Very little is known, however, of the functioning of these glands in wild birds as well as the proportion of dietary salt loads that are excreted by this extra-renal route. Green and Brothers's (1989) paper is thus interesting as, in it, they attempt to calculate whether the salt glands of the fairy prions and diving petrels could account for much of the excretion of the salt derived from their diet and the ingestion of sea water.

The sodium concentration of nasal secretions collected from the fairy prions was not very concentrated, compared with that of other seabirds (450 mmol l^{-1}), and they estimated that if 90% of sodium efflux were via the salt gland, this would require a secretory rate of 350 ml kg^{-1} d^{-1}. Secretory rates of up to 0.5 ml kg^{-1} min^{-1} have been measured in herring gulls (Schmidt-Nielsen, 1960) and this would mean that the nasal salt glands of the fairy prions would need to operate for about 12 h each day to maintain sodium balance. Green *et al.* (1988) found that the nasal secretions of the little penguin, *Eudyptula minor*, were, at 1175 mOsm kg^{-1}, much more concentrated than those of the fairy prion, and they estimated that the total daily influx of sodium could be eliminated via this route in as little as 4 h. This seems a reasonable possibility but further ecophysiological work is clearly needed to verify such theoretical estimates. Robertson and Newgrain (1992) tested experimentally the efficacy of both tritium and sodium-22 isotopes in estimating food and energy intake in emperor penguins (*Aptenodytes forsteri*) and concluded that tritium was the most reliable isotope for estimates of overall food intake, with sodium-22 being valuable for the measurement of sea water intake.

A number of other studies have also used radioactive isotopes and oxygen-18 to estimate rates of food and energy intake of other penguins, including little penguins (*Eudyptula minor*) (Green *et al.*, 1988; Gales, 1989; Gales and Green, 1990), jackass penguins (*Spheniscus demersus*) (Nagy *et al.*, 1984), and gentoo penguins (*Pygoscelis papua*) (Gales *et al.*, 1993). The primary aim of these studies has been to estimate feeding rates and the extent to which the severe population declines of these birds over the past 50 years have been the result of competition with commercial fisheries. Given that approximately 80% of all birds in sub-Antarctic waters are penguins and, in the Antarctic region, effectively all birds are penguins (Croxall, 1984), one can see the need for such data.

Gales *et al.* (1993) estimated that the 4700 pairs of gentoo penguins at Macquarie Island consume *c.* 2470 t of prey each year (fish and squid) and the 16 500 pairs at Heard Island consume some 9630 t annually. Nagy *et al.* (1984) estimated that an adult jackass penguin consumes approximately 138 kg of fish per year in meeting its energy requirements. They also calculated that 2.21×10^4 t of fish were consumed annually by penguins using the South African fishing grounds and that they would account for roughly 8% of the average commercial anchovy catch in that area. Studies such as this are clearly vital if we are to have any idea of the long-term impact of expanding fisheries on these native birds. Equally vital is the need for data on the abundance of krill, which is also such a vital food source for Antarctic birds and mammals; the proposals to harvest krill commercially, that have been floated from time to time, are indeed frightening. Almost a decade ago Gales *et al.* (1993) noted that

> Increased pressure is being exerted on the marine environment around Heard and Macquarie Islands by Australian and foreign fisheries interests. The quotas being set by the Australian Government have no regard to the requirements of the seabird and marine mammal populations and to date that information does not exist. (p.137.)

Little has changed in the interim since their paper was published.

Of all the penguins that inhabit the Antarctic desert, none is more fascinating than the emperor penguin (*Aptenodytes forsteri*) with its extraordinary life cycle. The male incubates on its feet the single egg, produced by the female, during some of the fiercest storms and gales known on the face of the Earth. The emperor penguin is unique among Antarctic birds in that its breeding season coincides with the Antarctic winter. In mid-March, as the southern winter approaches, emperor penguins leave the sea and walk over newly frozen sea ice to their rookery, which may be anywhere between 50 and 120 km distant. On reaching the rookery the males and females court and mate, and during early May the female lays a single large egg, which the male collects and places on his feet and then envelops in his brood pouch (Wilson, 1907; Prévost, 1961). The female is then free to return to the sea to feed. The male, by contrast, is totally responsible for the incubation of the egg, which lasts between 62 and 69 days and during which air temperatures may be as low as −30 °C. The male of course cannot feed during this period and loses up to 40% of his initial body mass during this prolonged fast, metabolising in the process all the reserves of lipids that he has built up prior to the breeding season (Le Maho, 1984).

The eggs hatch early in July and the female usually returns to the rookery at this time with a stomach filled with food for the chick but, if she is delayed, the male is capable of feeding the chick with its first few meals with an oesophageal secretion akin to pigeon crop milk (Prévost and Vilter, 1963). Once the female returns to the rookery, she takes over the chick and the male then walks back to the sea, which may be even further away as new ice develops during the winter. After feeding, the male returns to the rookery and then both parents begin a shuttle service back and forth between the sea and the rookery carrying food for the chick, which lasts for about six months until the chick is fledged in late December. Nagy *et al.* (2001) have recently measured the FMR of free-ranging emperor penguins in McMurdo Sound and found that, at 4.1 × BMR, this is the lowest cost of foraging estimated to date among the eight penguin species studied, as shown in Table 8.1. Ancel *et al.* (1992) used satellite tracking to locate the feeding grounds of the males and found that they fed in areas of open water known as 'polynias', which occurred at distances of between 296 and 895 km from the rookery. The mean foraging range of the king penguin (*A. patagonicus*) was also found to be 471 km when birds were monitored by satellite tracking (Jouventin *et al.*, 1994).

The physiological challenges faced by emperor penguins, especially the males, have been studied in some detail; these include thermoregulation (Le Maho *et al.*, 1976; Pinshow *et al.*, 1976) and the biochemical changes occurring during their prolonged fast (Le Maho, 1976; Cherel and Le Maho, 1985). Pinshow *et al.* (1976) estimated from treadmill studies *in situ* that walking 200 km to and from sea to rookery requires less than 15% of the bird's energy reserves but that thermoregulatory requirements should consume 85% of those same reserves. Figure 8.6 shows the relation between metabolic rate and ambient temperature and indicates that the emperor penguin's lower critical temperature is close to −10 °C, below which it must increase heat production to prevent a fall in body temperature. Male emperor penguins, however, diminish considerably their thermoregulatory costs by huddling together in large groups of up to 1000 birds, and in this way they protect themselves from the gale-force and freezing winds that blow throughout the winter period. As Le Maho *et al.* (1976) and Ancel *et al.* (1997) show, without this social behaviour, survival during the long fast at winter temperatures would be impossible.

Yvon Le Maho, from Strasbourg University in France, has done most to clarify our understanding of the extended fast of both emperor and king penguins in the Antarctic and how protein reserves are protected while lipids

Table 8.1. *A comparison of field metabolic rates (FMR), measured with doubly labelled water, and foraging costs of penguins*

Species	Period	Body mass (kg)	Overall FMR (kJ d^{-1})	×BMR	Foraging FMR (kJ h^{-1})	×BMR	Reference
Little penguin (*Eudyptula minor*)	Annual	1.09	1050	3.7	—	—	1,2
Little penguin	Non-breeding	1.13	1010	3.6	81	6.9	1,2
Little penguin	Breeding	1.05	1953	6.9	116	9.9	1,2
African penguin (*Spheniscus demersus*)	Breeding	3.17	1945	2.6	207	6.6	3
Chinstrap penguin (*Pygoscelis antarctica*)	Breeding	3.79	5600	4.7	247	5.0	4
Adélie penguin (*Pygoscelis adeliae*)	Breeding	3.99	3790	3.1	260	5.1	5,6
Macaroni penguin (*Eudyptes chrysolophus*)	Breeding	4.27	3085	4.1	241	7.7	7
Gentoo penguin (*Pygoscelis papua*)	Breeding	6.10	3740	3.1	362	7.2	7
King penguin (*Aptenodytes patagonicus*)	Breeding	12.9	7410	3.1	464	4.6	8
Emperor penguin (*Aptenodytes forsteri*)	Non-breeding	23.7	9546	2.9	590	4.1	9

Sources: [1]Costa *et al.* (1986), [2]Gales and Green (1990), [3]Nagy *et al.* (1984), [4]Moreno and Sanz (1996), [5]Nagy and Obst (1992), [6]Chappell *et al.* (1993), [7]Davis *et al.* (1989), [8]Kooyman *et al.* (1992), [9]Nagy *et al.* (2001). Table from Nagy *et al.* (2001).

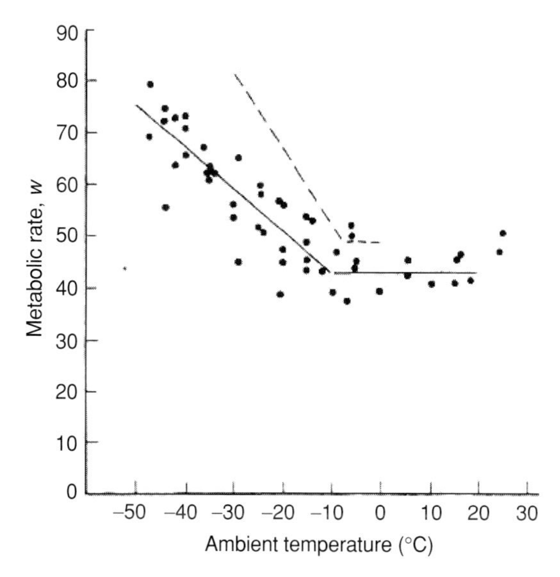

Figure 8.6. Variation in the metabolic rate of emperor penguins, *Aptenodytes forsteri*, as a function of ambient temperature. Note that the lower critical temperature for this species is approximately −10 °C. (From Pinshow *et al.* (1976) with permission.)

remain in the body (Le Maho, 1984). The period of fasting is characterised by three phases: (I) an initial phase lasting 10–14 d when mass loss is high and nitrogen excretion is elevated; (II) a second very prolonged phase where mass loss is constant at about 1% of the body mass per day and nitrogen excretion is minimal and, finally, (III) a third critical phase where the rate of mass loss doubles and nitrogen excretion increases as body proteins are catabolised (see Figure 8.7). Rates of uric acid excretion increase approximately three-fold at this time (Robin, 1984). The abrupt start of Phase III apparently coincides with the time when the bird's extensive lipid reserves are exhausted and blood glucose concentrations start to be maintained by gluconeogenesis. Whether this metabolic change is signalled by an increase in the rate of secretion of corticosterone from the adrenal gland is not known, cortisol and corticosterone being catabolic steroid hormones, but it is at this precise time that the male emperor penguin abandons its fast and heads to the sea if the female does not return. Under these circumstances, of course, the chick will die.

Diving abilities of penguins, seals and whales

Another field of great interest about marine birds and mammals is their diving ability. The South Georgian shag, *Phalacrocorax georgianus*, for example, is a

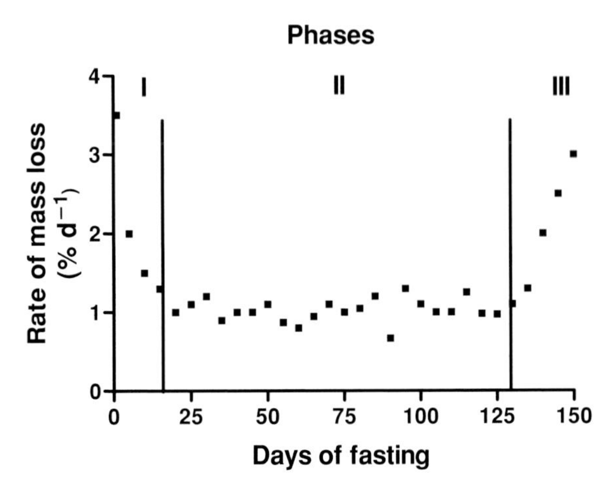

Figure 8.7. Rate of mass loss of fasting male emperor penguins (*Aptenodytes forsteri*) showing the three phases, with Phase III being the critical one where protein catabolism commences once the bird's fat reserves have been exhausted. See text for details. (Adapted from Le Maho (1984).)

foot-propelled pursuit diver that can remain submerged for up to 5.2 min and reach depths of 116 m (Croxall *et al.*, 1991). Bevan *et al.* (1997) recorded heart rate and stomach temperatures during dives on Bird Island, South Georgia, in the Antarctic and found that heart rate fell from 104 to 65 beats min^{-1} (bradycardia) during dives and the minimum heart rate was negatively correlated with both dive duration and dive depth (i.e. the lower the rate, the deeper and longer the dive). Although diving ability in birds is usually assumed to be fuelled aerobically, and calculations are made of the so-called aerobic dive limit, or ADL, Croxall *et al.* (1991) found evidence of anaerobic fuelling of dives in the South Georgian shag and similar results have been reported by Ponganis *et al.* (1997) in the emperor penguin and by Ponganis *et al.* (1999) for the king penguin, *Aptenodytes patagonicus*, which has been recorded as diving to a depth of 240 m by Kooyman *et al.* (1982). Kooyman *et al.* (1992), working with instrumented emperor penguins, also found that muscle oxygen depletion must occur frequently during prolonged dives and that a significant energy contribution from anaerobic glycolysis must occur. Ancel *et al.* (1992) recorded dives of over 400 m in depth by free-swimming emperor penguins in the Ross Sea. The gentoo penguin dives to depths of up to 210 m with dive durations of 7 min having been recorded (Stonehouse, 1975; Bost, 1994). It seems likely that future research will establish anaerobic fuelling as a common feature in these deep-diving birds. As Nagy *et al.* (2001) emphasise in their study of the emperor penguin, 'More research is required to determine ADLs with measurement of postdive lactate concentrations and to examine O_2 store depletion rates within the breath-hold period.'

Rather more research has been done on the diving physiology of marine mammals – primarily seals – but data on whales are accumulating steadily. Much of the earlier work with seals was carried out with captive or restrained, rather than free-ranging, animals, but recent exciting advances in telemetric techniques are rapidly changing this situation and providing vital information of a genuinely ecophysiological nature. The thinking of researchers has changed dramatically a number of times over the years with regard to the nature of the physiological and metabolic changes associated with diving in mammals. For over four decades, laboratory studies with aquatic animals have shown that simulated diving (i.e. simply forcing the animal's head under water) is accompanied by apnoea, bradycardia and peripheral vasoconstriction, a set of adjustments known loosely as the 'diving response', which flowed from the pioneering researches of Scholander and his colleagues in the 1940s (Scholander, 1940; Scholander *et al.*, 1942). This led to the accepted concept that seals performed long anaerobic dives interspersed with long recovery periods, during which accumulated lactate was burned off.

The first studies with freely diving animals in the 1970s found, on the other hand, that aquatic birds and mammals generally carried out short aerobic dives, interspersed with short recovery periods (see, for example, Kooyman and Campbell, 1972) and this led to the already mentioned concept of the 'aerobic dive limit', or ADL, which is defined as the diving duration beyond which lactate concentrations rise in the blood (Kooyman *et al.*, 1980, 1983; Guppy *et al.*, 1986). As more data have accumulated, however, with instrumented birds and freely diving seals, the situation has become more complex and rather less simple to interpret (Kooyman, 1998; Ponganis and Kooyman, 2000). Firstly, it is clear that a number of seals routinely carry out dives that exceed by far their calculated ADL and must thus be relying on anaerobic metabolism to fuel their underwater activity (Ponganis *et al.*, 1993). Thompson and Fedak (1993), for example, found that heart rates of grey seals (*Halichoerus grypus*) fell from 110 beats min^{-1} to an amazing 4 beats min^{-1} during long dives, the duration of which far exceeded estimated aerobic dive limits, even assuming resting metabolic rates. These data suggest that grey seals employ significant cellular energy-sparing mechanisms to prolong their dives and that these may affect in an important way their mode of prey capture.

Even more confusing, recent studies with elephant seals (*Mirounga angustirostris*) and grey seals have found that these species may spend as much as 90% of their time submerged, with some dives being longer than the predicted ADL, but without extended recovery periods at the surface, which one would have thought necessary to oxidise accumulated blood lactate (Le Boef *et al.*, 1989; Hindell *et al.*, 1992; Thompson and Fedak, 1993). Reed *et al.* (1994) worked with captive but freely diving grey seals and found that they exhibited high rates of gas exchange at the surface but, paradoxically, found that the oxygen concentration in the first breaths of air expired when the seal surfaced were not low, as expected, but quite high. This suggests that actively respiring tissue such as muscle is being under-perfused with blood during the dive and that blood oxygen stores are consequently not being used by the tissues. This would imply that there is little transfer of oxygen between blood and muscle during the dive and that the muscles use oxygen stored in myoglobin, which is depleted during the dive. As the dives recorded in this study were short, shallow dives, it suggests that grey seals routinely employ substantial physiological responses to conserve oxygen, and that these responses are not reserved solely for prolonged dives.

Reed *et al.* (1994) analysed the metabolic characteristics of the muscles of a number of seals and found no novel enzymatic pathways or muscle fibre types adapted for diving; the concentrations of myoglobin can adequately explain the differing diving capacities of the species. Castellini *et al.* (1985) report some very interesting early results following the injection of a bolus of radioactively labelled metabolites (glucose and palmitate) into the circulation of seals (species not specified!), which highlight the diametrically opposed metabolic demands placed on these animals when either exercising or diving. They found that processing of glucose and palmitate differed dramatically between the two energetic states and recommended that this approach be used when attempting to define the precise nature of the metabolic responses employed by freely diving seals.

Studies with species such as the southern elephant seal, *Mirounga leonina*, would be enormously valuable as this seal is a prodigious diver, spending 90% of its time at sea, diving to average depths of 300–600 m (Slip *et al.*, 1994) but with recorded dives down to 1500 m (McConnell *et al.*, 1992; McConnell and Fedak, 1996). Bennett *et al.* (2001) have recently studied diurnal and seasonal variations in the duration and depth of the longest dives by elephant seals at South Georgia and advance the very interesting speculation that the hormone melatonin may be involved in regulating diving behaviour and help protect the seals against high concentrations of damaging free radicals that are potentially absorbed at the termination of each dive.

Much less is known of the physiological adaptations that support diving in whales, even though their performance is impressive with known dives, for example, in excess of 600 m lasting for 18 min having been recorded for the white whale, *Delphinapterus leucas*, by Martin *et al.* (1993). Sperm whales have been reported to remain submerged for over 1 h (Lockyer, 1997). One of the few detailed studies is that of Shaffer *et al.* (1997), working with trained white whales diving to a platform that was positioned at depths between 5 and 300 m. They found that dives varied in duration from 2.2 to 13.3 min and that blood lactate concentrations were increased by as much as 4–5 times above resting after dives of 11 min. A large blood volume, high haemoglobin concentration and a high haematocrit are factors that give whales a high blood oxygen-carrying capacity (Ridgway *et al.*, 1984). Maximum breath-hold duration in the white whale was found to be 17 min, with an ADL of 9–10 min. To date, this is the only study of the physiological changes accompanying diving in a whale and suggests that they may conform to the anaerobic diving paradigm earlier advanced for seals by Kooyman *et al.* (1980). It is to be hoped that the formidable technical difficulties of working with free-diving whales, involving the repetitive sampling of blood and lung gases, will be resolved in the not too distant future and make possible comparisons with what we already know of diving birds and seals.

Gas exchange and heart rate changes have been studied in the harbour dolphin, *Phocoena phocoena*, by Reed *et al.* (2000). This species, at 80 kg, is at the low extreme of body size in cetaceans. They found that breath-hold durations in this laboratory study of two juveniles were lower than recorded from free-diving individuals, which may reach 3–5 min, but they concluded that this species was not well adapted physiologically for long duration dives, but instead for short, relatively shallow dives with rapid gas exchange at the surface.

Osmoregulation and kidney function

The problems posed by osmoregulation in sea water-living birds and mammals have already been alluded to. As we have seen, most marine birds possess nasal salt-secreting glands that are at least capable of coping with excessive salt loads from the diet and sea water drinking, even if the ecophysiological proof of this is yet to come. The situation is different with marine mammals, pinnipeds and cetaceans, which lack such extrarenal routes for the excretion of sodium chloride. In addition, from what little is known of their kidney morphology and function, they lack the capacity to produce very concentrated urine, unlike that of the desert rodents discussed in Chapter 6.

Beuchat (1990) has collated data from the literature on body size, medullary thickness and urinary concentrating ability from a wide range of mammals. She found that the maximum urine concentration actually declines with increasing body size, and derived the significant allometric relation of:

$$U_{\text{osm}} = 2564 M^{-0.097},$$

where U_{osm} is the osmolality of the urine in mOsm kg^{-1} and M is body mass in kg. Maximum urinary concentrations for the three dolphin species listed range from 1700 to 2658 mOsm kg^{-1} and those for the two seals were 2144 and 2298 mOsm kg^{-1}, which is about the concentration recorded for the white rat. I am not aware of any published figures from whales, but one might expect maximal urinary concentration to be even lower and, at 155 tonnes the above equation would predict a maximum urine concentration of only 804 mOsm kg^{-1} for the blue whale, which is lower than that of sea water!

Hedges *et al.* (1979) studied the structure of the kidney of the harbour dolphin, *Phocoena phocoena*, and concluded that its reniculate (lobulate) morphology is indicative of superior concentrating capacity and allows them to maintain osmotic balance while drinking sea water (known as 'mariposia'). Hui (1981) used tritiated water and sodium-22 to measure low rates of sea water consumption in common dolphins (*Delphinus delphis*) and estimated this at 12–13 ml kg^{-1} d^{-1}. Similar claims have been made for the Weddell seal by Kooyman and Draber (1968) and common seal (*Phoca vitulina*) by Tarasoff and Toews (1972) but full-scale studies of kidney function, and the role of adrenal and pituitary hormones, are lacking in both seals and cetaceans. Vardy and Bryden (1981) described the morphology of the kidney of three species of pinniped but concluded that they showed no evidence of specialisation to their marine existence. Reilly and Fedak (1991) have measured rates of water turnover and energy expenditure in free-living male common seals, using doubly labelled water, and found that water influx rates were slightly below allometric predictions.

Early studies by Malvin *et al.* (1978) suggested that the adrenal steroid hormone aldosterone and the kidney enzyme renin may play a significant role in electrolyte homeostasis in seals, although the action of vasopressin in cetaceans was questionable (Malvin *et al.*, 1971). Bradley *et al.* (1954), in a very early publication, reported an antidiuretic effect of pitressin (an extract of the neurohypophysis of pigs) on rates of urine flow in harbour seals and Hong *et al.* (1982) confirmed this in the harbour seal (*Phoca vitulina*) and ringed seal (*Pusa hispida*) by injecting arginine vasopressin (AVP). The physiological role of AVP as an antidiuretic hormone in marine mammals has not, however, been clearly established and requires studies in which changes in circulating concentrations of the hormone are monitored in different states of hydration and electrolyte loading.

Rudy Ortiz is working to bridge this gap and has recently published some interesting papers on osmoregulation in West Indian manatees (*Trichechus manatus*), which are members of the family Sirenia, and known as 'sea cows'. Ortiz *et al.* (1998) found that plasma AVP concentrations were higher in sea water-caught manatees than in fresh water (2.1 ± 0.48 *versus* 0.6 ± 0.04 pg ml^{-1}) whereas plasma aldosterone concentrations were lower (95.0 ± 12.6 *versus* 659.6 ± 103.3 pg ml^{-1}), thus suggesting classic osmoregulatory rôles for these two hormones in this species. Plasma AVP concentrations were also significantly correlated with plasma osmolality, which is supporting evidence for AVP functioning as a physiological antidiuretic hormone in the manatee.

Ortiz *et al.* (1996) also studied osmoregulation and the role of AVP in fasting elephant seal pups (*Mirounga angustirostris*) but were unable to conclude that AVP functioned to conserve body water in this species. In fact, although rates of urine production fell throughout the 8–12 week fast that normally follows their weaning, so did plasma AVP concentrations, and it would appear that water conservation was achieved through a reduction in renal blood flow and glomerular filtration rate (GFR), in a similar fashion to that already seen in desert rock wallabies in Chapter 6 (see Adams and Costa, 1993; Houser and Costa, 2001).

Central to this debate is the question of the extent of salt intake of sea water mammals, and particularly the amount of salt water that they may ingest with, and in addition to, their food. Sea water drinking has been questioned in dolphins (Telfer *et al.*, 1970) but is implied in both seals and dolphins from various turnover studies using radioisotopes (see Depocas *et al.*, 1971; Gentry, 1981; Hui, 1981). Hopefully, the coming years will see more work on water and electrolyte balance and kidney function of marine mammals that will answer these questions. Interestingly, $^{18}O : {}^{16}O$ isotope ratio studies with fossil whales suggest that pakicetid whales, which lived some seven million years ago, drank fresh water whereas the ratio in the bones of protocetid whales indicates that sea water drinking is a relatively recent innovation in whales (Thewissen *et al.*, 1996; Roe *et al.*, 1998).

Sea lions

Ecophysiology deals with reproduction as perhaps the most important event in an animal's life, when its genetic material is passed on to another individual, and it is perhaps fitting to conclude this chapter with some details of one of the most extraordinary stories to emerge concerning the reproductive cycle of a marine mammal. Pinnipeds are divided taxonomically into the Phocidae (true seals) and the Otariidae (sea lions and fur seals) and both have evolved life cycles where the young are born and weaned on ice or land. True seals have a brief lactation period in which growth of the young is rapid, fuelled by energy-dense milk, and weaning is abrupt. The sea lions on the other hand produce milk of lower lipid content, lactation is more protracted, growth of the young is slower and weaning is less abrupt. Despite these differences, seals and sea lions characteristically display a synchronous annual breeding season, which is cued in the different species to the seasonal availability of food resources (Campagna *et al.*, 2001; Crocker *et al.*, 2001). The one exception appears to be the Australian sea lion, *Neophoca cinerea.*

The present distribution of this species extends from the Abrolhos Islands on the central west coast of Australia to South Australia but, in the past, populations once extended to Tasmania and the east coast of Australia (Baudin, 1802; Flinders, 1814; Gales *et al.*, 1994). Breeding has been recorded on 51 islands, 28 of which are in Western Australia, and the total population is estimated at between 9000 and 12 000 individuals. The average size of Australian sea lion breeding populations is, however, the smallest of any otariid, with only five of the known 51 breeding populations producing more than 100 pups per season. Five colonies on South Australian islands account for almost 50% of the annual crop of young with rates varying between 120 and 300 pups per season. The other 46 colonies produce, on average, only 20–30 pups per season (Gales *et al.*, 1994).

Nick Gales and Dan Costa have recently reviewed the life history of the Australian sea lion and summarised evidence, originally established by Higgins (1993), to show that this species does not have an annual breeding cycle, breeding instead on a 17.6 month cycle (Gales and Costa, 1997). Since the cycle is just less than 18 months, this means that in any one population the time and season of breeding will drift gradually, with animals breeding in every month of the year over time. If this were not enough to make the species unusual, Gales and Costa (1997) also found that all populations are not synchronous, even when located on nearby islands, and populations may breed at times differing one from the other by as much as ten months. There are, of course, well-known cases of a species showing a great variation in timing of breeding in different parts of its geographic range (Bronson, 1989), and there are also opportunistic breeders such as the red kangaroo, *Macropus rufus,* and birds such as the budgerygah in Australia that can breed at any time of the year, depending on rainfall (Serventy, 1971; Newsome, 1975). So far as I am aware, however, the Australian sea lion is the only instance of a mammal that breeds supra-annually and thus completely independently of time and season.

The gestation period of the Australian sea lion is also 17.6 months, followed by a post-partum oestrus and mating. Delayed implantation was first reported in this species by Tedman (1991). Figure 8.8 shows measurements of plasma oestradiol-17β and progesterone concentrations throughout pregnancy by Gales *et al.* (1997). There is a significant increase in the level of both hormones

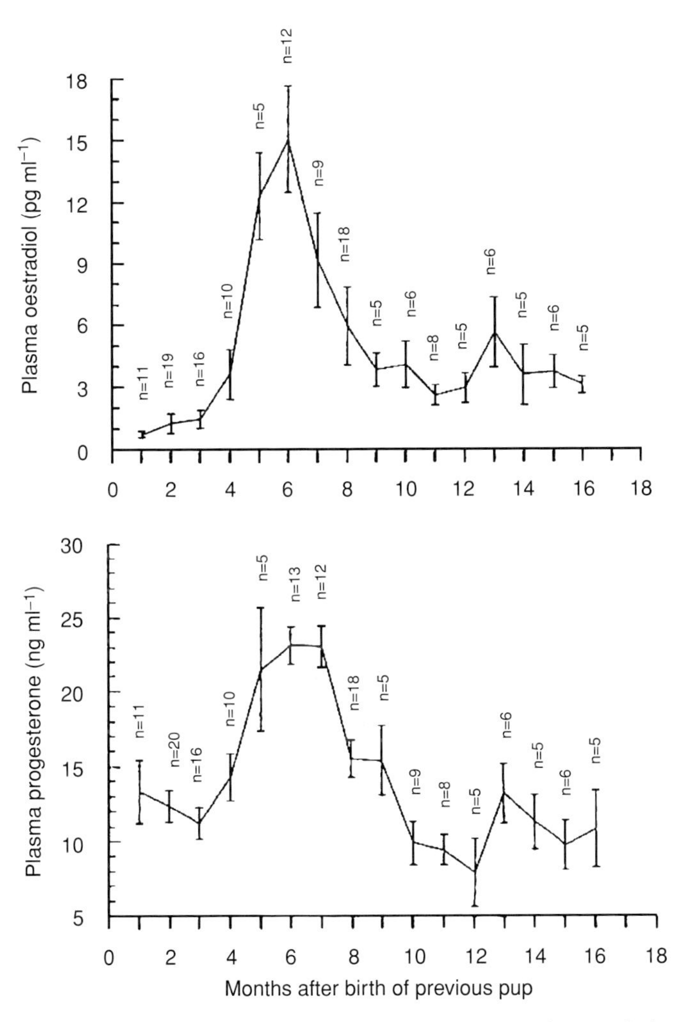

Figure 8.8. Changes in circulating concentrations of oestradiol-17β and progesterone in female Australian sea lions, *Neophoca cinerea*, during pregnancy. The sequential increase, then fall, of both hormones at 5–6 months into the pregnancy has been interpreted as signalling the time of blastocyst implantation, which is then followed by the longest post-implantation period of any sea lion. (From Gales *et al.* (1997) with permission.)

at 5–6 months of the pregnancy, which they interpret as the time of blastocyst reactivation and implantation, leaving a prolonged placental phase of up to 11–12 months, the longest yet recorded for any pinniped.

The interpretation advanced by Gales and Costa (1997) to explain the evolution of the quite unique life history of the Australian sea lion is based

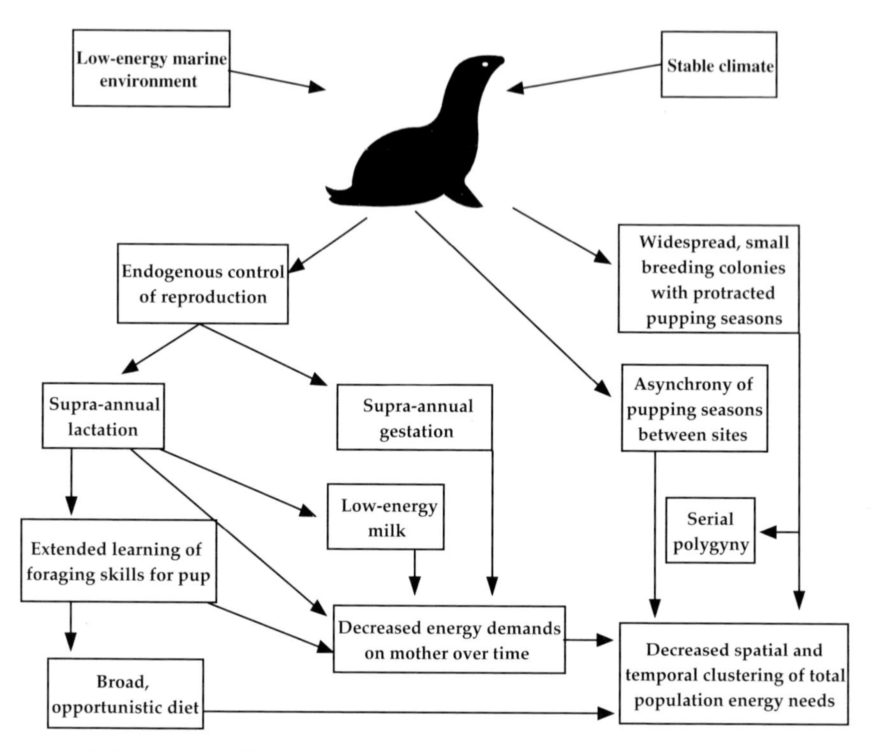

Figure 8.9. Schema illustrating the unusual life history of the Australian sea lion, which appears to have evolved an asynchronous supra-annual breeding cycle in response to the nutrient-poor waters of the southern coast of Australia. (Adapted from Gales and Costa (1997) with permission.)

on their opportunistic feeding behaviour and catholic tastes with respect to prey. Prey are not abundant in the oligotrophic waters along the west and south coasts of Australia, and energetic studies indicate that the sea lions need to work harder in catching their prey than do similar-sized California sea lions, with an FMR estimated at some 6.8 times their predicted BMR (Costa *et al.*, 1989). Taken together with a supra-annual life cycle, and a lack of temporal synchrony between breeding populations, these features suggest an animal adapted to a stable, low-energy environment. The waters frequented by the Australian sea lion are characterised by their low productivity (Pearce, 1991) and have been described as one of the most nutrient-deprived marine environments to be found in the world (Rochford, 1980).

The west coast of the Australian continent is atypical in not having a nutrient-rich cold current running northward along the coast, as part of the general ocean circulation, the result of which would be a coastal desert, as is seen in Africa and in North and South America. Instead, a warm current known as the Leeuwin current flows southwards (Pearce and Walker, 1991) and extends a tropical marine influence well into temperate latitudes, with corals, for example, proliferating on the Abrolhos Islands at approximately 28°S. This current has its origins in the nutrient-depleted waters of the Indonesian archipelago and flows down the west coast of Australia, preventing the normal, northward flow of cold, nutrient-rich water from the Antarctic.

Gales and Costa (1997) argue that this nutrient-poor, relatively stable marine environment explains the evolution of the unique reproductive pattern and life history of *Neophoca cinerea.* Some endogenous rhythms of reproduction that have been described in birds are thought to be adaptive in non-variable trophic environments (Murton and Westwood, 1977) and circannual reproductive patterns, linked to the warm Leeuwin current, have also been described in some Western Australian birds by Wooller *et al.* (1991). For the Australian sea lion, there is no nutritional advantage to be had by breeding at any one particular time of the year; by prolonging gestation and by spreading the energetic demands of lactation over a longer period, it is better able to garner the resources needed from a poor environment. Gales and Costa (1997) offer an interesting flow chart, shown in Figure 8.9, which summarises their interpretation of the determinants of life history traits of this fascinating and unique otariid.

9

Conclusion

The preservation of the Earth's fast-disappearing biodiversity is one of the major preoccupations of biologists today. One may take Australia as a typical example. The continent was colonised by the Europeans a little over 200 years ago, and 200 species have become extinct in that short time – one species per year – which is a rate greater than that recorded in any other country or habitat, including the Amazonian rain forests. The major reason for this has been the destruction and modification of the habitat such that the indigenous animals are no longer able to survive. In Australia, introduced predators (cat, fox, dog) and competitors (sheep, cow, rabbit) have also played a part in exterminating vulnerable marsupials, but whatever the agent, humans must shoulder the ultimate responsibility.

How can ecophysiology help to stem this tide of extinction? Hopefully, by providing us with precise details of how animals manage to survive in their diverse environments before they become extinct, and by pinpointing aspects of their biology where they are particularly vulnerable. Often when individuals of a species become so rare as to be judged 'threatened' or 'vulnerable to extinction' it is very difficult to reverse a trend that has perhaps been in action but gone unnoticed for decades. The western swamp tortoise, *Pseudemydura umbrina*, is a good example. This is a small fresh water tortoise first collected in Western Australia in 1839 and thought to be extinct until a schoolboy brought a live specimen into the WA Museum in 1953. The Museum curator, Glauert, did not recognise the tortoise and described it as a new species, *Emydura inspectata*. It was only when Ernest Williams, a chelonian expert in the United States, recognised in 1958 that it was identical with the single specimen held in the Museum in Vienna that the rediscovery was made (Williams, 1958). A search was made in the swamp where the boy collected the specimen and a small number of tortoises were found and taken into captivity in Perth Zoo. They subsequently laid eggs and a few young were hatched, but then they stopped breeding. Some 25 years later the situation had not changed and *Pseudemydura umbrina* was listed as the world's most endangered chelonian, and Australia's most threatened vertebrate, with an estimated 12–15 individuals left alive.

The few remaining tortoises were taken over by the State Department of Conservation and Land Management (CALM), but all efforts to induce them to breed were fruitless. It was only as a result of a chance encounter between two researchers that the problem was solved. Bruno Colomb, a French doctor who is passionate about birds, was working in my laboratory alongside Gerald Kuchling, who is equally passionate about tortoises. Bruno suggested using ultrasound to check the state of the ovaries of the females – a common medical procedure with which he was familiar – and, when Gerald tried this we were amazed to see developing follicles! These grew larger during early spring then mysteriously started to regress and finally disappear.

Ecophysiology played its role when we puzzled over why the eggs had not developed. We realised that the western swamp tortoise is an unusual chelonian that aestivates over summer, and only feeds in temporary ponds that form in winter and disappear at the end of spring. Just as the waters surrounding the Australian coast are impoverished in terms of essential nutrients, such as nitrogen and phosphate, so are the fresh water streams and ponds that form in late winter and spring. These fresh water pools are, however, rapidly colonised by a suite of invertebrate crustaceans and insects that provide a short-lived but rich source of food for birds, fish and tortoises.

We reasoned that perhaps the tortoises failed to breed because the amount of food that they were given was insufficient to fuel the growth of the follicles in the ovaries and what they needed, instead, was a huge burst of nutrients, as they would experience if living in the wild. In the next spring we therefore enormously increased the amount of food available to them and they went on to lay eggs, from which the first young hatched. Today some hundreds of western swamp tortoises have been bred successfully in captivity and released back into the wild and it is now one of Australia's most successful recovery programmes of an endangered species (Kuchling, 1999).

We have also seen, when discussing marine birds and mammals in Chapter 8, how the future of the Antarctic populations of these unique species is by no means secure. Longline fisheries have resulted in the deaths of huge numbers of albatrosses and the disastrous story of the world-wide decline in numbers of whales due to hunting, which is only now starting to be reversed, is well known. As a result of the efforts of researchers such as Rosemary Gales in Australia, John Croxall in the UK and Henri Weimerskirch and Pierre Jouventin in France, the plight of albatrosses has gained increasing attention world-wide and been the subject of numerous meetings, reports and publications (see, for example, Weimerskirch, 1987, 1998; Cherel *et al.*, 1996; Croxall and Gales, 1998) as well as the excellent book edited by Robertson and Gales (1998).

Weimerskirch *et al.* (1999) have also shown that the white-chinned petrel (*Procellaria aequinoctialis*) forms the majority of birds killed by longline fisheries in the Southern Ocean, even though this species breeds in Antarctic waters. This is because satellite tracking has revealed that this species has the longest mean foraging range ever recorded for a seabird during the incubation period, ranging from 2200 to 3495 km; Crozet Islands birds fly to South Africa, and South Georgia birds reach Patagonia to feed (see Figure 9.1). These data indicate that conservation measures limited to Antarctic waters are quite ineffectual in protecting seabirds with such extensive foraging ranges.

The long-term effects of global warming on Antarctic populations of birds and mammals are also unknown at this time, but predictions from what data are available are very disturbing. Barbraud and Weimerskirch (2001) have recently published population data on the extent of the emperor penguin colony at Dumont d'Urville Station in Terre Adélie over the period from 1952 until 2000, representing the longest time series available on demographic parameters of an Antarctic marine predator breeding on fast ice. They found that the population has declined by 50% and attribute this to a marked decrease in adult survival during the late 1970s, when there was a prolonged abnormally warm period with reduced sea-ice extent. Meteorological data show that, after a period of stability in the 1960s, with average temperatures of $-17.3\,^{\circ}$C, winter temperatures began to vary extensively and were high throughout the 1970s and remained so until the early 1980s, averaging almost three degrees warmer at $-14.7\,^{\circ}$C. During this warm phase, annual survival rates of both adult males and females, based on mark-and-return capture rates, were reduced, falling from 0.95 to as low as 0.65, and this coincided with a rapid contraction in the size of the breeding population.

Adult emperor penguins forage only at sea and feed preferentially on Antarctic krill (*Euphausia superba*), fish (primarily *Pleurogramma antarcticum*) and squid. There are no data for the extent of these resources during the 1970s but one can only speculate that the local warming of the sea and the reduction of the extent of the sea ice had a negative impact on food abundance, which translated into poor adult survival for the penguins and their young. Christophe Barbraud and Henri Weimerskirch were able to establish that the extent of the pack ice has important implications for the survival of the

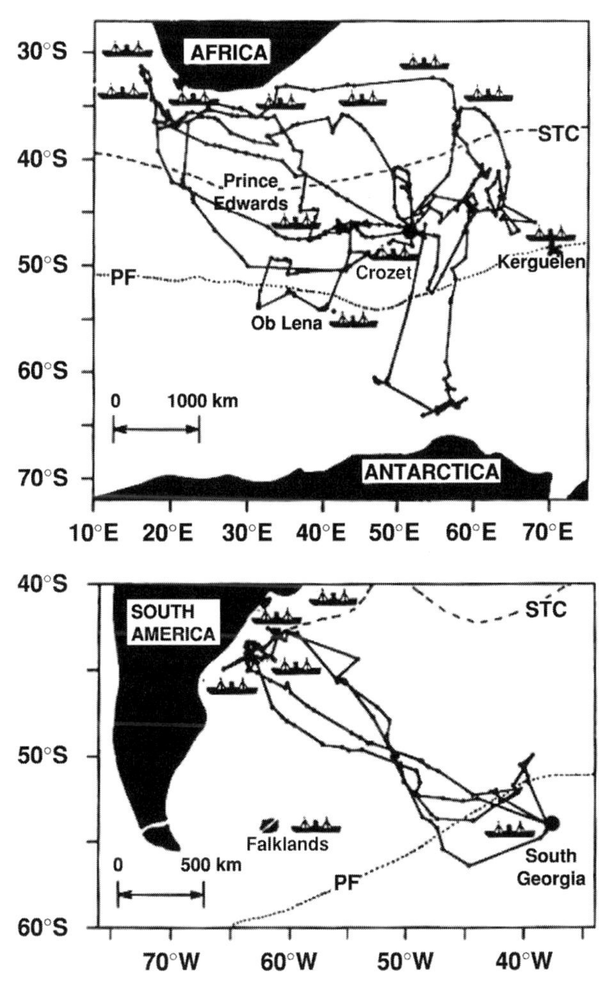

Figure 9.1. Foraging areas of white-chinned petrels, *Procellaria aequinoctialis*, which has the longest mean foraging range during the incubation period of any seabird at 2200–3495 km and forms the majority of the birds killed each year by longline fisheries. (From Weimerskirch *et al.* (1999) with permission.)

emperor penguins and that it operates via two opposing mechanisms. Annual sea-ice extent affects adult survival by increasing food availability, but sea-ice extent in winter negatively affects hatching success by increasing the distance between the colony and the feeding grounds that the adults must traverse. There must, thus, be a trade-off for the penguins in terms of the extent of winter pack ice that is most favourable for their long-term survival. Put in this way, it is obvious that if current predictions of global warming, with melting of the ice caps and changes in sea level, are correct (Easterling *et al.*, 2000), it could spell the end for these magnificent birds and many other marine species living in the Antarctic oceans.

Ecophysiological investigations on these, and other animals, are thus vital if conservation biologists are to have any chance of managing rapidly changing environments in such a way as to ensure the animals' long-term survival. Without a detailed knowledge of a species' requirements and its 'realised niche' in the sense of Hutchinson (1978), we are relegated to the category of spectators and helpless bystanders as species disappear at an ever-increasing rate. The case of the tiny marsupial honey possum, *Tarsipes rostratus*, discussed in Chapter 4, is typical. This is an ancient and highly specialised marsupial that was once widespread but is now restricted to a small fragment of its former range in Western Australia. It is currently not listed as 'threatened' or 'endangered' but its unique habitat – banksia woodland – is fast disappearing under pressure from farmers and land developers and from a fungal disease, *Phytophthora cinnammomi*, known locally as 'jarrah dieback,' as it also attacks certain species of banksia. Our knowledge of the nutritional requirements of the honey possum has grown rapidly in the past few years as a result of a concentrated research effort funded by the Australian Research Council. We now have reliable data on rates of intake of pollen and nectar of free-ranging animals and we are close to measuring rates of protein turnover of breeding females in the field. This species will probably join the 'threatened' list within a decade or two but, by then, we hope to know enough about critical aspects of its biology to be able to manage reservations to ensure its long-term persistence.

Ecophysiology is a new discipline with real relevance to conservation biologists as its essence is to discover the nature of the reality experienced by animals free-ranging in their natural environment, rather than in laboratory cages (Bronson, 1998). Armed with such knowledge, it should prove possible for environmental managers to ensure the long-term persistence of threatened species, despite the many deleterious changes occurring at the present time to the world's environments. One may only hope.

Appendix 1
Population estimation methods

Schnabel method

The following equations for estimating population size are based on Krebs (1999).

The equation for estimating the size of the population by using the Schnabel method is:

$$N = \frac{\Sigma(C_t M_t)}{\Sigma R_t},$$

where: N = estimated population size
C_t = total number of individuals caught in sample t
R_t = number of individuals already marked when caught in sample t
U_t = number of individuals marked for the first time and released in sample t
M_t = number of marked individuals in the population just before the tth sample is taken (e.g. $M_4 = U_1 + U_2 + U_3$).

To obtain confidence intervals for the Schnabel method, the following equations were applied:

$$\text{Variance } 1/N = \frac{\Sigma R_t}{(\Sigma C_t M_t)^2}.$$

$$\text{Standard error of } 1/N = \sqrt{\text{variance } 1/N}.$$

The standard error and a t-table were used to obtain confidence limits for $1/N$. These confidence limits were then inverted for N:

$$1/N \pm t_\alpha \text{ S.E.}$$

where S.E. = standard error of $1/N$

t_α = value from t-table for $(100 - \alpha)\%$ confidence limits.

Jolly–Seber method

The following equations for estimating population size are based on Krebs (1999).

The equation for estimating the size of the population by using the Jolly–Seber method is:

$$N_t = M_t/\alpha_t,$$

where: N_t = estimated population size just before sample time t

M_t = estimated size of the marked population just before sample time t

$= [(s_t + 1)Z_t/R_t + 1] + m_t$

α_t = proportion of animals marked

$= (m_t + 1)/(n_t + 1)$

R_t = number of the individuals released at sample t and caught again in some later sample

Z_t = number of individuals marked before sample t, not caught in sample t, but caught in some sample after sample t

m_t = number of marked animals caught in sample t

u_t = number of unmarked animals caught in sample t

n_t = total number of animals caught in sample $t = m_t + u_t$

s_t = total number of animals released after sample t

$= (n_t$ – accidental deaths or removals).

Calculations for Schnabel method of estimating abundance

Sample	No. caught C_t	No. recaptures R_t	No. newly marked (less deaths)	Marked animals at large M_t	$C_t{}^*M_t$	$C_t{}^*M_t^2$	$R_t{}^*M_t$	R_t^2/C_t	R_t/C_t
1	0	0	0	0	0	0	0	—	—
2	0	0	0	0	0	0	0	—	—
3	3	0	3	0	0	0	0	—	0
4	2	0	2	3	6	18	0	0	0
5	3	0	3	5	15	75	0	0	0
6	4	1	3	8	32	256	8	0.25	0.25
7	0	0	0	11	0	0	0	—	—
8	4	2	2	11	44	484	22	1	0.5
9	2	0	2	13	26	338	0	0	0
10	1	0	1	15	15	225	0	0	0
11	4	2	2	16	64	1024	32	1	0.5
12	3	2	1	18	54	972	36	1.33333	0.666667
13	6	5	1	19	114	2166	95	4.16667	0.833333
14	2	2	0	20	40	800	40	2	1
15	5	5	0	20	100	2000	100	5	1
16	1	0	1	20	20	400	0	0	0
17	5	2	3	21	105	2205	42	0.8	0.4
18	2	0	2	24	48	1152	0	0	0
19	5	5	0	26	130	3380	130	5	1
20	0	0	0	26	0	0	0	—	—
Total	52	26	26	276	813	15495	505	20.55	—

Schnabel

N = 31.3
variance $1/N$ = 0.00004
std error of $1/N$ = 0.01
lower 95% CI = 21.6
upper 95% CI = 48.5

Formulae for Schnabel method of estimating abundance

Sample	No. caught C_t	No. recaptures R_t	No. newly market (less deaths)	Marked animals at large M_t	$C_t{}^*M_t$	$C_t{}^*M_t^2$	$R_t{}^*M_t$	R_t^2/C_t	R_t/C_t
1	0	0	=B6-C6	0	=B6∗E6	=B6∗E6^2	=C6∗E6	=C6^2/B6	=C6/B6
2	0	0	=B7-C7	0	=B7∗E7	=B7∗E7^2	=C7∗E7	=C7^2/B7	=C7/B7
3	3	0	=B8-C8	0	=B8∗E8	=B8∗E8^2	=C8∗E8	=C8^2/B8	=C8/B8
4	2	0	=B9-C9	3	=B9∗E9	=B9∗E9^2	=C9∗E9	=C9^2/B9	=C9/B9
5	3	0	=B10-C10	5	=B10∗E10	=10B∗E10^2	=C10∗E10	=C10^2/B10	=C10/B10
6	4	1	=B11-C11	8	=B11∗E11	=B11∗E11^2	=C11∗E11	=C11^2/B11	=C11/B11
7	0	0	=B12-C12	11	=B12∗E12	=B12∗E12^2	=C12∗E12	=C12^2/B12	=C12/B12
8	4	2	=B13-C13	11	=B13∗E13	=B13∗E13^2	=C13∗E13	=C13^2/B13	=C13/B13
9	2	0	=B14-C14	13	=B14∗E14	=B14∗E14^2	=C14∗E14	=C14^2/B14	=C14/B14
10	1	0	=B15-C15	15	=B15∗E15	=B15∗E15^2	=C15∗E15	=C15^2/B15	=C15/B15
11	4	2	=B16-C16	16	=B16∗E16	=B16∗E16^2	=C16∗E16	=C16^2/B16	=C16/B16
12	3	2	=B17-C17	18	=B17∗E17	=B17∗E17^2	=C17∗E17	=C17^2/B17	=C17/B17
13	6	5	=B18-C18	19	=B18∗E18	=B18∗E18^2	=C18∗E18	=C18^2/B18	=C18/B18
14	2	2	=B19-C19	20	=B19∗E19	=B19∗E19^2	=C19∗E19	=C19^2/B19	=C19/B19
15	5	5	=B20-C20	20	=B20∗E20	=B20∗E20^2	=C20∗E20	=C20^2/B20	=C20/B20
16	1	0	=B21-C21	20	=B21∗E21	=B21∗E21^2	=C21∗E21	=C21^2/B21	=C21/B21
17	5	2	=B22-C22	21	=B22∗E22	=B22∗E22^2	=C22∗E22	=C22^2/B22	=C22/B22
18	2	0	=B23-C23	24	=B23∗E23	=B23∗E23^2	=C23∗E23	=C23^2/B23	=C23/B23
19	5	5	=B24-C24	26	=B24∗E24	=B24∗E24^2	=C24∗E24	=C24^2/B24	=C24/B24
20	0	0	=B25-C25	26	=B25∗E25	=B25∗E25^2	=C25∗E25	=C25^2/B25	=C25/B25
Total	=SUM(B6:B25)	=SUM(C6:C25)	=SUM(D6:D25)	=SUM(E6:E25)	=SUM(F6:F25)	=SUM(G6:G25)	=SUM(H6:H25)	=SUM(I9:I25)	—

Schnabel

N = F26/C26
variance $1/N$ = 26/(813*813)
std error of $1/N$ = B29^0.5
lower 95% CI = 813/37.67
upper 95% CI = 813/16.77

Appendix 2
Estimation of food intake in *Tiliqua rugosa*

Body mass (g)	Estimated FMR (kJ d^{-1})[1]	FMR (l CO_2 d^{-1})*	Estimated MWP (ml d^{-1})[2]	Water influx (ml d^{-1})	Dietary water intake (ml d^{-1})[3]	Water content of diet (ml 100 g^{-1})	Food intake (g d^{-1})
508.0	55.2	2.7	1.8	11.5	9.7	55.0	20.9
494.4	53.8	2.6	1.7	6.8	5.1	55.0	12.3
320.6	35.9	1.7	1.1	34.0	32.8	55.0	61.7
401.1	44.3	2.1	1.4	14.0	12.6	55.0	25.5
411.2	45.3	2.2	1.4	5.8	4.4	55.0	10.6
418.6	46.1	2.2	1.5	8.0	6.6	55.0	14.6
337.2	37.7	1.8	1.2	3.5	2.3	55.0	6.3
496.5	54.0	2.6	1.7	16.1	14.4	55.0	29.3
423.5	***46.6***	***2.2***	***1.5***	***12.5***	***11.0***	***55.0***	***22.7***

* 20.8 kJ = 1 l of CO_2 with a high-carbohydrate diet

[1] FMR in kJ d^{-1} = 0.166 $BM^{0.932}$ where BM is in g

[2] MWP = 0.662 ml per litre of CO_2 produced

[3] Water influx − MWP

Appendix 3
Simultaneous measurement of sodium and potassium concentration in plasma or urine using the IL143 digital flame photometer

All solutions are provided with the instrument from Instrumentation Laboratory de Puerto Rico Ltd., Cidra, Puerto Rico 00639.

Lithium diluent (15 mmol l^{-1})

Dilute 10 ml of 1500 STOCK solution in 1 l of distilled water.

Standard ($Na^+ = 140$, $K^+ = 5$)

Take 0.1 ml and add 20 ml of lithium diluent. Also set up a standard curve with 50, 100, 150 and 200 mmol l^{-1} NaCl standards and run these samples at the beginning and at the end of the run. Set up a similar series of KCl standards ranging from 1 to 10 mmol l^{-1} for plasma and 50 to 200 mmol l^{-1} with urine. Calculate a regression between the readings and the true concentrations for each ion and use these regressions to calculate the values for your samples.

Samples

Add 10 μl of sample to 2.0 ml of the lithium diluent.

Operation

- Turn on gas tap (value set at 35 lb in^{-2}, high-purity propane).
- Turn on IL143 electrical switch.
- Turn on air (set at 27 lb in^{-2}).
- Machine should ignite – if not, ensure that syphon tube is filled with water.
- Check that aspiration rate is between 30 and 50 s ml^{-1}.
- Set zero with lithium diluent and 140/5 standard using balance control.
- Adjust lithium control if necessary.

- If cannot set the 140, reduce the aspiration rate.
- After running samples, aspirate cleaning solution for 30 s followed by distilled water.

Turning off

- Turn off main gas tap (do not touch valve setting).
- Wait until NO GAS light comes on.
- Turn off air.
- Wait until automatic striker is activated.
- Turn off IL143 electrical switch.

N.B. Lithium control should not be adjusted unless pointer moves outside the limits. Potassium read 0–20 mmol l^{-1} or 0–200 mmol l^{-1} depending on range switch. When running samples, check zero and 140/5 standard after every ten samples to ensure stability.

Appendix 4
Determination of plasma urea nitrogen

Urea nitrogen

Urea is synthesised in the liver from ammonia produced as a result of deamination of amino acids and is the chief means of excreting nitrogen from the animal. It is customary in most laboratories to express urea as urea nitrogen. This came about from the desire to compare the quantity of nitrogen excreted in the form of urea with other nitrogenous compounds. Hence, the standardisation of the method is in **units of urea nitrogen**. Since the molecular mass of urea is 60 Da and it contains two nitrogen atoms, with a combined mass of 28 Da, an urea nitrogen value can be converted to urea by multiplying by 60/28 or 2.14.

Method principle

Urea is hydrolysed to ammonium carbonate by urease, and the ammonia that is released from the carbonate by alkali reacts with phenol and sodium hypochlorite in an alkaline medium to form a blue indophenol. Sodium nitroprusside serves as a catalyst. The intensity of the blue colour is proportional to the quantity of urea in the specimen.

Reagents

Phenol nitroprusside solution

Dissolve 5 g phenol and 25 mg sodium nitroprusside in 500 ml double-distilled (d-d) water. Store at 4 °C. Stable for two months.

Alkaline hypochlorite

Dissolve 2.5 g NaOH in 250 ml d-d water in 500 ml volumetric flask. Cool, add 0.21 g sodium hypochlorite. Mix well and make up to

500 ml with d-d water. Store in an amber bottle at 4 °C. Stable for 2 months.

EDTA buffer, pH 6.5

Dissolve 5 g disodium salt EDTA in 450 ml d-d water. Adjust pH to 6.5 with 1 M NaOH. Make up to 500 ml with d-d water.

Urease working solution (0.4 U ml^{-1} for urine; 2.4 U ml^{-1} for plasma)

Stock urease (560 U ml^{-1}) is diluted as follows for plasma: 0.5 ml in 100 ml EDTA buffer. Store at 4 °C. Stable for 3 weeks.

Urea standards

Stock standard (500 mg urea nitrogen per 100 ml): 1.0717 g urea and 0.1 g NaN_3 are dissolved in 100 ml d-d water. Store at 4 °C. Stable for 6 months.

Working standards (10, 50, 100, 500 μg urea N ml^{-1}):

100 μl STOCK	⟶	50 ml d-d H_2O	(10 μg urea N ml^{-1})
500 μl STOCK	⟶	50 ml d-d H_2O	(50 μg urea N ml^{-1})
1 ml STOCK	⟶	50 ml d-d H_2O	(100 μg urea N ml^{-1})
5 ml STOCK	⟶	50 ml d-d H_2O	(500 μg urea N ml^{-1})

Method

1. Label large glass test-tubes (20 ml), one tube as a blank, four tubes as standards and appropriate number for samples.
2. Pipette 1 ml urease working solution into each tube.
3. Pipette 10 μl plasma sample into sample tubes.
4. Pipette 10 μl of each working standard into appropriate tubes.
5. Mix all tubes and incubate 15 min at 37 °C.
6. Add rapidly, and successively, mixing after each addition, 5 ml phenol nitroprusside solution, followed by 5 ml alkaline hypochlorite.
7. Place the tubes in a water bath at 37 °C for 20 min.
8. Measure absorbance at 560 nm, using a spectrophotometer and setting the zero with the blank.

Calculations

Use curve-fitting software to construct the standard curve (should be a straight line). Read off unknowns as μg urea nitrogen ml^{-1}.

To convert to urea (μg ml^{-1}), multiply urea nitrogen by 60/28.

Appendix 5
Radioimmunoassay of testosterone in plasma

Extraction

- Pipette 100 μl assay buffer into glass stoppered extraction tubes.
- Pipette 10 μl plasma into each aliquot of buffer. Vortex briefly.
- Pipette 2.5 ml diethyl ether into each sample, stopper, and vortex for 1 min. Allow to settle.
- Using a Pasteur pipette, remove supernatant carefully to a clean assay tube.
- Repeat, using a further 2.5 ml diethyl ether, and combine the supernatants.
- Dry down under compressed air at 37 °C.

Assay

Prepare standards

- Pipette duplicate 500 μl of each working standard into glass assay tubes.
- Pipette duplicate 500 μl ethanol into tubes for 0 pg of steroid.
- Dry down under compressed air at 37 °C.

Prepare assay and solvent blanks

- Set up an assay tube as the assay blank. Antibody will be omitted from this tube, thus it is an estimate of any non-specific binding.
- Pipette 5 ml diethyl ether into an assay tube as a solvent blank. Dry down.

Addition of antiserum

- Pipette 200 μl of working antibody solution (WAS) into tubes containing dried solvent (solvent blank), sample extracts, and standards.
- Vortex thoroughly for 5 s, but avoid frothing.

- Pipette 200 μl (×2) WAS into scintillation vials, add 5 ml scintillation fluid, cap, label and set aside to count . . . . Input Counts 1 (WAS)
- Pipette 200 μl working antibody blank solution (WABS) into assay blank and vortex.
- Pipette 200 μl WABS into scintillation vials, cap as above and set aside to count . . . . Input Counts 2 (WABS)
- Cover all tubes with Alfoil and incubate at 37 °C for 45 min, followed by room temperature for 3 h.

Separation of free and bound fraction

- Place tubes in ice bath for 15 min.
- Add 1 ml cold (0 °C) dextran-coated charcoal (stirred on automatic stirrer) to each tube. Vortex briefly after each addition.
- Incubate all tubes at 0 °C for 10 min exactly.
- Centrifuge at 4000 rpm for 10 min at 0 °C.
- Pipette 600 μl from supernatant into scintillation vials. Add 5 ml scintillation fluid, cap and label.
- Mix all scintillation tubes thoroughly by vortexing, and count each sample 3 times for 10 min.

Reagents

Buffer 0.05 M borate, pH 8.0

Dissolve 2 g AR grade boric acid crystals in approximately 480 ml distilled water. Adjust pH to 8.0 with drop-wise addition of 10 N NaOH, and dilute to 500 ml with distilled water. Store at 4 °C.

[^{3}H]testosterone

Using [1,2,6,7,-^{3}H]testosterone (NET-370, Amersham), dilute 250 μCi (9.25 MBq) in 25 ml absolute ethanol (Merck grade) to provide stock: 10 μCi ml^{-1}.

Prepare a working solution of [^{3}H]testosterone in assay buffer by diluting stock 1 : 22 as follows:

Pipette 115 μl stock into a 5 ml glass vial. Dry down, and take up in 2.5 ml assay buffer. Make freshly each 2–3 weeks. Count 50 μl (×2) before use: *c.* 50 000 dpm per 50 μl.

Testosterone standard

- Rinse clean glassware in methanol before use and dry thoroughly. Use only freshly opened ethanol (Merck or BDH HyperSolv).

- Prepare 1 mg ml^{-1} stock as follows. Weigh approximately 1 mg chromatographically pure testosterone accurately. Dissolve in the equivalent volume of absolute ethanol.
- Dilute 100 μl of stock to 100 ml with absolute ethanol to give working stock A (1 μg ml^{-1}).
- Dilute 1 ml of working stock A to 100 ml with absolute ethanol to give working stock B (10 ng ml^{-1}).
- Dilute working stock B to obtain working standards as follows:

125 μl	diluted in 50 ml ethanol	12.5 pg/500 μl
250 μl	" "	25 pg/500 μl
500 μl	" "	50 pg/500 μl
1.0 ml	" "	100 pg/500 μl
1.5 ml	" "	150 pg/500 μl
2.0 ml	" "	200 pg/500 μl
3.0 ml	" "	300 pg/500 μl

Keep all stocks and working standards sealed with parafilm at 4 °C.

Bovine gamma globulin

Dissolve 250 mg bovine serum globulin (Commonwealth Serum Laboratories) in 10 ml 1 N NaCl containing 0.1% NaN_3 (sodium azide). Store at 4 °C.

Bovine serum albumin

Dissolve 1 g bovine serum albumin (Sigma Fraction V) in 10 ml distilled water containing 0.1% NaN_3. Store at 4 °C.

Testosterone antiserum

Reconstitute the lyophilised antiserum with the volume of distilled water indicated on the label of the vial (usually 1 ml). Store in 100 μl volumes in micro-Eppendorf tubes at −10 °C. This antiserum is used at 1/100 dilution in the assay tube.

Working antibody solution (WAS)

Made up freshly with each assay, and with the reagents in the following order:

Borate buffer	90 parts
[^{3}H]testosterone working solution	5 parts
Bovine serum albumin	2 parts
Bovine gamma globulin	2 parts
Testosterone antiserum	1 part

Working antibody blank solution (WABS)

Made up in the same proportions as the working antibody solution, except that antiserum is omitted and the volume is compensated with buffer (i.e. 91 parts).

Dextran-coated charcoal suspension

500 mg Norit A acid-washed charcoal and 50 mg Dextran T_{40} made up to 100 ml with 0.1% gelatin. Store at 4 °C.

Gelatin 0.1%

100 mg gelatin dissolved in 100 ml 0.05 M borate buffer (assay buffer). Store at 4 °C.

Scintillation fluid

Ultima Gold from Packard United Technologies. Use 3 ml in plastic screw-top mini-vials.

Calculations

Count samples in a liquid scintillation counter and calculate dpm for each tube, averaging counts for three cycles.

Calculation of percentage bound

Calculate the percentage of [^{3}H]testosterone bound to the antibody in each standard and each sample as follows.

$$\frac{\text{dpm in STD (or sample)} \times \text{dilution factor } 1.2/0.6}{\text{Input counts } 1(\text{WAS})} \times 100.$$

Standard curve

Plot the standard curve using a 4-parameter logistic curve with pg testosterone as X and % radioactivity bound as Y. This curve uses the following equation:

$$Y = \frac{A - D}{1 + (X/C)^B} + D,$$

where A is the upper asymptote and D is the lower asymptote of the curve, C is the value of the steroid concentration at the centre or inflection-point of the sigmoid curve and B is related to the slope (about 0.15).

Samples

Calculate each sample, using the equation, and to express in pg/ml plasma, multiply by 100.

Solvent blank

From the % radioactivity bound, using the above equation, calculate the amount of 'testosterone equivalents' in the solvent blank. This should be less than 5 pg per sample. If greater, solvent should be purified by redistillation.

Non-specific binding

The non-specific binding (NSB) is calculated as:

$$\frac{\text{dpm in assay blank}}{\text{Input Counts 2 (WABS)}} \times 100$$

and should always be less than 5%. It may be subtracted from each sample calculation, but is usually regarded as an assay control. If greater than 5%, the conditions of the assay should be investigated. For example, the concentration of dextran-coated charcoal may be too low, or the cleanliness of the glassware may be in question.

Appendix 6
Preparation of 'stripped plasma'

1. Weigh out 50 mg Norit A acid-washed charcoal for each 1 ml plasma to be stripped in a 12 ml centrifuge tube.
2. To remove 'fines' in the charcoal, add about 5 ml of distilled water, vortex, centrifuge and remove supernatant with a Pasteur pipette.
3. Add plasma to the charcoal, vortex well, and incubate at 37 °C for 30 min.
4. Remove plasma with a Pasteur pipette to a clean tube and centrifuge again.
5. Remove plasma, as before, to an airtight container and store at –15 °C.

Appendix 7
Radioimmunoassay of testosterone in faeces

Extraction efficiency

It is important to establish the solvent and its method of use that will optimise the extraction of testosterone from faeces of the particular animal species under study. This will require a minimum of six faecal samples that have each received the same amount of radioactive testosterone, or tracer (approximately 10 000 dpm), which are then exposed to the different solvents. After the extraction procedure, an exact fraction of the solvent is measured for radioactivity, in order to calculate the recovery of the tracer added to the medium. The most common solvents used are ethanol or diethyl ether. The extraction method for diethyl ether is to add ten volumes of solvent, mix by vortexing for 1 min, and count an aliquot (calculated exactly) of the supernatant for radioactivity. The other extraction method, using ethanol, usually requires ten volumes of 90% ethanol added to the faeces, followed by boiling for 20 min, but the concentration of water in the ethanol may need to be varied between 80 and 90% in order to improve the extraction of steroid. Again, this depends on the species under study. The protocol below outlines an ethanolic extraction procedure, which recovered 90% of testosterone from our study animal.

Extraction

- Dry faeces in oven at 40 °C for 24 h and macerate well in a coffee grinder (Braun).
- To a weighed sample (30 mg) in a glass test-tube, add sufficient distilled water to enable homogenisation (300 μl). Homogenise thoroughly by vortexing for 30 s.

- Pipette 50 μl recovery solution (5000 dpm) to faecal sample and vortex 20 s to mix sample thoroughly.
- Pipette 50 μl recovery solution to scintillation vials, add 3 ml scintillation fluid, cap, label and set aside for counting . . . . . . Input Counts 1 (Recovery)
- Add 2.7 ml 100% ethanol (BDH, HyperSolv grade for HPLC) and place tube in 75–80 °C water bath. Boil sample for 20 min. (N.B. Solvent volume must be calculated in order to arrive at ten faecal volumes of 90% ethanol.)
- Centrifuge, and remove supernatant to clean conical glass tube.
- Repeat extraction procedure by resuspending the pellet in ten volumes 90% ethanol (3 ml) , boil for 20 min and combine the supernatants.
- Evaporate to dryness under compressed air in a water bath at 40 °C.

Chromatography

The system used in this assay is a lipophilic, hydrophobic gel (Lipidex 5000™ from Packard Technologies). Glass columns (0.6 cm i.d. and 0.9 cm o.d. × 13 cm) are packed by gravity flow with gel suspended in column solvent (*n*-hexane : chloroform, 9 : 1). Calibrate the columns with *c.* 20 000 dpm radioactive steroid, together with *c.* 100 pg cold steroid, loaded onto the column with column solvent. Elute 1 ml volumes and count radioactivity to establish the elution profile of the steroid under investigation. In this assay:

- To dried ethanolic extract, add 0.5 ml column solvent, *n*-hexane : CH_3Cl (9 : 1), and vortex 5 s.
- Using a Pasteur pipette, load carefully onto Lipidex 5000™ column (5 ml bed volume, equilibrated in column solvent, 9 : 1).
- Allow to run in and repeat the load process with two more volumes of 0.5 ml column solvent.
- Discard load volume each time.
- Pipette 6 ml column solvent (9 : 1) to column and discard.
- Collect next 8 ml into a graduated glass centrifuge tube.
- Make up to 8 ml, if necessary, with column solvent (9 : 1) and mix fraction with Pasteur pipette.
- Pipette 1 ml of fraction to scintillation vial, dry down, add 5 ml scintillation fluid and set aside to count . . . . Recovery dpm
- Dry down remaining 7 ml, cover and store for assay.

Assay

Prepare standards

- Pipette 500 μl of each working standard into glass centrifuge tubes.
- Duplicate 500 μl ethanol (Merck grade) in tubes to serve as 0 pg of steroid.

- Pipette 50 μl compensatory [^{3}H]testosterone (2500 dpm) to each tube.
- Dry down all tubes.
- Pipette duplicate 50 μl compensatory counts into two scintillation vials, add 5 ml scintillation fluid, cap, label, and set aside to count . . . Input Counts 2 (compensatory dpm).

Prepare assay blank

- Set up a tube as the assay blank. Antibody will be omitted from this tube, thus, it is an estimate of any non-specific binding.
- Pipette 50 μl compensatory [^{3}H]testosterone to assay blank and dry down.

Solvent blank

- Collect 9 ml column solvent (9 : 1). Dry down and add 50 μl compensatory [^{3}H]testosterone. Dry down.

Addition of antiserum

- Pipette 200 μl of working antibody solution (WAS) into tubes containing dried column solvent, sample extracts, and standards. Vortex 3–4 s.
- Pipette 200 μl (×2) WAS into scintillation vials, add 5 ml scintillation fluid, cap, label and set aside to count . . . . Input Counts 3 (WAS)
- Pipette 200 μl working antibody blank solution (WABS) into assay blank. Vortex.
- Pipette 200 μl WABS into scintillation vials, cap as above and set aside to count . . . . Input Counts 4 (WABS)
- Cover all tubes with Alfoil and incubate at 37 °C for 45 min, followed by room temperature for 3 h.

Separation of free and bound fraction

- Place tubes in ice-bath for 15 min.
- Add 1ml cold (0 °C) dextran-coated charcoal (stirred on automatic stirrer) to each tube. Vortex briefly after each addition.
- Incubate all tubes at 0 °C for 10 min **exactly**.
- Centrifuge at 4000 rpm for 10 min at 0 °C.
- Pipette 600 μl from supernatant into scintillation vials. Add 5 ml scintillation fluid, cap and label.
- Mix all scintillation tubes thoroughly by vortexing, and count each sample three times for 10 min.

Reagents

Testosterone standard

Prepare 1 mg ml^{-1} stock as follows:

Weigh approximately 1 mg chromatographically pure testosterone (Cal Biochem) accurately. Dissolve in the equivalent volume of absolute ethanol (Merck or BDH HyperSolv).

Dilute 100 μl of STOCK to 100 ml with absolute ethanol to give working stock A (1 μg ml^{-1})

Dilute 1 ml of working stock A to 100 ml with absolute ethanol to give working stock B (10 ng ml^{-1}).

Dilute working stock B to obtain working standards as follows:

125 μl	working stock B in 50 ml ethanol	12.5 pg/500 μl
250 μl	" "	25 pg/500 μl
500 μl	" "	50 pg/500 μl
750 μl	" "	75 pg/500 μl
1.0 ml	" "	100 pg/500 μl
1.5 ml	" "	150 pg/500 μl
2.0 ml	" "	200 pg/500 μl

Keep all stocks and working standards sealed with Parafilm at 4 °C.

Buffer

Dissolve 2 g AR grade boric acid crystals in approximately 480 ml distilled water. Adjust pH to 8.0 with drop-wise addition of 10 N NaOH, and dilute to 500 ml with distilled water. Store at 4 °C.

[^{3}H]testosterone

Using [2,4,6,7-^{3}H]testosterone (Amersham TRK 402), dilute 250 μCi (9.25 MBq) in 25 ml absolute ethanol (Merck grade) to provide stock, 10 μCi (370 kBq) ml^{-1}.

Prepare the working solution of [^{3}H]testosterone in assay buffer by diluting stock 1 : 22 as follows.

Pipette 115 μl stock into a 5 ml glass vial. Dry down, and take up in 2.5 ml assay buffer. Make freshly each 2–3 weeks.

Count 50 μl (×2) before use .... *c.* 50 000 dpm per 50 μl

Prepare a recovery solution by diluting the working solution 1 in 10 with assay buffer. Count 50 μl before use.................. *c.* 5000 dpm per 50 μl

Prepare a compensatory solution as follows:
Stock is diluted 1 in 300 with absolute ethanol . . . . . . *c.* 2500 dpm per 50 μl
Compensatory counts are added to the standards prior to the working antibody solution in order to compensate for the radioactive label added to the sample tubes in the recovery step.

Bovine gamma globulin

Dissolve 250 mg bovine serum globulin (Commonwealth Serum Laboratories) in 10 ml 1 N NaCl containing 0.1% NaN_3 (sodium azide). Store at 4 °C.

Bovine serum albumin

Dissolve 1 g bovine serum albumin (Sigma Fraction V) in 10 ml distilled water containing 0.1% NaN_3. Store at 4 °C.

Dextran-coated charcoal suspension

500 mg Norit A acid-washed charcoal and 50 mg Dextran T_{40} made up to 100 ml with 0.1% gelatin. Store at 4 °C.

Gelatin 0.1%

100 mg gelatin dissolved in 100 ml 0.05 M borate buffer (assay buffer). Store at 4 °C.

Testosterone antiserum

Reconstitute the lyophilised antiserum with the volume of distilled water indicated on the label of the vial (usually 1 ml). Store in 100 μl volumes in micro-Eppendorf tubes at −10 °C. Antiserum is used at 1/100 dilution in the assay tube.

Working antibody solution (WAS)

Made up freshly with each assay, and with the reagents in the following order:

Borate buffer . . . . 90 parts
[^{3}H]oestradiol working solution . . . . 5 parts
Bovine serum albumin . . . . 2 parts
Bovine gamma globulin . . . . 2 parts
Testosterone antiserum . . . . 1 part

Working antibody blank solution (WABS)

Made up in the same proportions as the working antibody solution, except that antiserum is omitted and the volume is compensated with buffer (i.e. 91 parts).

Scintillation fluid

Ultima Gold from Canberra Packard. Use 3 ml in plastic screw-top mini-vials.

Lipidex

Hydroxyalkoxypropyl-Sephadex from Packard United Technologies.

Calculations

Count samples in a liquid scintillation counter and calculate dpm for each tube, averaging counts for three cycles.

Standard curve

The percentage of $[^3H]$testosterone bound to the antibody in each standard is expressed as:

$$\frac{\text{dpm in STD} \times \text{dilution factor } 1.2/0.6}{\text{Input Counts 2 (compensatory)} + \text{Input Counts 3 (WAS)}} \times 100.$$

Using a 4-parameter logistic curve plot pg testosterone as X and % bound as Y. This curve uses the following equation:

$$Y = \frac{A - D}{1 + (X/C)^B} + D,$$

where A is the upper asymptote and D is the lower asymptote of the curve, C is the value of the steroid concentration at the centre or inflection-point of the sigmoid curve and B is related to the slope (about 0.15).

Samples

To account for procedural loss during the assay, radioactivity is added in step 4. The % recovery of steroid in the assay can then be expressed as:

$$\frac{\text{dpm in recovery} \times \text{dilution factor } (8/1)}{\text{Input Counts 1 (recovery)}} \times 100 = (A).$$

The amount of radioactivity bound in the sample is expressed as a percentage of the total amount of radioactivity presented to the assay **in each sample**. The total amount consists of the radioactivity in the WAS (Input Counts 3), together with the radioactivity from the Recovery (Input Counts 1). However, only a part of the Input Counts 1 enters the assay and this portion can be calculated as:

$$\text{Input Counts 1} \times 7/8 \times (A)/100 = (B).$$

Thus, % radioactivity bound can be expressed as:

$$\frac{\text{dpm in sample} \times 1.2/0.6}{(B) + \text{Input Counts 3 (WAS)}} \times 100.$$

The amount of testosterone in pg in the sample can then be derived from the standard curve.

To calculate testosterone concentrations in the original faecal sample, procedural loss must be corrected for, by the percentage recovery in each sample:

$$\text{testosterone (pg) in sample} = \text{pg testosterone} \times 100/(A) = (C).$$

To express in pg per faecal sample, $(C) \times 8/7$, as only 7/8 of the extract was taken through the assay $= (D)$.

Testosterone per g dried faeces is calculated as: (D)/mass of faecal sample in g.

Appendix 8
The comparative method

In this very simple worked example, some imaginary data from two species will be analysed to see whether they show a correlation, in two ways: with and without taking cognisance of the phylogenetic relationship between the two species. Imagine that we have measured rates of evaporative water loss of four species belonging to the same genus (A, B, C and D), which vary in body size, and we wish to see if rates of water are simply a function of differences in body mass between the species. The primary data are shown in the following table.

Species	Body mass (g)	Rate of evaporative water loss (mg h^{-1})
A	12	32
B	15	36
C	24	48
D	30	58

Graphing these mean data for each of the four species, as seen in Figure A1, results in a highly significant correlation with $r^2 = 0.997$.

If we now imagine that the four species are related in a single phylogeny, shown in Figure A2, we may then align our two sets of values for body mass and EWL for the four species as in Figure A3, with nodal (ancestral) values being determined by simply averaging the values for the extant species.

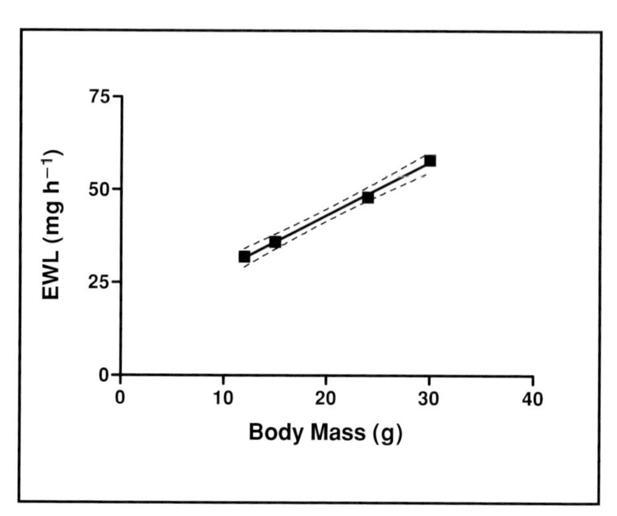

Figure A1.

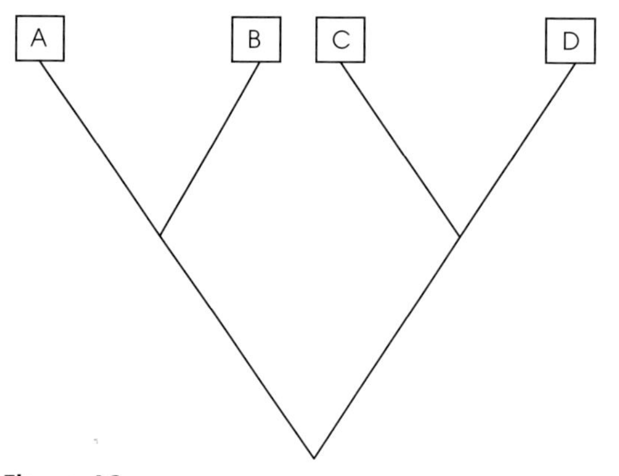

Figure A2.

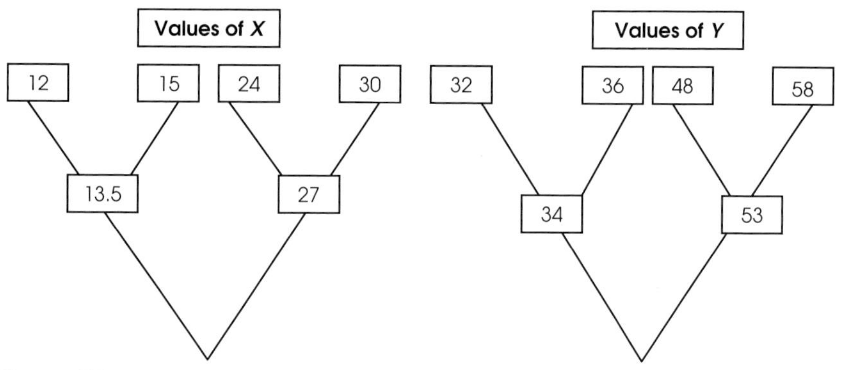

Figure A3.

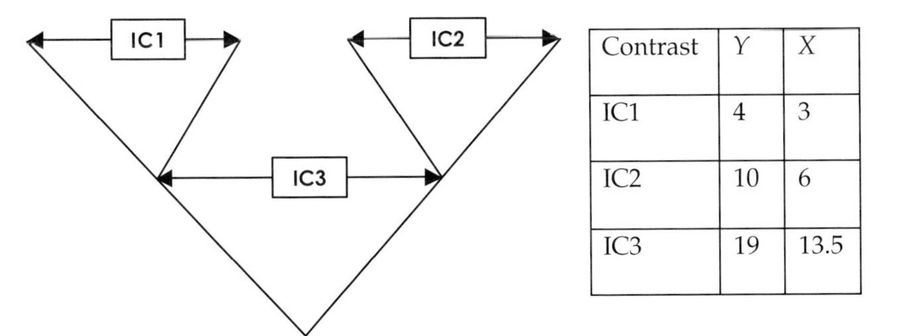

Contrast	Y	X
IC1	4	3
IC2	10	6
IC3	19	13.5

Figure A4.

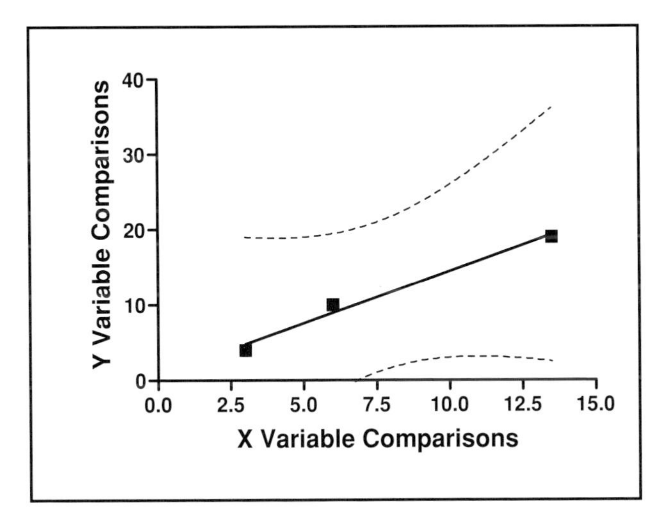

Figure A5.

Pairwise independent contrasts (IC) between the species for the X and Y values are then computed as follows by subtraction as shown in Figure A4.

The resulting regression of the Y variable comparisons against the X variable comparisons (Figure A5) has only three points instead of four and much wider 95% confidence limits, but the coefficient of determination (r^2) is still very high at 0.983 (i.e. we can accept the relation between body mass and EWL).

This method of independent contrasts is based on a model proposed by Felsenstein (1985) and assumes a Brownian motion model of evolution (i.e. changes in the character sets of the species are random and independent of each other). This is because the difference between species A and B reflects only the evolutionary changes that have occurred since they split from their common ancestor and any similarity between the species that is the result of their common ancestry is effectively subtracted out. The same argument applies to the contrasts between species C and D and the two ancestral nodes.

The advantage of the independent contrasts method is that, by partitioning the variation in an appropriate fashion, it can all be used to assess the strength of the comparative relationship. This method is perhaps the most simple and no account is taken of branch lengths in the phylogeny which are all considered to be equal and values for higher nodes are calculated as the average of lower nodes. Reference should be made to Harvey and Pagel (1993), Garland *et al.* (1992) and Martins *et al.* (2002) for more complex and perhaps realistic procedures, which differ in their assumptions and statistical treatment of the data.

Appendix 9
Basic turnover equations

Fractional turnover rate (k)

$$k = \frac{\ln(C_1/C_2)}{t}, \tag{A1}$$

where C_1 and C_2 are the initial and final *specific activities* of the isotope in body water and t is the elapsed time.

Half-life ($T_{1/2}$) of the isotope in the body

$$T_{1/2} = \frac{0.693}{t}, \tag{A2}$$

where $0.693 = \ln(C_1/C_2)$ when C_2 is exactly 50% of C_1.

Total water flux

$$\frac{\text{ml } H_2O \text{ flux}}{\text{kg d}} = \frac{1000W \ln(C_1/C_2)}{Mt}, \tag{A3}$$

where W = total body water content in millilitres and M = body mass in grams.

Water efflux with linear change in body water

$$\frac{\text{ml } H_2O \text{ efflux}}{\text{kg d}} = \frac{2000(W_2 - W_1) \ln(C_1 W_1/C_2 W_2)}{(M_1 + M_2)\ln(W_2/W_1)t}. \tag{A4}$$

Water efflux with exponential change in body water

$$\frac{\text{ml H}_2\text{O efflux}}{\text{kg d}} = \frac{2000\,W_1 \ln (W_2/W_1) \ln (C_1 W_1/C_2 W_2)}{(M_1 + M_2) \ln (W_1/W_2)t}. \tag{A5}$$

Water influx

$$\frac{\text{ml H}_2\text{O influx}}{\text{kg d}} = \frac{\text{ml H}_2\text{O efflux}}{\text{kg d}} + \frac{2000(W_2 - W_1)}{t(M_1 + M_2)}. \tag{A6}$$

References

Acher, R. and Chauvet, J. (1997) Principles in protein evolution: composite domain evolution of neurohyphopysial prehormones. In *Advances in Comparative Endocrinology* (ed. S. Kawashima and S. Kikuyama), pp. 1191–9. Monduzzi Editore, Bologna, Italy.

Adams, N. J., Brown, C. R. and Nagy, K. A. (1986) Energy expenditure of free-ranging Wandering albatrosses, *Diomedea exulans. Physiol. Zool.*, **56:** 583–91.

Adams, S. and Costa, D. P. (1993) Water conservation and protein metabolism in Northern elephant seal pups during the postweaning fast. *J. Comp. Physiol.* B, **163:** 367–73.

Algar, D. (1986) An ecological study of macropod marsupial species on a reserve. PhD thesis, The University of Western Australia, Perth.

Alkon, P. U., Pinshow, B. and Degen, A. A. (1984) Seasonal water turnover rates and body water volumes in desert chukars. *Condor*, **84:** 332–7.

Ambrose, S. J. and Bradshaw, S. D. (1988a) The water and electrolyte metabolism of free-ranging and captive White-Browed scrubwrens, *Sericornis frontalis* (Acanthizidae) from arid, semi-arid and mesic environments. *Aust. J. Zool.*, **36:** 29–51.

Ambrose, S. J. and Bradshaw, S. D. (1988b) Seasonal changes in standard metabolic rates in the White-Browed scrubwren, *Sericornis frontalis* (Aves: Acanthizidae) in arid, semi-arid and mesic environments. *Comp. Biochem. Physiol.*, **89A:** 79–83.

Ambrose, S. J., Bradshaw, S. D., Withers, P. C. and Murphy, D. P. (1996) Water and energy balance of captive and free-ranging Spinifexbirds (*Eremiornis carteri*) North (Aves: Sylviidae) on Barrow Island, Western Australia. *Aust. J. Zool.*, **44:** 107–17.

Amos, W., Worthington Wilmer, J., Fullard, K., Burg, T. M., Croxall, J. P., Bloch, D. and Coulson, T. (2001) The influence of parental relatedness on reproductive success. *Proc. R. Soc. Lond.* B, **268:** 2021–7.

Ancel, A., Kooyman, G. L., Ponganis, E. P., Gendner, J.-P., Lignon, J., Mestre, X., Huin, N., Thorson, P. H., Robisson, P. and Le Maho, Y. (1992) Foraging behaviour of Emperor penguins as a resource detector in winter and summer. *Nature (Lond.)*, **360:** 336–8.

Ancel, A., Visser, H., Handrich, Y., Masman, D. and Le Maho, Y. (1997) Energy saving in huddling penguins. *Nature (Lond.)*, **385:** 304–5.

Andrewartha, H. G. and Birch, L. C. (1954) *The Distribution and Abundance of Animals*. University of Chicago Press, Chicago. 506pp.

Anstee, S. D. and Needham, D. J. (1998) The use of a new field anaesthesia technique to allow the correct fitting of radio collars on the Western Pebble Mound mouse, *Pseudomys chapmani. Aust. Mammal.*, **20:** 99–101.

Arena, P. C., Richardson, K. C. and Cullen, L. K. (1988) Anaesthesia in two species of large Australian skinks. *Vet. Rec.*, **123:** 155–8.

Arnould, J. P. Y., Briggs, D. R., Croxall, J. P., Prince, P. A. and Wood, A. G. (1996) The foraging behaviour and energetics of Wandering albatrosses, brooding chicks. *Antarctic Sci.*, **8:** 229–36.

Aschoff, J. and Pohl, H. (1970) Rhythmic variations in energy metabolism. *Fed. Proc.*, **29:** 1541–52.

Astheimer, L. B., Buttemer, W. A. and Wingfield, J. (1992) Interactions of corticosterone with feeding, activity and metabolism in passerine birds. *Ornis Scand.*, **23:** 355–65.

Astheimer, L. B., Buttemer, W. A. and Wingfield, J. C. (1994) Gender differences in the adrenocortical response to ACTH challenge in an arctic passerine, *Zonotrichia leucophrys gambelii. Gen. Comp. Endocrinol.*, **94:** 33–43.

Astheimer, L. B., Buttemer, W. A. and Wingfield, J. C. (1995) Seasonal and acute changes in adrenocortical responsiveness in an Arctic-breeding bird. *Horm. Behav.*, **29:** 442–57.

Atchley, W. R. and Anderson, D. (1978) Ratios and the statistical analysis of biological data. *Syst. Zool.*, **27:** 71–8.

Augee, M. L. and Gooden, B. A. (1992) Monotreme hibernation – some afterthoughts. In *Platypus and Echidnas* (ed. M. L. Augee), pp. 174–6. The Royal Zoological Society of New South Wales, Mosman, New South Wales.

Avery, R. A. (1982) Field studies of body temperature temperatures and thermoregulation. In *Biology of the Reptilia* (ed. C. Gans and F. H. Pough), pp. 93–166. Academic Press, New York.

Baddouri, K., Butlen, D., Imbert-Teboul, M., Le Bouffant, F., Marchetti, J., Chabardes, D. and Morel, F. (1984) Plasma antidiuretic hormone levels and kidney responsiveness to vasopressin in the jerboa *Jaculus orientalis. Gen. Comp. Endocrinol.*, **54:** 203–15.

Baddouri, K., El Hilali, M., Marchetti, J. and Menard, J. (1987) Renal excretion capacity in hydrated desert rodents (*Jaculus orientalis* and *Jaculus deserti*). *J. Comp. Physiol.* B, **157:** 237–40.

Baddouri, K., Marchetti, J., Hilali, M. and Menard, J. (1981) Mesure de l'hormone antidiurétique et de l'activité rénine plasmatique chez les Rongeurs désertiques (*Jaculus orientalis* et *Jaculus deserti*). *C.R. Acad. Sci. (Paris)*, **292:** 1143–6.

Bailey, N. J. (1952) Improvements in the interpretation of recapture data. *J. Anim. Ecol.*, **21:** 120–7.

Baker, J. R., Anderson, S. S. and Fedak, M. A. (1988) The use of a ketamine–diazepam mixture to immobilise wild grey seals (*Halichoerus grypus*) and southern elephant seals (*Mirounga leonina*). *Vet. Rec.*, **123:** 287–9.

Baker, J. R., Fedak, M. A., Anderson, S. S., Arnbom, T. and Baker, R. (1990) Use of a tiletamine–zolazepam mixture to immobilise wild grey seals and southern elephant seals. *Vet. Rec.*, **126:** 75–7.

Bakker, H. R. and Bradshaw, S. D. (1977) The effect of hypothalamic lesions on water metabolism of the toad *Bufo marinus*. *J. Endocrinol.*, **75:** 161–72.

Bakker, H. R. and Bradshaw, S. D. (1983) Renal function in the spectacled Hare Wallaby (*Lagorchestes conspicillatus*): Effects of dehydration and protein deficiency. *Aust. J. Zool.*, **31:** 101–8.

Bakker, H. R. and Bradshaw, S. D. (1989) Water turnover and electrolyte balance of the Spectacled Hare Wallaby (*Lagorchestes conspicillatus*) on Barrow Island. *Comp. Biochem. Physiol.*, **92A:** 521–9.

Bakker, H. R. and Main, A. R. (1980) Condition, body composition and total body water estimation in the quokka, *Setonix brachyurus* (Macropodidae). *Aust. J. Zool.*, **28:** 395–406.

Bankir, L. and de Rouffignac, C. (1985) Urinary concentrating ability: insights from comparative anatomy. *Am. J. Physiol.*, **249:** R643–66.

Banta, M. R. and Holcombe, D. W. (2002) The effects of thyroxine on metabolism and water balance in a desert-dwelling rodent, Merriam's kangaroo rat (*Dipodomys merriami*). *J. Comp. Physiol.* B, **172:** 17–25.

Barbraud, C. and Weimerskirch, H. (2001) Emperor penguins and climate change. *Nature (Lond.)*, **411:** 183–6.

Barker-Jorgensen, C. (1997) Urea and amphibian water economy. *Comp. Biochem. Physiol.*, **117A:** 161–70.

Barnett, J. L. (1973) A stress response in *Antechinus stuartii* (Macleay). *Aust. J. Zool.*, **21:** 501–13.

Bartholomew, G. A. (1972) The water economy of seed-eating birds that survive without drinking. *Proc. 15th Int. Ornithol. Congr.*, pp. 237–54. W. Junk, The Hague.

Bartholomew, G. A. (1982) The diversity of temporal heterothermy. In *Living in the Cold: Physiological and Biochemical Adaptations* (ed. H. C. Heller, X. J. Musacchia and L. C. H. Wang), pp. 1–12. Elsevier, New York.

Baudin, F. (1802/2001) *Mon Voyage aux Terres Australes* (ed. J. Bonnemains, J.-M. Argentin and M. Marin). Museum d'Histoire Naturelle, Le Havre, Le Havre. 467pp.

Baudinette, R. and Schmidt-Nielsen, K. (1974) Energy cost of gliding flight in Herring gulls. *Nature (Lond.)*, **248:** 83–4.

Bayomy, M. F., Shalan, A. G., Bradshaw, S. D., Withers, P. C., Stewart, T. and Thompson, G. (2002) Water content, body weight and acid mucopolysaccharides, hyaluronidase and beta-glucuronidase in response to aestivation in Australian desert frogs. *Comp. Biochem. Physiol.* (A), **131:** 881–92.

Bell, D. T., Moredount, J. C. and Loneragan, W. A. (1987) Grazing pressure by the Tammar (*Macropus eugenii*) on the vegetation of Garden Island, Western Australia, and the potential impact on food reserves of a controlled burning regime. *J. Roy. Soc. W. A.*, **69:** 89–94.

Ben Chaouacha-Chekir, R., Lachiver, F. and Cheniti, T. (1983) Données préliminaires sur le taux de renouvellement d'eau chez un Gerbillidé désertique, *Psammomys obesus*, étudié dans son environnement naturel en Tunisie. *Mammalia*, **47:** 543–7.

Bennett, A. F. (1982) The energetics of reptilian activity. In *Biology of the Reptilia* (ed. C. Gans and F. H. Pough), pp. 155–99. Academic Press, New York.

Bennett, A. F. (1987) Interindividual variability: an underutilized resource. In *New Directions in Ecological Physiology* (ed. M. E. Feder, A. F. Bennett, W. W. Burggren and R. B. Huey), pp. 147–69. Cambridge University Press, Cambridge.

Bennett, A. F. and Dawson, W. R. (1976) Metabolism. In *Biology of the Reptilia* (ed. C. Gans and W. R. Dawson), pp. 127–223. Academic Press, New York.

Bennett, A. F. and Licht, P. (1972) Anaerobic metabolism during activity in lizards. *J. Comp. Physiol.* B, **81:** 277–88.

Bennett, K. A., McConnell, B. J. and Fedak, M. A. (2001) Diurnal and seasonal variations in the duration and depth of the longest dives in southern elephant seals (*Mirounga leonina*): possible physiological and behavioural constraints. *J. Exp. Biol.*, **204:** 649–62.

Bennett, P. M. and Harvey, P. H. (1987) Active and resting metabolism in birds: allometry, phylogeny and ecology. *J. Zool., Lond.*, **213:** 327–63.

Bentley, P. J. (1971) *Endocrines and Osmoregulation*. Springer-Verlag, Berlin. 300pp.

Berk, M. L. and Heath, J. E. (1975) An analysis of behavioural thermoregulation in the lizard *Dipsosaurus dorsalis*. *J. Thermal Biol.*, **1:** 15–22.

Bernard, C. (1878) *Leçons sur les Phénomènes de la Vie communs aux Animaux et aux Végétaux*. J.-B. Baillière et Fils, Paris.

Berry, K. H. (1984) The status of the desert tortoise (*Gopherus agassizii*) in the United States. Report of Desert Tortoise Council, U.S. Fish & Wildlife Service, Sacramento, California.

Bertolino, S., Viano, C. and Currado, I. (2001) Population dynamics, breeding patterns and spatial use of the garden dormouse (*Eliomys quercinus*) in an Alpine habitat. *J. Zool., Lond.*, **253:** 513–21.

Beuchat, C. A. (1990) Body size, medullary thickness, and urine concentrating ability in mammals. *Am. J. Physiol.*, **258:** R298–308.

Bevan, R., Boyd, I. L., Butler, P., Reid, K., Woakes, A. and Croxall, J. (1997) Heart rates and abdominal temperatures of free-ranging South Georgian shags, *Phalacrocorax georgianus*. *J. Exp. Biol.*, **200:** 661–75.

Bevan, R. M., Butler, P. J., Woakes, A. J. and Prince, P. A. (1995) The energy expenditure of free-ranging Black-browed albatrosses. *Phil. Trans. R. Soc. Lond.* B, **350:** 119–31.

Billiards, S. S., King, J. M. and Agar, N. S. (1999) Comparative erythrocyte metabolism in three species of marsupials from Western Australia. *Comp. Haem. Int.*, **9:** 86–91.

Bishop, C. M. and Hall, M. R. (1991) Non-invasive monitoring of avian reproduction by simplified faecal steroid analysis. *J. Zool., Lond.*, **224:** 649–68.

Block, W. and Vannier, G. (1994) What is ecophysiology? Two perspectives. *Acta Oecol. (Berl.)*, **15:** 5–12.

Bonnet, X., Bradshaw, S.D., Shine, R. and Pearson, D. (1999) Why do snakes have eyes? The (non-)effect of blindness in island tiger snakes (*Notechis scutatus*). *Behav. Ecol. Sociobiol.*, **46:** 267–72.

Bonnet, X., Lagarde, F., Henen, B. T., Corbin, J., Nagy, K. A., Naulleau, G., Balhoul, K., Chastel, O. and Legrand, A. (2001) Sexual dimorphism in steppe tortoises (*Testudo horsfieldii*); influence of the environment and sexual selection on body shape and mobility. *Biol. J. Linn. Soc.*, **72:** 357–72.

Bonnet, X. and Naulleau, G. (1994) Utilisation d'un indice de condition corporelle (BCI) pour l'étude de la reproduction chez les serpents. *C.R. Acad. Sci. (Paris)*, **317:** 34–41.

Borut, A., Horowitz, M. and Castel, A. (1972) Blood volume regulation in the spiny mouse: capillary changes due to dehydration. *Symp. Zool. Soc. Lond.*, **31:** 175–89.

Bost, C. A. (1994) Maximum diving depth and diving patterns of the Gentoo penguin, *Pygoscelis papua* at the Crozet Islands. *Mar. Ornithol.*, **22:** 237–44.

Bradley, A. J. (1987) Stress and mortality in the Red-Tailed Phascogale, *Phascogale calura* (Marsupialia: Dasyuridae). *Gen. Comp. Endocrinol.*, **67:** 85–100.

Bradley, A. J. (1997) Reproduction and life history in the red-tailed phascogale, *Phascogale calura* (Marsupialia: Dasyuridae): the adaptive-stress senescence hypothesis. *J. Zool., Lond.*, **241:** 739–55.

Bradley, A. J., McDonald, I. R. and Lee, A. K. (1980) Stress and mortality in a small marsupial (*Antechinus stuartii* Macleay). *Gen. Comp. Endocrinol.*, **40:** 188–200.

Bradley, S. E., Mudge, G. H. and Blake, W. D. (1954) The renal excretion of sodium, potassium and water by the Harbour seal (*Phoca vitulina* L.): effect of apnea, sodium, potassium, and water loading; pitressin; and mercurial diuresis. *J. Cell. Comp. Physiol.*, **43:** 1–22.

Bradshaw, F. J. and Bradshaw, S. D. (2001) Maintenance nitrogen requirement of an obligate nectarivore, the Honey possum, *Tarsipes rostratus*. *J. Comp. Physiol.* B, **171:** 59–67.

Bradshaw, S. D. (1970) Seasonal changes in the water and electrolyte metabolism of *Amphibolurus* lizards in the field. *Comp. Biochem. Physiol.*, **36:** 689–718.

Bradshaw, S. D. (1978a) Aspects of hormonal control of osmoregulation in desert reptiles. In *Comparative Endocrinology* (ed. P. J. Gaillard and H. H. Boer), pp. 213–16. Elsevier North-Holland Biomedical Press, Amsterdam.

Bradshaw, S. D. (1978b) Volume regulation in desert reptiles and its control by pituitary and adrenal hormones. In *Osmotic and Volume Regulation* (ed. C. B. Jorgensen and E. Skadhauge), pp. 38–59. Munksgaard, Cϕpenhagen.

Bradshaw, S. D. (1981) Ecophysiology of Australian desert lizards: studies on the genus *Amphibolurus*. In *Biogeography and Ecology in Australia* (ed. A. Keast), pp. 1395–434. Junk, The Hague.

Bradshaw, S. D. (1983) Recent endocrinological research on the Rottnest Island quokka (*Setonix brachyurus*). *J. R. Soc. W. A.*, **66:** 55–61.

Bradshaw, S. D. (1986) *Ecophysiology of Desert Reptiles*. Academic Press, Sydney. 324pp.

Bradshaw, S. D. (1988) Desert reptiles: a case of adaptation or pre-adaptation? *J. Arid Envts*, **14:** 155–74.

Bradshaw, S. D. (1990) Aspects of hormonal control of osmoregulation in desert marsupials. In *Progress in Comparative Endocrinology* (ed. A. Epple, C. G. Scanes and M. H. Stetson), pp. 516–21. Wiley-Liss, New York.

Bradshaw, S. D. (1992a) Le problème du stress dans les études écophysiologiques: Stratégies de mesure et de contrôle. *Bull. Soc. Ecophysiol. (Paris)*, **17:** 69–76.

Bradshaw, S. D. (1992b) L'Ecophysiologie d'une île désertique en Australie: études de bilans énergétiques et homéostasie de vertébrés terrestres dans un milieu aride. *Bull. Soc. Ecophysiol. (Paris)*, **17:** 83–92.

Bradshaw, S. D. (1996) Hormones, stress and their relevance to problems of conservation. In *Environmental and Conservation Endocrinology*, 3rd Congress of the Asia & Oceania Society for Comparative Endocrinology (ed. J. Joss), pp. 1–4. Macquarie University, Sydney, Sydney.

Bradshaw, S. D. (1997a) *Homeostasis of Desert Reptiles*. Springer-Verlag, Heidelberg. 213pp.

Bradshaw, S. D. (1997b) Water metabolism of endangered Marsupial species. In *Advances in Comparative Endocrinology* (ed. S. Kawashima and S. Kikuyama), pp. 1701–5. Monduzzi Editore, Bologna, Italy.

Bradshaw, S. D. (1999) Ecophysiological studies on desert mammals: insights from stress physiology. *Aust. Mammal.*, **21:** 55–65.

Bradshaw, S. D. (2000) Field studies of the nutrition of Australian native animals. *Proc. Nutr. Soc. Aust.*, **24:** 155–84.

Bradshaw, S. D. and Bradshaw, F. J. (1999) Field energetics and the estimation of pollen and nectar intake in the marsupial honey possum, *Tarsipes rostratus*, in heathland habitats of south-western Australia. *J. Comp. Physiol.* B, **169:** 569–80.

Bradshaw, S. D. and Bradshaw, F. J. (2002) Short-term movements and habitat utilisation of the marsupial Honey possum, *Tarsipes rostratus*. *J. Zool., Lond.*, **258:** 343–8.

Bradshaw, S. D., Cheniti, T. and Lachiver, F. (1976) Echanges hydriques chez deux rongeurs désertiques, *Meriones shawii* et *Meriones libycus* étudiés dans leur environnement naturel en Tunisie. *Bull. Soc. Ecophysiol. (Paris)*, **1:** 30–1.

Bradshaw, S. D., Cohen, D., Katsaros, A., Tom, J. and Owen, F. J. (1987a) Determination of ^{18}O by prompt nuclear reaction analysis: application for measurement of microsamples. *J. Appl. Physiol.*, **63:** 1296–302.

Bradshaw, S. D. and De'ath, G. (1991) Variation in condition indices due to climatic and seasonal factors in an Australian desert lizard, *Amphibolurus nuchalis. Aust. J. Zool.*, **39:** 373–85.

Bradshaw, S. D. and Main, A. R. (1968) Behavioural attitudes and regulation of temperature in *Amphibolurus lizards. J. Zool., Lond.*, **154:** 193–221.

Bradshaw, S. D., Morris, K. D. and Bradshaw, F. J. (2001) Water and electrolyte homeostasis and kidney function of desert-dwelling marsupial wallabies in Western Australia. *J. Comp. Physiol.* B, **171:** 23–32.

Bradshaw, S. D., Morris, K. D., Dickman, C. R., Withers, P. C. and Murphy, D. (1994) Field metabolism and turnover in the Golden Bandicoot (*Isoodon auratus*) and other small mammals from Barrow Island, Western Australia. *Aust. J. Zool.*, **42:** 29–41.

Bradshaw, S. D., Saint Girons, H. and Bradshaw, F. J. (1991) Patterns of breeding in two species of agamid lizards in the arid sub-tropical Pilbara region of Western Australia. *Gen. Comp. Endocrinol.*, **82:** 407–24.

Bradshaw, S. D., Saint Girons, H., Naulleau, G. and Nagy, K. A. (1987b) Material and energy balance of some captive and free-ranging reptiles in western France. *Amphib. Rept.*, **8:** 129–42.

Bradshaw, S. D. and Shoemaker, V. H. (1967) Aspects of water and electrolyte changes in a field population of *Amphibolurus lizards. Comp. Biochem. Physiol.*, **20:** 855–65.

Braithwaite, R. W. and Lee, A. K. (1979) A mammalian example of semelparity. *Am. Nat.*, **113:** 151–5.

Brand, M. D., Couture, P., Else, P. L., Withers, K. W. and Hulbert, A. J. (1991) Evolution of energy metabolism. *Biochem. J.*, **275:** 81–6.

Braun, E. (1985) Comparative aspects of the urimary concentrating process. *Renal Physiol.*, **8:** 249–69.

Brett, J. R. (1958) Implications and assessments of environmental stress. In *Investigations of Fish-Power Problems* (ed. P. A. Larkin), pp. 69–83. University of British Columbia, Vancouver.

Bronson, F. H. (1989) *Mammalian Reproductive Biology*. The University of Chicago Press, Chicago. 325pp.

Bronson, F. H. (1998) Energy balance and ovulation: small cages versus natural habitats. *Reprod. Fert. Dev.*, **10:** 127–37.

Brooker, B. and Withers, P. C. (1994) Kidney structure and renal indices of dasyurid marsupials. *Aust. J. Zool.*, **42:** 163–76.

Brookhyser, K. M., Aulerich, R. J. and Vomachka, A. J. (1977) Adaptation of the orbital sinus bleeding technique to the chinchilla (*Chinchilla laniger*). *Lab. Anim. Sci.*, **27:** 251–4.

Brothers, N. (1991) Albatross mortality and associated bait loss in the Japanese long-line fishery in the Southern Ocean. *Biol. Cons.*, **55:** 255–68.

Brown, J. L., Wasser, S. K., Wildt, D. E., Graham, L. H. and Monfort, S. L. (1997) Faecal steroid analysis for monitoring ovarian and testicular function in diverse wild carnivore, primate and ungulate species. *Int. J. Mamm. Biol.*, **62:** 27–31.

Brown, K. (2000) Ecologists spar over population counts of threatened Desert Tortoise. *Sciences, N.Y.*, **290:** 36.

Brown, M. B., Berry, K. H., Schumacher, I. M., Nagy, K. A., Christopher, M. M. and Klein, P. A. (1999) Seroepidemiology of upper respiratory tract disease in the desert tortoise in the western Mojave desert of California. *J. Wildl. Dis.*, **35:** 716–27.

Brownfield, M. S. and Wunder, W. (1976) Relative medullary area: a new structural index for estimating urinary concentrating capacity of mammals. *Comp. Biochem. Physiol.*, **55A:** 69–75.

Burt, W. H. (1943) Territoriality and home range concepts as applied to mammals. *J. Mammal.*, **24:** 346–52.

Burton, T. M. and Likens, G. E. (1975) Energy flow and nutrient cycling in salamander populations in the Hubbard Brook Experiment Forest, New Hampshire. *Ecology*, **58:** 1068–80.

Buscarlet, L. A. (1974) The use of ^{22}Na for determining the food intake of the migratory locust. *Oikos*, **25:** 204–8.

Cade, T. J., Tobin, C. A. and Gold, A. (1965) Water economy and metabolism of estrildine finches. *Physiol. Zool.*, **38:** 9–33.

Campagna, C., Werner, R., Karesh, W., Marin, M., Koontz, F., Cook, R. and Koontz, C. (2001) Movements and location at sea of South American sea lions (*Otaria flavescens*). *J. Zool., Lond.*, **257:** 205–20.

Cannon, W. B. (1929) Organization for homeostasis. *Physiol. Rev.*, **9:** 399–429.

Cannon, W. B. (1939) *The Wisdom of the Body*. W.W. Norton & Co. Inc., New York. 350pp.

Carey, H. V. and Martin, S. L. (1996) Hibernation and the stress response. In *Adaptations to the Cold* (ed. F. Geiser, A. J. Hulbert and S. C. Nicol), pp. 319–25. The University of New England Press, Armidale, NSW.

Carey, H. V., Frank, C. L. and Yee Aw, T. (2000) Cellular response to metabolic stress in hibernating mammals. In *Life in the Cold: Eleventh International Hibernation Symposium* (ed. G. Heldmaier and M. Klingenspor), pp. 339–46. Springer-Verlag, Berlin.

Carnegie, D. W. (1898) *Spinifex and Sand*. Arthur Pearson, London.

Case, T. J. (1976) Seasonal aspects of thermoregulatory behavior in the chuckwalla, *Sauromalus obesus* (Reptilia, Lacertilia, Iguanidae). *J. Herp.*, **10:** 85–95.

Castellini, M. A., Murphy, B. J., Fedak, M., Ronald, K., Gofton, N. and Hochachka, P. W. (1985) Potentially conflicting metabolic demands of diving and exercise in seals. *J. Appl. Physiol.*, **58:** 392–9.

Caughley, G. (1977) *Analysis of Vertebrate Populations*. Tom Wiley & Sons, Inc., London. 234pp.

Caughley, G. (1994) Directions in conservation biology. *J. Anim. Ecol.*, **63:** 215–44.
Caughley, G. and Gunn, A. (1996) *Conservation Biology in Theory and Practice.* Blackwell Science, Cambridge, Massachusetts, USA.
Caughley, G. and Sinclair, A. R. E. (1994) *Wildlife Ecology and Management.* Blackwell Science, Cambridge, Massachusetts, USA. 334pp.
Cavigelli, S. A. (1999) Behavioral patterns associated with fecal cortisol levels in free-ranging ring-tailed lemurs, *Lemur catta. Anim. Behav.*, **57:** 935–44.
Chapman, T. E. and McFarland, L. Z. (1971) Water turnover of coturnix quail with individual observations on a burrowing owl, Petz conure and vulturine fish eagle. *Comp. Biochem. Physiol.*, **39A:** 653–6.
Chappell, M. A., Shoemaker, V. H., Janes, D. N., Maloney, S. K. and Butcher, T. L. (1993) Energetics of foraging in breeding Adélie penguins. *Ecology*, **74:** 2450–61.
Chard, T. (1978) *An Introduction to Radioimmunoassay and Related Techniques*, North Holland Publishing Company, Amsterdam, New York and Oxford. pp. 293–534.
Chauvet, M. T., Coln, T., Hurpet, D., Chauvet, J. and Acher, R. (1983) A multigene family of vasopressin-like hormones? Identification of mesotocin, lysopressin and phenypressin in Australian macropods. *Biochem. Biophys. Res. Commun.*, **116:** 258–63.
Chauvet, M. T., Hurpet, D., Chauvet, J. and Acher, R. (1980) Phenypressin (Phe^2-Arg^8-vasopressin), a new neurohypophysial peptide found in marsupials. *Nature (Lond.)*, **287:** 640–2.
Cherel, Y. and Le Maho, Y. (1985) Five months of fasting in King penguin chicks: body mass loss and fuel metabolism. *Am. J. Physiol.*, **249:** R387–92.
Cherel, Y., Weimerskirch, H. and Duhamel, G. (1996) Interactions between longline vessels and seabirds in Kerguelen waters and a method to reduce seabird mortality. *Biol. Cons.*, **75:** 63–70.
Christopher, M. M., Berry, K. H., Wallis, I. R., Nagy, K. A., Henen, B. T. and Peterson, C. C. (1999) Reference intervals and physiologic alterations in haematologic and biochemical values of free-ranging tortoises in the Mojave desert. *J. Wildl. Dis.*, **35:** 212–38.
Clarke, B. C. and Nicolson, S. W. (1994) Water, energy, and electrolyte balance in captive Namib sand-dune lizards (*Angolosaurus skoogi*). *Copeia*, **1994:** 962–74.
Claus, R., Hoppen, H. O. and Karg, H. (1981) The secret of truffles: a steroidal pheromone. *Experientia*, **37:** 1178–9.
Cockburn, A. and Lazenby-Cohen, K. A. (1992) Use of nest trees by *Antechinus stuartii*, a semelparous lekking marsupial. *J. Zool., Lond.*, **226:** 657–80.
Cogger, H. G. (1974) Thermal relations of the mallee dragon, *Amphibolurus fordii* (Lacertilia: Agamidae). *Aust. J. Zool.*, **22:** 319–39.
Cohen, D. D., Katsaros, A. and Garton, D. (1986) Some characteristics of the $^{18}O(p, \alpha)^{15}N$ reaction. *Nucl. Instrum. Methods Phys. Res.*, **B15:** 555–8.
Cole, L. C. (1954) The population consequences of life history phenomena. *Quart. Rev. Biol.*, **29:** 103–37.

Colwell, E. K. (1974) Predictability, constancy, and contingency of periodic phenomena. *Ecology*, **55:** 1148–53.

Corbett, J. L., Farrell, D. J., Leng, R. A., McClymont, G. L. and Young, B. A. (1971) Determination of the energy expenditure of penned and grazed sheep from estimates of carbon dioxide entry rate. *Br. J. Nutr.*, **21:** 277–86.

Costa, D. P., Dann, P. and Disher, W. (1986) Energy requirements of free-ranging Little penguins, *Eudyptula minor. Comp. Biochem. Physiol.*, **85A:** 135–8.

Costa, D. P., Kretzmann, M. and Thorson, P. H. (1989) Diving pattern and energetics of the Australian sea lion, *Neophoca cinerea. Am. Zool.*, **29:** 71A (Abstr.).

Costa, W. R., Nagy, K. A. and Shoemaker, V. H. (1976) Observations of the behavior of jackrabbits (*Lepus californicus*) in the Mojave Desert. *J. Mammal.*, **57:** 399–402.

Coward, W. A. and Cole, T. J. (1991) The doubly labelled water method for measurement of energy expenditure in humans: risks and benefits. In *The Doubly Labelled Water Method: Technical Recommendations for Use in Humans* (ed. R. G. Whitehead and A. Prentice), pp. 294–7. Report of an IDECG Working Group. Vienna, Austria.

Cowles, R. B. and Bogert, C. M. (1944) A preliminary study of the thermal requirements of desert reptiles. *Bull. Am. Mus. Nat. Hist.*, **83:** 265–96.

Crawford, E. C. and Lasiewski, R. C. (1968) Oxygen consumption and respiratory evaporation of the emu and rhea. *Condor*, **70:** 333–9.

Crocker, D. E., Gales, N. J. and Costa, D. P. (2001) Swimming speed and foraging strategies of New Zealand sea lions (*Phocatros hookeri*). *J. Zool., Lond.*, **254:** 267–77.

Croxall, J. and Gales, R. P. (1998) An assessment of the conservation status of albatrosses. In *Ecology and Conservation of Albatrosses* (ed. G. C. Robertson and R. P. Gales), pp. 46–66. Surrey Beatty and Sons, Chipping Norton, Australia.

Croxall, J. P. (1984) Seabirds. In *Antarctic Ecology* (ed. R. M. Laws), pp. 531–618. Academic Press, New York.

Croxall, J. P., Naito, Y., Kato, A., Rothery, P. and Briggs, D. R. (1991) Diving patterns and performance in the Antarctic blue-eyed shag *Phalacrocorax atriceps. J. Zool., Lond.*, **225:** 177–99.

Croxall, J. P. and Ricketts, C. (1983) Energy costs of incubation in the Wandering albatross *Diomedea exulans. Ibis*, **125:** 33–9.

Crum, B. G., Williams, J. B. and Nagy, K. A. (1985) Can tritiated water-dilution space accurately predict total body water in chukar partridges. *J. Appl. Physiol.*, **59:** 1383–8.

Czekala, N. M. and Lasley, B. L. (1977) A technical note on sex determination in monomorphic birds using faecal steroid analysis. *Int. Zool. Ybk*, **17:** 209–11.

Daan, S., Masman, D. and Groenewold, A. (1990) Avian basal metabolic rates; their association with body composition and energy expenditure. *Am. J. Physiol.*, **259:** R333–40.

Dameron, G. W., Weingand, K. W., Duderstadt, J. M., Odioso, L. W., Dierckman, T. A., Schwecke, W. and Baran, K. (1992) Effect of bleeding site on clinical

laboratory testing of rats: orbital venous plexus *versus* posterior vena cava. *Lab. Anim. Sci.*, **42:** 299–301.
Dantzler, W. H. (1989) *Comparative Physiology of the Vertebrate Kidney*. Springer-Verlag, Heidelberg. 198pp.
Dantzler, W. H. and Schmidt-Nielsen, B. (1966) Excretion in fresh-water turtle (*Pseudemys scripta*) and desert tortoise (*Gopherus agassizii*). *Am. J. Physiol.*, **210:** 198–210.
Davies, S. J. J. F. (1984) Nomadism as a response to desert conditions in Australia. *J. Arid Envts*, **7:** 183–96.
Davis, D. E. (1982) *CRC Handbook of Census Methods for Terrestrial Vertebrates*. CRC Press, Boca Raton, Florida. 397 pp.
Davis, R. W., Croxall, J. P. and O'Connell, M. J. (1989) The reproductive energetics of Gentoo (*Pygoscelis papua*) and Macaroni (*Eudyptes chrysolophus*) penguins at South Georgia. *J. Anim. Ecol.*, **58:** 59–74.
Dawson, T. J. and Schmidt-Nielsen, K. (1966) Effect of thermal conductance on water economy in the antelope jackrabbit, *Lepus alleni. J. Cell Physiol.*, **67:** 463–72.
Dawson, W. R. (1984) Physiological studies of desert birds: present and future considerations. *J. Arid Envts*, **7:** 133–56.
de Castri, F. (1981) Mediterranean shrublands of the world. In *Ecosystems of the World: II. Mediterranean-Type Shrublands* (ed. F. di Castri, D. W. Goodall and R. L. Specht), pp. 1–52. Elsevier, Amsterdam.
de Rouffignac, C., Bankir, L. and Roinel, N. (1981) Renal function and concentrating ability in a desert rodent: the gundi (*Ctenodactylus vali*). *Pflügers Arch.*, **390:** 138–44.
de Rouffignac, C. and Morel, F. (1969) Micropuncture study of water, electrolytes and urea movements along the loop of Henle in *Psammomys. J. Clin. Invest.*, **48:** 474–86.
de Rouffignac, C. and Morel, F. (1973) Étude comparée du renouvellement de l'eau chez quatre espèces de rongeurs, dont deux espèces d'habitat désertique. *J. Physiol. (Paris)*, **58:** 309–22.
de Witt, C. B. (1967) Behavioral thermoregulation in the desert iguana. *Sciences, N. Y.*, **158:** 809–10.
Degen, A. A. (1994) Field metabolic rates of *Acomys russatus* and *Acomys cahirinus* and a comparison with other rodents. *Isr. J. Zool.*, **40:** R60–5.
Degen, A. A. (1997) *Ecophysiology of Small Desert Mammals*. Springer, Berlin, New York. xii + 296pp.
Degen, A. A., Hazan, A., Kam, M. and Nagy, K. A. (1991) Seasonal water influx and energy expenditure of free-living sand rats. *J. Mammal.*, **72:** 652–7.
Degen, A. A., Kam, M., Hazan, A. and Nagy, K. A. (1986) Energy expenditure and water flux in three sympatric desert rodents. *J. Anim. Ecol.*, **55:** 421–9.
Degen, A. A., Kam, M. and Jurgrau, D. (1988) Energy requirements of fat sand rats (*Psammomys obesus*) and their efficiency of utilization of the saltbush *Atriplex halmus* for maintenance. *J. Zool., Lond.*, **215:** 443–52.

Degen, A. A., Khokhlova, I. S., Kam, M. and Nagy, K. A. (1997) Body size, granivory and seasonal dietary shifts in desert gerbilline rodents. *Funct. Ecol.*, **11:** 53–9.
Degen, A. A., Pinshow, B. and Alkon, P. U. (1982) Water flux in Chukar partridges (*Alectoris chukar*) and a comparison with other birds. *Physiol. Zool.*, **55:** 64–71.
Degen, A. A., Pinshow, B. and Alkon, P. U. (1985) Summer water turnover rates in free-living Chukars and Sand partridges in the Negev Desert. *Condor*, **85:** 333–7.
Degen, A. A., Pinshow, B. and Ilan, M. (1990) Seasonal water flux, urine and plasma osmotic concentrations in free-living sand rats feeding solely on saltbush. *J. Arid Envts*, **18:** 59–66.
Degen, A. A., Pinshow, B. and Kam, M. (1992) Field metabolic rates and water influxes of two sympatric Gerbillidae: *Gerbillus allenbyi* and *G. pyramidum. Oecologia*, **90:** 586–90.
Demeneix, B. A. and Henderson, N. E. (1978a) Serum T_4 and T_3 in active and torpid ground squirrels, *Spermophilus richardsoni. Gen. Comp. Endocrinol.*, **35:** 77–85.
Demeneix, B. A. and Henderson, N. E. (1978b) Thyroxine metabolism in active and torpid ground squirrels, *Spermophilus richardsoni. Gen. Comp. Endocrinol.*, **35:** 86–92.
Depocas, F., Hart, J. S. and Fisher, H. D. (1971) Sea water drinking and water flux in starved and fed Harbour seals. *Can. J. Physiol. Pharmacol.*, **49:** 53–62.
Diamond, J. M. (1982) Big-bang reproduction and ageing in male marsupial mice. *Nature (Lond.)*, **298:** 115–16.
Diaz, G. B., Ojeda, R. A. and Dacar, M. (2001) Water conservation in the South American desert mouse opossum, *Thylamys pusilla* (Didelphimorphia, Didelphidae). *Comp. Biochem. Physiol.*, **130A:** 323–30.
Dickman, C. R. (1993) Evolution of semelparity in male dasyurid marsupials: a critique and an hypothesis of sperm competition. In *Biology and Management of Australasian Carnivorous Marsupials* (ed. M. Roberts, J. Carnio, G. Crawshaw and M. Hutchins), pp. 25–38. Metropolitan Toronto Zoo, Toronto.
Dickman, C. R. and Braithwaite, R. W. (1992) Postmating mortality of males in dasyurid marsupials, *Dasyurus* and *Parantechinus. J. Mammal.*, **73:** 143–7.
Dickman, C. R., Predavec, M. and Downey, F. J. (1995) Long-range movements of small mammals in arid Australia: implications for land management. *J. Arid Envts*, **31:** 441–52.
Dobzhansky, T. (1953) Biology and evolution. *Am. Biol. Teacher*, **25:** 125–9.
Dodson, P. (1978) On the use of ratios in growth studies. *Syst. Zool.*, **27:** 62–7.
Downs, C. T. and Perrin, M. R. (1990) Field water-turnover rates of three *Gerbillurus* species. *J. Arid Envts.*, **19:** 199–208.
Dunlap, K. D. and Wingfield, J. C. (1995) External and internal influences on indices of physiological stress. I. Seasonal and population variation in adrenocortical secretion of free-living lizards, *Sceloporus occidentalis. J. Exp. Zool.*, **271:** 36–46.
Eales, J. G. (1997) Iodine metabolism and thyroid-related functions in organisms lacking thyroid follicles: are thyroid hormones also vitamins? *Proc. Soc. Exp. Biol. Med.*, **214:** 302–17.

Easterling, D. R., Meehl, G. A., Parmesan, C., Changnon, S. A., Karl, T. R. and Mearns, L. O. (2000) Climate extremes: Observations, modeling and impacts. *Sciences, N. Y.*, **289:** 2068–74.

Ekins, R. P. (1974) Basic principles and theory. In *Radioimmunoassay and Saturation Analysis* (ed. P. H. Sönksen), pp. 3–11. British Medical Bulletin vol. 30, no. 1. The British Council, London.

Eldridge, M. D. B., King, J. M., Loupis, A. K., Spencer, P. B. S., Taylor, A. C., Pope, L. C. and Hall, G. P. (1999) Unprecedented low levels of genetic variation and inbreeding depression in an island population of the Black-Footed rock-wallaby. *Cons. Biol.*, **13:** 531–41.

Else, P. L. and Hulbert, A. J. (1987) Evolution of mammalian endothermic metabolism: "leaky" membranes as a source of heat. *Am. J. Physiol.*, **253:** R1–7.

Emerson, S. B. and Hess, D. L. (1996) The role of androgens in opportunistic breeding, tropical frogs. *Gen. Comp. Endocrinol.*, **103:** 220–30.

Fairclough, R. J., Rabjohns, M. A. and Peterson, A. J. (1977) Chromatographic separation of androgens, estrogens and progestogens on hydroxyalkoxypropyl-Sephadex (Lipidex®). *J. Chromatography*, **133:** 412–14.

Fancy, S. G., Blanchard, J. M. and Holleman, D. F. (1986) Validation of doubly-labeled water method using a ruminant. *Am. J. Physiol.*, **251:** R143–9.

Fänge, R., Schmidt-Nielsen, K. and Robinson, M. (1958) Control of secretion from the avian salt gland. *Am. J. Physiol.*, **195:** 321–6.

Fänge, R. and Schmidt-Nielsen, K. O., H. (1958) The salt gland of the Herring Gull. *Biol. Bull.*, **115:** 162–71.

Feder, M. E. and Burggren, W. W. (eds) (1992) *Environmental Physiology of Amphibians*. University of Chicago Press, Chicago.

Felsenstein, J. (1985) Phylogenies and the comparative method. *Am. Nat.*, **125:** 1–15.

Felsenstein, J. (1988) Phylogenies and the comparative method. *A. Rev. Ecol. Syst.*, **19:** 445–71.

Fernandez, P., Anderson, D. J., Sievert, P. R. and Huyvaert, K. P. (2001) Foraging destinations of three low-latitude albatross (*Phoebastria*) species. *J. Zool., Lond.*, **254:** 391–404.

Fisher, C. D., Lindgren, E. and Dawson, W. R. (1972) Drinking patterns and behaviour of Australian desert birds in relation to their ecology and abundance. *Condor*, **74:** 111–36.

Flinders, M. (1814) *A Voyage to Terra Australis*. W. Nicol, London.

Flint, E. N. and Nagy, K. A. (1984) Flight energetics of free-living Sooty Terns. *Auk*, **101:** 288–94.

Frankel, A. I., Cook, B., Graber, J. and Nalbandov, A. V. (1967) Determination of corticosterone in plasma by fluorimetric techniques. *Endocrinology*, **80:** 181–94.

Friend, G. R. (1984) Relative efficiency of two pitfall-drift fence systems for sampling small vertebrates. *Aust. Zool.*, **21:** 423–33.

Friend, G. R., Smith, G. T., Mitchell, D. S. and Dickman, C. R. (1989) Influence of pitfall and drift fence design on capture rates of small vertebrates in semi-arid habitats of Western Australia. *Aust. Wildl. Res.*, **16:** 1–10.

Gabe, M., Agid, R., Martoja, M., Saint Girons, M.-C. and Saint Girons, H. (1964) Données histophysiologiques et biochimiques sur l'hibernation et le cycle annuel chez *Eliomys quercinus* L. *Arch. Biol. Liège*, **75:** 1–87.

Gabrielsen, G. W., Mehlum, F. and Nagy, K. A. (1987) Daily energy expenditure and energy utilisation of free-ranging Black-legged kittiwakes. *Condor*, **89:** 126–32.

Gadgil, M. and Bossert, W. H. (1970) Life historical consequences of natural selection. *Am. Nat.*, **104:** 1–24.

Gales, N. J. and Costa, D. P. (1997) The Australian sea lion: a review of an unusual life history. In *Marine Mammal Research in the Southern Hemisphere* (ed. M. A. Hindell and C. Kemper), vol. 1, pp. 78–87. Surrey Beatty & Sons, Chipping Norton, Sydney.

Gales, N. J., Shaughnessy, P. D. and Dennis, T. E. (1994) Distribution, abundance and breeding cycle of the Australian sea lion *Neophoca cinerea* (Mammalia: Pinnipedia). *J. Zool., Lond.*, **234:** 353–70.

Gales, N. J., Williamson, P., Higgins, L. V., Blackberry, M. A. and James, I. (1997) Evidence for a postimplantation period in the Australian sea lion (*Neophoca cinerea*). *J. Reprod. Fert.*, **111:** 159–63.

Gales, R. (1989) Validation of the use of tritiated water, doubly labelled water, and ^{22}Na for estimating food, energy, and water intake in little penguins, *Eudyptula minor*. *Physiol. Zool.*, **62:** 147–69.

Gales, R. and Green, B. (1990) The annual energetics cycle of little penguins (*Eudyptula minor*). *Ecology*, **71:** 2297–312.

Gales, R., Green, B., Libke, J., Newgrain, K. and Pemberton, D. (1993) Breeding energetics and food requirements of gentoo penguins (*Pygoscelis papua*) at Heard and Macquarie Islands. *J. Zool., Lond.*, **231:** 125–39.

Gallagher, K. J., Morrison, D. A., Shine, R. and Grigg, G. C. (1983) Validation and use of ^{22}Na turnover to measure food intake in free-ranging lizards. *Oecologia*, **60:** 76–82.

Gans, C. (1979) Momentarily excessive construction as the basis for protoadaptation. *Evolution*, **33:** 227–33.

Garavanta, C. A. M., Wooller, R. D. and Richardson, K. C. (2000) Movement patterns of honey possums, *Tarsipes rostratus*, in the Fitzgerald River National Park, Western Australia. *Wildl. Res.*, **27:** 179–83.

Garland, T., Jr., Harvey, P. H. and Ives, A. R. (1992) Procedures for the analysis of comparative data using phylogenetically independent contrasts. *Syst. Biol.*, **41:** 18–32.

Gauthier, M. and Thomas, D. W. (1990) Evaluation of the accuracy of ^{22}Na and tritiated water for the estimation of food consumption and fat reserves in passerine birds. *Can. J. Zool.*, **68:** 1590–4.

Geiser, F. (1990) Influence of polyunsaturated and saturated dietary lipids on adipose tissue, brain and mitochondrial fatty acid composition of a mammalian hibernator. *Biochim. Biophys. Acta*, **1046:** 159–66.

Geiser, F. (1991) The effect of unsaturated and saturated dietary lipids on the pattern of daily torpor and the fatty acid composition of tissues and membranes of the deer mouse *Peromyscus maniculatus. J. Comp. Physiol.* B, **161:** 590–7.
Geiser, F. (1994) Hibernation and daily torpor in marsupials: a review. *Aust. J. Zool.*, **42:** 1–16.
Geiser, F. and Broome, L. S. (1991) Hibernation in the pygmy possum *Burramys parvus* (Marsupialia). *J. Zool., Lond.*, **223:** 593–602.
Geiser, F., Firth, B. T. and Seymour, R. S. (1992) Polyunsaturated dietary lipids lower the selected body temperature of a lizard. *J. Comp. Physiol.* B, **162:** 1–4.
Geiser, F., Hulbert, A. J. and Nicol, S. C. (eds) (1996) *Adaptations to the Cold.* University of New England Press, Armidale, NSW. 404 pp.
Geiser, F. and Kenagy, G. J. (1987) Polyunsaturated lipid diet lengthens torpor and reduces body temperature in a hibernator. *Am. J. Physiol.*, **252:** R897–901.
Geiser, F. and Kenagy, G. J. (1988) Torpor duration in relation to temperature and metabolism in hibernating ground squirrels. *Physiol. Zool.*, **61:** 442–9.
Geiser, F. and Ruf, T. (1995) Hibernation *versus* daily torpor in mammals and birds: physiological variables and classification of torpor patterns. *Physiol. Zool.*, **68:** 935–66.
Geiser, F., Stahl, B. and Leramonth, R. P. (1992) The effect of dietary fatty acids in the pattern of torpor in a marsupial. *Physiol. Zool.*, **65:** 1236–45.
Gentry, R. L. (1981) Seawater drinking in eared seals. *Comp. Biochem. Physiol.*, **8A:** 81–6.
Gettinger, R. D. (1983) Use of doubly labeled water ($^3H_2{}^{18}O$) for determination of H_2O flux and CO_2 production by a mammal in a humid environment. *Oecologia*, **59:** 54–7.
Gilligan, D. M., Woodworth, L. M., Montgomery, M. E., Nurthen, R. K., Briscoe, D. A. and Frankham, R. (2000) Can fluctuating asymmetry be used to detect inbreeding and loss of genetic diversity in endangered populations? *Anim. Cons.*, **3:** 97–104.
Goldstein, D. L. (1993) Renal glomerular and tubular responses to saline infusion in a marine bird, Leach's storm petrel. *J. Comp. Physiol.* B, **163:** 167–73.
Goldstein, D. L. and Bradshaw, S. D. (1998) Water and sodium regulation of water and sodium balance in the field by Australian honeyeaters (Aves: Meliphagidae). *Physiol. Zool.*, **71:** 214–25.
Goldstein, D. L. and Nagy, K. A. (1985) Resource utilization by desert quail: time and energy, food and water. *Ecology*, **66:** 378–87.
Goldstein, D. L. and Rothschild, E. L. (1993) Daily rhythms in renal and cloacal excretion by captive and wild song sparrows. *Physiol. Zool.*, **66:** 708–19.
Gould, S. J. and Lewontin, R. C. (1979) The spandrels of San Marco and the Panglossian paradigm: a critique of the adaptationist programme. *Proc. R. Soc. Lond.* B, **205:** 581–98.
Gould, S. J. and Vrba, E. S. (1982) Exaptation – a missing term in the science of form. *Paleobiology*, **8:** 4–15.

Gower, D. B. and Ruparelia, B. A. (1993) Olfaction in humans with special reference to odorous 16-androstenes: their occurrence, perception and possible social, psychological and sexual impact. *J. Endocrinol.*, **137:** 167–87.

Goymann, W., Mostl, E., Van't Hof, T., East, M. L. and Hofer, H. (1999) Noninvasive fecal monitoring of glucocorticoids in spotted hyenas, *Crocuta crocuta. Gen. Comp. Endocrinol.*, **114:** 340–8.

Grant, B. W. (1990) Trade-offs in activity time and physiological performance for thermoregulating desert lizards, *Sceloporus merriami. Ecology*, **71:** 2323–33.

Grant, B. W. and Dunham, A. E. (1998) Thermally imposed time constraints on the activity of the desert lizard *Sceloporus merriami. Ecology*, **69:** 167–76.

Gray, D. A. and Simon, E. (1983) Mammalian and avian antidiuretic hormone: studies related to possible species variation in osmoregulatory systems. *J. Comp. Physiol.* B, **151:** 241–6.

Gray, J. M., Yarian, D. and Remenofsky, M. (1990) Corticosterone, foraging behavior and metabolism in dark-eyed juncos, *Junco hyemalis. Gen. Comp. Endocrinol.*, **79:** 375–84.

Green, B. and Brothers, N. (1989) Water and sodium turnover and estimated food consumption rates in free-living Fairy Prions (*Pachyptila turtur*) and common diving petrels (*Pelecanoides urinatrix*). *Physiol. Zool.*, **62:** 702–15.

Green, B., Brothers, N. and Gales, R. (1988) Water, sodium and energy turnover in free-living little penguins, *Eudyptula minor. Aust. J. Zool.*, **36:** 429–40.

Green, B. and Eberhard, I. (1983) Water and sodium intake, and estimated food consumption, in free-living Eastern quolls, *Dasyurus viverrinus. Aust. J. Zool.*, **31:** 871–80.

Green, B., Griffiths, M. and Newgrain, K. (1992) Seasonal patterns in water, sodium, and energy turnover in free-living echidnas, *Tachyglossus aculeatus* (Mammalia: Monotremata). *J. Zool., Lond.*, **227:** 351–66.

Green, B., King, D. and Butler, W. H. (1986) Water, sodium and energy turnover in free-living Perenties, *Varanus giganteus. Aust. Wildl. Res.*, **13:** 589–95.

Green, B., King, D. and Bradley, A. J. (1989) Water and energy metabolism and estimated food consumption rates of free-living Wambengers, *Phascogale calura* (Marsupialia: Dasyuridae). *Aust. Wildl. Res.*, **16:** 501–7.

Greenwald, L. (1989) The significance of renal medullary thickness. *Physiol. Zool.*, **62:** 1005–14.

Greer, A. E. (1990) *The Biology and Evolution of Australian Lizards*. Surrey Beatty & Sons, Sydney.

Gregory, P. T. (1982) Reptilian hibernation. In *The Biology of the Reptilia* (ed. C. Gans and H. Pough), pp. 53–154. Academic Press, New York.

Grémillet, J. H. and Plos, A. L. (1994) The use of stomach temperature records for the calculation of daily food intake in cormorants. *J. Exp. Biol.*, **189:** 105–15.

Grenot, C. J. (1967) Observations physio-écologiques sur la régulation thermique chez le lézard agamide *Uromastix acanthinurus* Bell. *Bull. Soc. Zool. France*, **92:** 51–66.

Grenot, C. J. (1973) Sur la biologie d'un rongeur héliophile du Sahara, le "Goundi" (Ctenodactylidae). *Acta Tropica*, **30:** 237–50.

Grenot, C. J., Garcin, L. and Tsèrè-Pagès, H. (1996) Cold hardiness and behaviour of the European common lizard, from French populations in winter. In *Adaptations to the Cold* (ed. F. Geiser, A. J. Hulbert and S. C. Nicol), pp. 115–21. University of New England Press, Armidale, NSW.

Grigg, G. and Beard, L. (2000) Hibernation by echidnas in mild climates: hints about the evolution of endothermy. In *Life in the Cold: Eleventh International Hibernation Symposium* (ed. G. Heldmaier and M. Klingenspor), pp. 5–19. Springer-Verlag, Berlin.

Grigg, G., Beard, L. A. and Augee, M. L. (1989) Hibernation in a montreme, the echidna (*Tachyglossus auculeatus*). *Comp. Biochem. Physiol.*, **92A:** 609–12.

Grossman, C. J. (1984) Regulation of the immune system by sex steroids. *Endocr. Rev.*, **5:** 435–55.

Guppy, M., Bradshaw, S. D., Fergusson, B., Hansen, I. A. and Atwood, C. (1987) Metabolism in lizards: low lactate turnover and advantages of heterothermy. *Am. J. Physiol.*, **22:** R77–82.

Guppy, M., Hill, R. D., Schneider, R. C., Qvist, J., Liggins, G. C., Zapol, W. M. and Hochachka, P. W. (1986) Microcomputer-assisted metabolic studies of voluntary diving of Weddell seals. *Am. J. Physiol.*, **250:** R175–87.

Guppy, M. and Withers, P. C. (1999) Metabolic depression in animals: physiological perspectives and biochemical generalizations. *Biol. Rev.*, **74:** 1–40.

Hackel, D. B., Schmidt-Nielsen, K., Haines, H. and Mikat, E. (1965) Diabetes mellitus in the sand rat (*Psammomys obesus*). *Lab. Exper.*, **14:** 200–7.

Haines, H., Hackel, D. B. and Schmidt-Nielsen, K. (1965) Experimental diabetes mellitus induced by diet in the sand rat. *Am. J. Physiol.*, **208:** 297–300.

Halpern, B. N. and Pacaud, A. (1951) Technique de prélèvement d'échantillons de sang chez les petits animaux de laboratoire par ponction de plexus opthalmique. *C.R. Soc. Biol. (Paris).*, **141:** 1465–7.

Hamilton, W. J. I. and Coetzee, C. G. (1969) Thermoregulatory behaviour of the vegetarian lizard *Angolosaurus skoogi* on the vegetationless northern Namib desert dunes. *Scientific Papers of the Namib Desert Research Station*, no. **47**, pp. 95–103.

Harper, J. M. and Austad, S. N. (2001) Effect of capture and season on fecal glucocorticoid levels in Deer Mice (*Permoyscus maniculatus*) and Red-Backed Voles (*Clethrionomys gapperi*). *Gen. Comp. Endocrinol.*, **123:** 337–44.

Harris, S., Cresswell, W. J., Forde, P. G., Trewhella, W. J., Woollard, T. and Wray, S. (1990) Home range analysis using radio-tracking data – a review of problems and techniques particularly applied to the study of mammals. *Mamm. Rev.*, **20:** 97–123.

Harvey, L. A., Propper, C. R., Woodley, S. K. and Moore, M. C. (1997) Reproductive endocrinology of the explosively breeding Desert Spadefoot Toad, *Scaphiopus couchii*. *Gen. Comp. Endocrinol.*, **105:** 102–13.

Harvey, P. H. and Pagel, M. D. (1991) *The Comparative Method in Evolutionary Biology*. Oxford University Press, London. 239pp.

Hayden, P. (1966) Seasonal occurrence of jackrabbits on Jackass Flat, Nevada. *J. Wildl. Man.*, **30:** 835–8.

Hazon, N. and Balemont, R. J. (1998) Endocrinology. In *The Physiology of Fishes* (ed. D. H. Evans), pp. 441–88. CRC Press, Boca Raton, Florida.

Heatwole, H. (1970) Thermal ecology of the desert dragon, *Amphibolurus inermis. Ecol. Monogr.*, **40:** 425–57.

Heatwole, H. (1976) *Reptile Ecology.* University of Queensland Press, Queensland, Australia. 178pp.

Heatwole, H. (1984) Adaptations of amphibians to aridity. In *Arid Australia* (ed. H. G. Cogger and E. E. Cameron), pp. 177–222. Australian Museum, Sydney.

Hedenström, A. (1993) Migration and soaring or flapping flight in birds: the relative importance of energy cost and speed. *Phil. Trans. R. Soc. Lond.* B, **342:** 353–61.

Hedges, N. A., Gaskin, D. E. and Smith, G. J. D. (1979) Rencular morphology and renal vascular system of the Harbour porpoise *Phocoena phocoena* (L.). *Can. J. Zool.*, **57:** 868–75.

Heisinger, J. F. and Breitenbach, R. P. (1969) Renal structural characteristics as indexes of renal adaptation for water conservation in the genus *Sylvilagus. Physiol. Zool.*, **42:** 160–72.

Heldmaier, G. and Klingenspoor, M. (eds) (2000) *Life in the Cold.* Springer-Verlag, Berlin.

Heldmaier, G. and Ruf, T. (1992) Body temperature and metabolic rate during natural hypothermia in endotherms. *J. Comp. Physiol.* B, **162:** 696–706.

Henen, B. T. (1997) Seasonal and annual energy budgets of female desert tortoises (*Gopherus agassizii*). *Ecology*, **78:** 283–96.

Henen, B. T., Peterson, C. C., Wallis, I. R., Berry, K. H. and Nagy, K. A. (1998) Effects of climatic variation on field metabolism and water relations of desert tortoises. *Oecologia*, **117:** 365–73.

Henle, K. (1992) Predation pressure, food availability, thermal environment, and precision of thermoregulation in a desert population of the skink *Morethia boulengeri*, with a comment on measuring thermoregulatory precision. *Zool. Jahrb. Abt. Syst.*, **119:** 405–12.

Henzell, R. P. (1972) Adaptation to aridity in lizards of the *Egernia whitei* species group. PhD thesis, University of Adelaide.

Herd, R. M. (1985) Estimating food intake by captive emus, *Dromaius novaehollandiae*, by means of sodium-22 turnover. *Aust. Wildl. Res.*, **12:** 455–60.

Hertz, P. E., Huey, R. B. and Stevenson, R. D. (1993) Evaluating temperature regulation by field-active ectotherms: the fallacy of the inappropriate question. *Am. Nat.*, **142:** 796–818.

Hewitt, S. (1981) Plasticity of renal function in the Australian desert rodent, *Notomys alexis. Comp. Biochem. Physiol.*, **69A:** 297–304.

Hiebert, S. M., Remenofsky, M., Salvante, K., Wingfield, J. and Gass, C. L. (2000) Noninvasive methods for measuring and manipulating corticosterone in hummingbirds. *Gen. Comp. Endocrinol.*, **120:** 235–47.

Higgins, L. V. (1993) The nonannual, nonseasonal breeding cycle of the Australian sea lion, *Neophoca cinerea. J. Mammal.*, **74:** 270–4.
Hillyard, S. D. (1975) The role of antidiuretic hormones in the water economy of the Spadefoot toad, *Scaphiopus couchi. Physiol. Zool.*, **46:** 242–51.
Hillyard, S. D. (1976) The movement of soil water across the isolated amphibian skin. *Copeia*, **1976:** 314–20.
Hindell, M. A., Slip, D. J., Burton, H. R. and Bryden, M. M. (1992) Physiological implications of continuous and deep dives of the Southern elephant seal (*Mirounga leonina*). *Can. J. Zool.*, **70:** 370–9.
Hoff, K. S. and Hillyard, S. D. (1993) Inhibition of cutaneous absorption in dehydrated toads by Saralasin is associated with changes in barometric pressure. *Physiol. Zool.*, **66:** 89–98.
Hoffman, K. H., Neuhäuser, T., Gerstenlauer, B., Weidner, K. and Lenz, M. (1993) Thermal control of endocrine events in cricket reproduction (*Gryllus bimaculatus de Geer*). In *Perspectives in Comparative Endocrinology* (ed. K. G. Davey, R. E. Peter and S. S. Tobe), pp. 561–7. National Resources Council of Canada, Ottawa.
Holloway, J. C. and Geiser, F. (1995) Influence of torpor on daily energy expenditure of the dasyurid marsupial *Sminthopsis crassicaudata. Comp. Biochem. Physiol.*, **112A:** 59–66.
Holloway, J. C. and Geiser, F. (2001) Seasonal changes in the thermoenergetics of the marsupial sugar glider, *Petaurus breviceps. J. Comp. Physiol.* B, **171:** 643–50.
Holmes, W. N. and Phillips, J. G. (1985) The avian salt gland. *Biol. Rev.*, **60:** 213–56.
Hong, S. K., Elsner, R., Claybaugh, J. R. and Ronald, K. (1982) Renal functions of the Baikal seal *Pusa sibirica* and Ringed seal *Pusa hispida. Physiol. Zool.*, **55:** 289–99.
Horowitz, M. and Borut, A. (1970) Effect of acute dehydration on body fluid compartments in three rodent species: *Rattus norvegicus*, *Acomys cahirinus* and *Meriones crassus. Comp. Biochem. Physiol.*, **35:** 283–90.
Horowitz, M. and Borut, A. (1973) Blood volume regulation in dehydrated rodents: plasma colloid osmotic pressure, total osmotic pressure and electrolytes. *Comp. Biochem. Physiol.*, **44A:** 1261–5.
Horowitz, M. and Borut, A. (1975) Blood volume regulation in dehydrated rodents: plasma protein turnover and sedimentation coefficients. *Comp. Biochem. Physiol.*, **51A:** 827–31.
Horowitz, M. and Borut, A. (1994) The Spiny mouse (*Acomys cahirinus*) – a rodent prototype for studying plasma volume regulation during thermal dehydration. *Isr. J. Zool.*, **40:** 117–25.
Houser, D. S. and Costa, D. P. (2001) Protein catabolism in suckling and fasting northern elephant seal pups (*Microunga angustirostris*). *J. Comp. Physiol.* B, **171:** 635–42.
Hudson, J. W. (1981) Role of the endocrine glands in hibernation with special reference to the thyroid gland. In *Survival in the Cold: Hibernation and Other Adaptations* (ed. X. J. Musacchia and L. Jansky), pp. 33–54. Elsevier/North-Holland, New York.

Huey, R. B. (1982) Temperature, physiology, and the ecology of reptiles. In *Biology of the Reptilia* (ed. C. Gans and F. H. Pough), pp. 25–91. Academic Press, New York.

Huey, R. B. and Kingsolver, J. G. (1993) Evolution of resistance to high temperature in ectotherms. *Am. Nat.*, **142:** S21–46.

Huey, R. B., Pianka, E. R. and Schoener, T. W. (eds) (1983) *Lizard Ecology*. Harvard University Press, Cambridge, Massachusetts. 501pp.

Hughes, M. R. (1995) Responses of gull kidneys and salt glands to NaCl loading. *Can. J. Physiol. Pharmacol.*, **73:** 1727–32.

Hughes, M. R., Roberts, J. R. and Thomas, B. R. (1987) Total body water and its turnover in free-living nestling glaucous-winged gulls with a comparison of body water and water flux in avian species with and without salt glands. *Physiol. Zool.*, **60:** 481–91.

Hui, C. (1981) Seawater consumption and water flux in the common dolphin *Delphinus delphis*. *Physiol. Zool.*, **54:** 430–40.

Hulbert, A. J. (1987) Thyroid hormones, membranes and the evolution of endothermy. In *Advances in Physiological Research* (ed. H. McLennan, J. R. Ledsome, C. H. S. McIntosh and D. R. Jones), pp. 305–19. Plenum Publishing Corp, New York.

Hulbert, A. J. (2000) Thyroid hormones and their effects: a new perspective. *Biol. Rev.*, **75:** 519–631.

Hulbert, A. J. and Else, P. L. (1981) Comparison of the "mammal machine" and the "reptile machine": energy use and thyroid activity. *Am. J. Physiol.*, **241:** R350–6.

Hulbert, A. J. and Else, P. L. (1989) Evolution of mammalian endothermic metabolism: mitochondrial activity and cell composition. *Am. J. Physiol.*, **256:** R63–9.

Hulbert, A. J. and Else, P. L. (1990) The cellular basis of endothermic metabolism: a role for "leaky" membranes. *News Physiol. Sci.*, **5:** 25–8.

Hulbert, A. J. and Else, P. L. (1999) Membranes as possible pacemakers of metabolism. *J. Theor. Biol.*, **199:** 257–74.

Hulbert, A. J. and Else, P. L. (2000) Mechanisms underlying the cost of living in animals. *A. Rev. Physiol.*, **62:** 207–35.

Hume, I. D. (1999) *Marsupial Nutrition*. Cambridge University Press, Cambridge. 434pp.

Hutchinson, G. E. (1978) *An Introduction to Population Ecology*. Yale University Press, New Haven and London. 260pp.

Hytten, F. E. (1976) Is viviparity the best means of reproduction? *Acta Paed. Acad. Sci. Hung.*, **17:** 1–8.

Imbert, M. and de Rouffignac, C. (1976) Role of sodium and urea in the renal concentrating mechanism in *Psammomys obesus*. *Pflügers Arch.*, **361:** 107–14.

Izumi, Y., Sugiyama, F., Sugiyama, Y. and Yagami, K. (1993) Comparison between the blood from orbital sinus and heart in analyzing plasma biochemical values – increase of plasma enzyme values in the blood from orbital sinus. *Jikken Dobutsu*, **42:** 99–102.

Jakob, E. M., Marshall, S. D. and Uetz, G. W. (1996) Estimating fitness: a comparison of body condition indices. *Oikos*, **77:** 61–7.

Jallageas, M. and Assenmacher, I. (1986) Endocrine correlates of hibernation in the edible dormouse (*Glis glis*). In *Living in the Cold* (ed. H. C. Heller, X. J. Musaccia and L. C. H. Wang), pp. 265–72. Elsevier, New York.

Jamison, R. L., Roinel, N. and de Rouffignac, C. (1979) Urinary concentrating mechanism in the desert rodent *Psammomys obesus*. *Am. J. Physiol.*, **236:** F448–53.

Jennings, D. H., Moore, M. C., Knapp, R., Matthews, L. and Orchinick, M. (2000) Plasma steroid-binding globulin mediation of differences in stress reactivity in alternate male phenotypes in tree lizards, *Urosaurus ornatus*. *Gen. Comp. Endocrinol.*, **120:** 289–99.

Jensen, E. V. (1991) Overview of the nuclear receptor family. In *Nuclear Hormone Receptors* (ed. E. V. Jensen), pp. 1–13. Academic Press, New York.

Jolly, G. M. (1965) Explicit estimates for capture-recapture data with both death and dilution – stochastic model. *Biometry*, **52:** 225–47.

Jones, M. E. E., Bradshaw, S. D., Fergusson, B. and Watts, R. (1990) The effect of available surface water on levels of anti-diuretic hormone (lysine vasopressin) and water and electrolyte metabolism of the Rottnest Island quokka (*Setonix brachyurus*). *Gen. Comp. Endocrinol.*, **77:** 75–87.

Jones, R. M. (1980) Metabolic consequences of accelerated urea synthesis during seasonal dormancy of spadefoot toads, *Scaphiopus couchi* and *Scaphiopus multiplicatus*. *J. Exp. Zool.*, **212:** 255–67.

Jones, C. M. and Geiser, F. (1992) Prolonged and daily torpor in the feathertail glider, *Acrobates pygmaeus* (Marsupialia: Acrobatidae). *J. Zool., Lond.*, **227:** 101–8.

Jørgensen, B. C. C. B. P. (1992) Effects of arginine vasotocin, cortisol and adrenergic factors on water balance in the toad *Bufo bufo*: physiology or pharmacology. *Comp. Biochem. Physiol.*, **101A:** 709–16.

Jouventin, P., Capdeville, D., Cuenot-Chaillet, F. and Boiteau, C. (1994) Exploitation of pelagic resources by a non-flying seabird: satellite tracking of the King penguin throughout the breeding cycle. *Mar. Ecol. Prog. Ser.*, **106:** 11–19.

Jouventin, P. and Weimerskirch, H. (1990) Satellite tracking of Wandering albatrosses. *Nature (Lond.)*, **343:** 746–8.

Kam, M. and Degen, A. A. (1988) Water, electrolyte and nitrogen balances of fat sand rats (*Psammomys obesus*) when consuming the saltbush *Atriplex halimus*. *J. Zool., Lond.*, **215:** 453–62.

Kayser, C. (1961) *The Physiology of Natural Hibernation*. Pergamon Press, Oxford and London. 323pp.

Keast, A. (1959) Australian birds: their zoogeography and adaptations to an arid environment. In *Biogeography and Ecology in Australia* (ed. A. Keast, R. L. Crocker and C. S. Christian), pp. 89–114. Junk, The Hague.

Kenagy, G. J., Sharbaugh, S. M. and Nagy, K. A. (1989) Annual cycle of energy and time expenditure in a Golden-mantled ground squirrel population. *Oecologia*, **78:** 269–82.

Kinnear, J. E., Bromilow, R. M., Onus, M. L. and Sokolowski, R. E. S. (1988) The Bromilow trap: a new risk-free soft trap suitable for small to medium-sized macropodids. *Aust. Wildl. Res.*, **15:** 235–7.

Klewen, R. and Winter, H. G. (1987) Contribution to the deep cooling marking of amphibians in the field. *Salamandra*, **23:** 159–65.

Koehn, R. K. and Bayne, B. L. (1989) Towards a physiological and genetical understanding of the energetics of the stress response. *Biol. J. Linn. Soc.*, **37:** 157–71.

Kohel, K. A., MacKenzie, D. S., Rostal, D. C., Grumbles, J. S. and Lance, V. A. (2001) Seasonality in plasma thyroxine in the desert tortoise, *Gopherus agassizii. Gen. Comp. Endocrinol.*, **121:** 214–22.

Kooyman, G. L. (1998) The physiological basis of diving to depth: birds and mammals. *A. Rev. Physiol.*, **60:** 19–32.

Kooyman, G. L. and Campbell, W. B. (1972) Heart rates in freely diving Weddell seals. *Comp. Biochem. Physiol.*, **43A:** 31–6.

Kooyman, G. L., Castellini, M. A., Davis, R. W. and Maue, R. A. (1983) Aerobic dive limits of immature Weddell seals. *J. Comp. Physiol.* B, **151:** 171–4.

Kooyman, G. L., Davis, R. W., Croxall, J. P. and Costa, D. P. (1982) Diving depths and energy requirements of king penguins. *Science*, **217:** 726–7.

Kooyman, G. L. and Draber, C. M. (1968) Observations on milk, blood and urine constituents of the Weddell seal. *Physiol. Zool.*, **41:** 187–94.

Kooyman, G. L., Ponganis, P. J., Castellini, M. A., Ponganis, E. P., Ponganis, K. V., Thorson, P. H., Eckert, S. A. and LeMaho, Y. (1992) Heart rates and swim speeds of Emperor penguins diving under sea ice. *J. Exp. Biol.*, **165:** 161–80.

Kooyman, G. L., Wahrenbrock, E. A., Castellini, M. A., Davis, R. W. and Sinnett, E. E. (1980) Aerobic and anaerobic metabolism during voluntary diving in Weddell seals: evidence for preferred pathways from blood chemistry and behaviour. *J. Comp. Physiol.* B, **138:** 335–46.

Kotiaho, J. S. (1999) Estimating fitness: comparison of body condition indices revisited. *Oikos*, **87:** 399–400.

Krebs, C. J. (1999) *Ecological Methodology*. Adddison-Wesley Educational Publishers, San Francisco, California.

Krebs, C. J. (2001) *Ecology: The Experimental Analysis of Distribution and Abundance*. Addison-Wesley Educational Publishers, San Francisco, California. 695pp.

Krebs, C. J. and Singelton, G. R. (1993) Indices of condition for small mammals. *Aust. J. Zool.*, **41:** 317–23.

Krogh, A. (1939) *Osmotic Regulation in Aquatic Animals*. Cambridge University Press, Cambridge. 242pp.

Kuchling, G. (1999) *The Reproductive Biology of the Chelonia.* Springer-Verlag, Heidelberg. 223pp.

Kuchling, G., Burbidge, A. A., Bradshaw, S. D. and DeJose, J. (1988) News of *Pseudemydura umbrina. IUCN Tortoise and Fresh-Water Turtle Specialist Group Newsletter*, vol. 3, pp. 2–3.

Kuchling, G., DeJose, J., Burbidge, A. A. and Bradshaw, S. D. (1992) Beyond captive breeding: the Western Swamp Tortoise, *Pseudemydura umbrina*, recovery programme. *Int. Zool. Ybk*, **31:** 37–41.

Kuhn, T. S. (1962) *The Structure of Scientific Revolutions.* The University of Chicago Press, Chicago. 210pp.

Kuhn, T. S. (1976) *The Copernican Revolution: Planetary Astronomy in the Development of Western Thought.* Havard University Press, Cambridge, Massachusetts. 297pp.

Lacas, S., Allevard, A. M., Ag'Atteinine, S., Gallo-Bona, N., Gauquelin-Koch, G., Hardin-Pouzet, H., Gharib, C., Sicard, B. and Maurel, D. (2000) Cardiac Natriuretic Peptide respones to water restriction in the hormonal adaptation of two semidesert rodents from West Africa (*Steatomys caurinus, Taterillus gracilis). Gen. Comp. Endocrinol.*, **120:** 176–89.

Lachiver, F. (1964) Thyroid activity in the Garden dormouse (*Eliomys quercinus* L.) studied from June to November. *Ann. Acad. Sci. Fenn.*, **71:** 285–94.

Lachiver, F., Cheniti, T., Bradshaw, S. D., Berthier, J. L. and Petter, F. (1978) Field studies in southern Tunisia on water turnover and thyroid activity in two species of *Meriones.* In *Environmental Endocrinology* (ed. I. Assenmacher and D. S. Farner), pp. 81–4. Springer-Verlag, Berlin.

Lasley, B. L. and Kirkpatrick, J. F. (1991) Monitoring ovarian function in captive and free-ranging wildlife by means of urinary and faecal steroids. *J. Zoo Wildl.*, **22:** 23–31.

Laurance, W. F. (1992) Abundance estimates of small mammals in Australian tropical rainforest: a comparison of four trapping methods. *Wildl. Res.*, **19:** 651–5.

Le Boef, B. J., Naito, Y., Huntley, A. C. and Asaga, T. (1989) Prolonged, continuous, deep diving by Northern elephant seals. *Cana. J. Zool.*, **67:** 2514–19.

Le Maho, Y. (1976) Thermorégulation et jeûne chez le Manchot Empereur et le Manchot Royal. Thèse de Doctorat, Laboratoire de Thermorégulation du CNRS, Université Claude-Bernard, Lyon. 109pp.

Le Maho, Y. (1984) Adaptations métaboliques au jeûne prolongé chez les oiseaux et les mammifères. *Bull. Soc. Ecophysiol. (Paris)*, **9:** 129–48.

Le Maho, Y., Delclitte, P. and Chatonnet, J. (1976) Thermoregulation in fasting Emperor penguins under natural conditions. *Am. J. Physiol.*, **231:** 913–22.

Lee, A. K. (1968) Water economy of the burrowing frog, *Heleioporus eyrei* (Gray). *Copeia*, **1968:** 741–5.

Lee, A. K., Bradley, A. J. and Braithwaite, R. W. (1977) Corticosteroid levels and male mortality in *Antechinus stuartii.* In *The Biology of Marsupials* (ed. B. Stonehouse and B. Gilmour), pp. 209–20. Macmillan, London.

Lee, A. K. and Cockburn, A. (1985) *Evolutionary Ecology of Marsupials.* Cambridge University Press, Cambridge. 274pp.

Lee, A. K. and Mercer, E. H. (1967) Cocoon surrounding desert-dwelling frogs. *Sciences, N. Y.*, **157:** 87–8.

Lee, A. K. and McDonald, I. R. (1985) Stress and population regulation in small mammals. *Oxford Rev. Reprod. Biol.*, pp. 261–304. Oxford University Press. Oxford.

Lee, A. K., Woolley, P. and Braithwaite, R. W. (1982) Life history strategies of dasyurid marsupials. In *Carnivorous Marsupials* (ed. M. Archer), pp. 1–11. Royal Zoological Society of New South Wales, Sydney, Australia.

Lee, J. C. (1980) Comparative thermal ecology of two lizards. *Oecologia*, **44:** 171–6.

Lee, P. and Schmidt-Nelsen, K. (1971) Respiratory and cutaneous evaporation in the Zebra finch: effect on water balance. *Am. J. Physiol.*, **220:** 1598–605.

Leslie, P. H. (1952) The estimation of population parameters from data obtained by means of the capture-recapture method. *Biometry*, **39:** 363–88.

Levitt, J. (1980) *Responses of Plants to Environmental Stress.* Academic Press, New York.

Licht, P. (1964) A comparative study of the thermal dependence of contractility in saurian skeletal muscle. *Comp. Biochem. Physiol.*, **13:** 27–34.

Licht, P., Denver, R. J. and Herrera, B. E. (1991) Comparative survey of blood thyroxine binding proteins in turtles. *J. Exp. Zool.*, **259:** 43–52.

Lifson, N., Gordon, G. B. and McClintock, R. (1955) Measurement of total carbon dioxide production by means of $D_2{}^{18}O$. *J. Appl. Physiol.*, **7:** 704–10.

Lifson, N. and McClintock, R. (1966) Theory of use of the turnover rates of body water for measuring energy and material balance. *J. Theor. Biol.*, **12:** 46–74.

Lincoln, F. C. (1930) Calculating waterfowl abundance on the basis of banding returns. *U.S.D.A. Circular*, **118:** 1–4.

Lockyer, C. (1997) Diving behaviour of the Sperm whale in relation to feeding. In *Sperm Whale Deaths in the North Sea* (ed. T. G. Jacques and R. H. Lambertson) (*Bull. Inst. R. Sci. Nat. Belg. Biol.*, **67**), pp. 47–52.

Louw, G. N. and Seely, M. K. (1982) *Ecology of Desert Organisms.* Longman, London and New York. 194pp.

Lovegrove, B. G. (2000) Daily heterothermy in mammals: coping with unpredictable environments. In *Life in the Cold: Eleventh International Hibernation Symposium* (ed. G. Heldmaier and M. Klingenspor), pp. 29–40. Springer-Verlag, Berlin.

Lovegrove, B. G., Raman, J. and Perrin, M. R. (2001) Daily torpor in elephant shrews (Macroscelidae: *Elephantulus* spp.) in response to food deprivation. *J. Comp. Physiol.* B, **171:** 11–21.

Low, B. S. (1978) Environmental uncertainty and the parental strategies of marsupials and placentals. *Am. Nat.*, **112:** 197–213.

Lyman, C. P. (1978) Natural torpidity, problems and perspectives. In *Strategies in Cold: Natural Torpidity and Thermogenesis* (ed. L. C. H. Wang and J. W. Hudson), pp. 9–19. Academic Press, New York.

Macfarlane, W. V., Kinne, R., Walmsley, C. M., Siebert, B. D. and Peter, D. (1967) Vasopressins and the increase of water and electrolyte excretion by sheep, cattle and camels. *Nature (Lond.)*, **214:** 979–81.

Macfarlane, W. V., Morris, R. J. H. and Howard, B. (1962) Water metabolism of merino sheep and camels. *Aust. J. Sci.*, **25:** 112–16.

Macfarlane, W. V., Morris, R. J. H. and Howard, B. (1963) Turnover and distribution of water in desert camels, sheep, cattle and kangaroos. *Nature (Lond.)*, **197:** 270–1.

Maclean, G. L. (1996) *The Ecophysiology of Desert Birds*. Springer, Berlin, New York. xi + 181pp.

MacLean, G. S., Lee, A. K. and Wilson, K. J. (1973) A simple method of obtaining blood from lizards. *Copeia*, **1973:** 338–9.

MacMillen, R. E. and Lee, A. K. (1967) Australian desert mice: independence of exogenous water. *Sciences, N.Y.*, **158:** 383–5.

Maddison, W. P. (1991) Squared-change parsimony reconstructions of ancestral states for continuous valued characters in a phylogenetic tree. *Syst. Zool.*, **40:** 304–14.

Maetz, J. (1971) Fish gills: mechanisms of salt transfer in fresh water and sea water. *Phil. Trans. R. Soc. Lond.* B, **262:** 209–49.

Magnusson, W. E. (1989) Ratios, statistics, and physiological models: comment on Packard and Boardman. *Physiol. Zool.*, **62:** 997–1000.

Main, A. R. (1961) The occurrence of Macropodidae on islands and its climatic and ecological implications. *J. R. Soc. W. A.*, **44:** 84–9.

Main, A. R. and Bakker, H. R. (1981) Adaptation of macropod marsupials to aridity. In *Ecological Biogeography of Australia* (ed. A. Keast), pp. 1490–519. Dr. W. Junk, The Hague.

Malan, A. (1993) Temperature regulation, enzyme kinetics, and metabolic depression in mammalian hibernation. In *Life in the Cold: Ecological, Physiological and Molecular Mechanisms* (ed. H. V. Carey), pp. 241–51. Westview Press, Boulder, Colorado.

Malan, A. (1996) The origins of hibernation: a reappraisal. In *Adaptations to the Cold* (ed. F. Geiser, A. J. Hulbert and S. C. Nicol), pp. 1–6. University of New England Press, Armidale, NSW.

Malvin, R. L., Bonjour, J. P. and Ridgway, S. (1971) Antidiuretic hormone levels in some cetaceans. *Proc. Soc. Exp. Biol. Med.*, **136:** 1203–5.

Malvin, R. L., Ridgway, S. and Cornell, L. H. (1978) Renin and aldosterone levels in dolphins and sea lions. *Proc. Soc. Exp. Biol. Med.*, **157:** 665–8.

Manley, B. J. F. and Parr, M. J. (1968) A new method of estimating population size, survivorship and birth rate from capture-recapture data. *Trans. Soc. Brit. Ent.*, **18:** 81–9.

Marder, J. and Gavrieli-Levin, I. (1987) The heat-acclimated pigeon: an ideal physiological model for a desert bird. *J. Appl. Physiol.*, **62:** 952–8.

Marder, J., Withers, P. C. and Philpot, G. (2003) Patterns of cutaneous water evaporation in Australian pigeons. *Isr. J. Zool.*, (in press).

Martin, A. R., Smith, T. G. and Cox, O. P. (1993) Studying the behaviour and movements of high Arctic belugas with satellite telemetry. In *Marine Mammals: Advances in Behavioural and Population Biology* (ed. I. L. Boyd), pp. 195–210. Clarendon Press, Oxford, UK.

Martins, E. P., Diniz-Filho, J. A. F. and Housworth, E. A. (2002) Adapative constraints and the phylogenetic comparative method: a computer simulation test. *Evolution*, **56:** 1–13.

Mason, J. W. (1975) A historical view of the stress field. *J. Hum. Stress.*, **1:** 6–12.

McCarron, H. C. K., Buffenstein, R., Fanning, F. D. and Dawson, T. (2001) Free-ranging heart rate, body temperature and energy metabolism in eastern grey kangaroos (*Macropus giganteus*) and red kangaroos (*Macropus rufus*) in the arid regions of South East Australia. *J. Comp. Physiol.* B, **171:** 401–11.

McClanahan, L. L. (1964) Osmotic tolerance of the muscles of two desert inhabiting toads, *Bufo cognatus* and *Scaphiopus couchii. Cell Physiol. Biochem.*, 12: 501.

McClanahan, L. L. (1967) Adaptations of the spadefoot toad, *Scaphiopus couchii*, to desert environments. *Comp. Biochem. Physiol.*, **20:** 73–99.

McClanahan, L. L. (1972) Changes in body fluids of burrowed spadefoot toads as a function of soil water potential. *Copeia*, **1972:** 209–16.

McClanahan, L. L. and Baldwin, R. (1969) Rate of water uptake through the integument of the desert toad, *Bufo punctatus. Cell Physiol. Biochem.*, **28:** 381–9.

McClanahan, L. L., Shoemaker, V. H. and Ruibal, R. (1976) Structure and function of the cocoon of a ceratophryd frog. *Copeia*, **1976:** 179–85.

McConnell, B. J., Chambers, C. and Fedak, M. A. (1992) Foraging ecology of southern Elephant seals in relation to bathymetry and productivity of the Southern Ocean. *Antarctic Sci.*, **4:** 393–8.

McConnell, B. J. and Fedak, M. A. (1996) Movements of southern Elephant seals. *Can. J. Zool.*, **74:** 1485–96.

McDonald, I. R., Lee, A. K., Than, K. A. and Martin, R. W. (1986) Failure of glucocorticoid feedback in males of a population of small marsupials (*Antechinus swaisonii*) during the period of mating. *J. Endocrinol.*, **108:** 63–8.

McDonald, I. R. and Bradshaw, S. D. (1993) Adrenalectomy and steroid replacement in a small macropodid marsupial, the quokka (*Setonix brachyurus*): metabolic and renal effects. *Gen. Comp. Endocrinol.*, **90:** 64–77.

McDonald, L., Bradshaw, S. D. and Gardner, A. (2003) Legal protection of fauna habitat in Western Australia. *Environmental and Planning Law Journal.* (submitted).

McEwen, B. S. (1998a) Protecting and damaging effects of stress mediators. *New Engl. J. Med.*, **338:** 171–9.

McEwen, B. S. (1998b) Stress, adaptation, and disease: allostasis and allostatic load. *Ann. N. Y. Acad. Sci.*, **840:** 33–44.

McGinnis, S. M. and Dickson, L. L. (1967) Thermoregulation in the desert iguana *Dipsosaurus dorsalis. Sciences, N.Y.*, **156:** 1757–9.

McGinnis, S. M. and Falkenstein, M. (1971) Thermoregulatory behavior in three sympatric species of iguanid lizards. *Copeia*, **1971:** 552–4.

Medica, P. A., Bury, R. B. and Luckenbach, R. (1980) Drinking and construction of water catchments by the desert tortoise *Gopherus agassizii*, in the Mojave desert. *Herpetologica*, **36:** 301–4.

Miller, T. and Bradshaw, S. D. (1979) Adrenocortical function in a field population of a macropodid marsupial (*Setonix brachyurus*, Quoy & Gaimard). *J. Endocrinol.*, **82:** 159–70.

Mills, H. and Bencini, R. (2000) New evidence for facultative male die-off in island populations of dibblers, *Parantechinus apicalis. Aust. J. Zool.*, **48:** 501–10.

Minnich, J. E. (1977) Adaptive responses in the water and electrolyte budgets of native and captive desert tortorises,*Gopherus agassizi*, to chronic drought. *Desert Tortoise Council Symp. Proc.*, 1977, pp. 102–29.

Minnich, J. E. and Ziegler, M. R. (1976) Comparison of field water budgets in the tortoises *Gopherus agassizzii* and *Gopherus polyphemus. Am. Zool.*, **16:** 219.

Moberly, W. R. (1968) The metabolic responses of the common iguana, *Iguana iguana* to activity and restraint. *Comp. Biochem. Physiol.*, **27:** 1–20.

Monod, T. (1973) *Les Déserts.* Editions Horizons de France, Paris. 247pp.

Morafka, D. J. (1994) Neonates: missing links in the life history of North American tortoises. In *Biology of North American Tortoises* (ed. R. B. Bury and D. J. Germano), pp. 161–73. National Biological Survey: Fisheries and Wildlife Research, no. 13. Washington, DC.

Moreno, J. and Sanz, J. J. (1996) Field metabolic rates of breeding Chinstrap penguins (*Pygoscelis antarctica*) in the South Shetlands. *Physiol. Zool.*, **69:** 586–98.

Moro, D. (2000) Kidney structure in two species of arid zone rodent: Lakeland Downs Short-Tailed mouse, *Leggadina lakedownensis*, and the House mouse, *Mus domesticus. Aust. Mammal.*, **21:** 251–5.

Moro, D. and Bradshaw, S. D. (1999) Comparative water and sodium balance, and metabolic physiology, of House mice and Short-tailed mice under laboratory conditions. *J. Comp. Physiol.* B, **169:** 538–48.

Moro, D. and Morris, K. D. (2000) Movements and refugia of Lakeland Downs short-tailed mice, *Leggadina lakedownensis*, and house mice, *Mus domesticus*, on Thevanard Island, Western Australia. *Wildl. Res.*, **27:** 11–20.

Morton, S. R. and MacMillen, R. E. (1982) Seeds as sources of preformed water for desert-dwelling granivores. *J. Arid Envts*, **8:** 235–43.

Motais, R. (1967) Les mécanismes d'échanges d'ioniques branchiaux chez les teleostéens. *Ann. Inst. Oceanogr. Monaco*, **45:** 1083.

Muchlinski, A. E., Hogan, J. M. and Stoutenburgh, R. J. (1990) Body temperature regulation in a desert lizard, *Sauromalus obesus*, under undisturbed field conditions. *Comp. Biochem. Physiol.*, **95A:** 579–83.

Muchlinski, A. E., Stoutenburgh, R. J. and Hogan, J. M. (1989) Fever response in laboratory maintained and free-ranging chuckwallas (*Sauromalus obesus*). *Am. J. Physiol.*, **257:** R150–5.

Mullen, R. K. (1970) Respiratory metabolism and body water turnover rates of *Perognathus formosus* in its natural environment. *Comp. Biochem. Physiol.*, **32:** 259–65.

Mullen, R. K. (1971) Energy metabolism and body water turnover rates of two species of free-living kangaroo rats, *Dipodomys merriami* and *Dipodomys microps. Comp. Biochem. Physiol.*, **39A:** 379–90.

Munck, A. U., Guyre, P. M. and Holbrook, N. J. (1984) Physiological functions of glucocorticoids in stress and their relation to pharmacological actions. *Endocr. Rev.*, **5:** 25–44.

Murphy, R. C. (1936) *Oceanic Birds of South America*. Macmillan, New York.

Murton, R. K. and Westwood, N. J. (1977) *Avian Breeding Cycles*. Clarendon Press, Oxford, UK.

Myers, R. D., Veale, W., L. and Yaksh, T. L. (1971) Change in body temperature of the unanaesthetized monkey produced by sodium and calcium ions perfused through the cerebral ventricles. *J. Physiol. (Lond.)*, **217:** 381–92.

Nagy, K. A. (1972) Water and electrolyte budgets of a free-living desert lizard, *Sauromalus obesus. J. Comp. Physiol.* B., **79:** 39–62.

Nagy, K. A. (1973) Behavior, diet and reproduction in a desert lizard, *Sauromalus obesus. Copeia*, **1973:** 93–102.

Nagy, K. A. (1980) CO_2 production in animals: analysis of potential errors in the doubly labeled water method. *Am. J. Physiol.*, **238:** R466–73.

Nagy, K. A. (1982a) Field studies of water relations. In *Biology of the Reptilia* (ed. C. Gans and H. Pough), pp. 483–501. Academic Press, New York.

Nagy, K. A. (1982b) Energy requirements of free-living iguanid lizards. In *Iguanas of the World: Their Behavior, Ecology, and Conservation* (ed. G. M. Burghardt and A. S. Rand), pp. 45–59. Noyes Publications, Park Ridge, New Jersey.

Nagy, K. A. (1983) *The Doubly Labeled Water Method: A Guide to its Use.* (*Misc. Pub. Univ. Calif.*) University of California Press. 45pp.

Nagy, K. A. (1987a) How do desert animals get enough water? In *Progress in Desert Research* (ed. L. Berkofsky and M. G. Wurtele), pp. 89–98. Rowman & Littlefield, Totowa, New Jersey.

Nagy, K. A. (1987b) Field metabolic rate and food requirement scaling in mammals and birds. *Ecol. Monogr.*, **57:** 111–28.

Nagy, K. A. (1988a) Seasonal patterns of water and energy balance in desert vertebrates. *J. Arid Envts*, **14:** 201–10.

Nagy, K. A. (1988b) Energetics of desert reptiles. In *Ecophysiology of Desert Vertebrates* (ed. P. K. Gosh and I. Prakash), pp. 166–86. Scientific Publishers, Jodhpur, India.

Nagy, K. A. (1992) The doubly labeled water method in ecological energetics studies of terrestrial vertebrates. *Bull. Soc. Ecophysiol. (Paris)*, Suppl. **17:** 9–14.

Nagy, K. A. (1994a) Seasonal water, energy and food use by free-living, arid-habitat mammals. *Aust. J. Zool.*, **42:** 55–63.

Nagy, K. A. (1994b) Field bioenergetics of mammals: what determines field metabolic rates? *Aust. J. Zool.*, **42:** 43–53.

Nagy, K. A. (2000) Energy costs of growth in neonate reptiles. *Herpet. Monogr.*, **14:** 378–87.

Nagy, K. A. (2001) Food requirements of wild animals: predictive equations for free-living mammals, reptiles and birds. *Nutr. Abstr. Rev.*, **B71:** 21R–31R.

Nagy, K. A. and Bradshaw, S. D. (1995) Energetics, osmoregulation and food consumption by free-living desert lizards, *Ctenophorus (=Amphibolurus) nuchalis. Amphib. Rept.*, **16:** 25–35.

Nagy, K. A. and Bradshaw, S. D. (2000) Scaling of energy and water fluxes in free-living arid-zone Australian marsupials. *J. Mammal.*, **81:** 962–70.

Nagy, K. A., Clarke, B. C., Seely, M. K., Mitchell, D. and Lighton, J. R. B. (1991) Water and energy balance in Namibian desert sand-dune lizards *Anglosaurus skoogi* (Anderson, 1916). *Funct. Ecol.*, **5:** 731–9.

Nagy, K. A. and Costa, D. P. (1980) Water flux in animals: analysis of potential errors in the tritiated water method. *Am. J. Physiol.*, **238:** R454–65.

Nagy, K. A., Girard, I. A. and Brown, T. K. (1999) Energetics of free-ranging mammals, reptiles, and birds. *A. Rev. Nutrit.*, **19:** 247–77.

Nagy, K. A. and Gruchacz, M. J. (1994) Seasonal water and energy metabolism of the desert-dwelling kangaroo rat (*Dipodomys merriami*). *Physiol. Zool.*, **67:** 1461–78.

Nagy, K. A., Kooyman, G. L. and Ponganis, P. J. (2001) Energetic cost of foraging in free-diving Emperor penguins. *Physiol. Biochem. Zool.*, **74:** 541–7.

Nagy, K. A., Lee, A. K., Martin, R. W. and Fleming, M. R. (1988) Field metabolic rate and food requirement of a small dasyurid marsupial, *Sminthopsis crassicaudata. Aust. J. Zool.*, **36:** 293–9.

Nagy, K. A. and Medica, P. A. (1986) Physiological ecology of desert tortoises in southern Nevada. *Herpetologica*, **42:** 73–92.

Nagy, K. A., Meienberger, C., Bradshaw, S. D. and Wooller, R. D. (1995) Field metabolic rate of a small marsupial mammal, the Honey Possum (*Tarsipes rostratus*). *J. Mammal.*, **76:** 862–6.

Nagy, K. A., Morafka, D. J. and Yates, R. A. (1997) Young desert tortoise survival: energy, water, and food requirements in the field. *Chelon. Cons. Biol.*, **2:** 396–404.

Nagy, K. A. and Obst, B. S. (1992) Food and energy requirements of Adélie penguins (*Pygoscelis adelie*) on the Antarctic peninsula. *Physiol. Zool.*, **65:** 1271–84.

Nagy, K. A. and Peterson, C. C. (1988) Scaling of water flux rate in animals. *Univ. Calif. Publ. Zool.*, **120:** 1–172.

Nagy, K. A., Seymour, R. S., Lee, A. K. and Braithwaite, R. (1978) Energy and water budgets in free-living *Antechinus stuartii* (Marsupialia: Dasyuridae). *J. Mammal.*, **59:** 60–8.

Nagy, K. A., Shoemaker, V. H. and Costa, W. R. (1976) Water, electrolyte and nitrogen budgets of jackrabbits (*Lepus californicus*) in the Mojave Desert. *Physiol. Zool.*, **49:** 351–63.

Nagy, K. A., Siegfried, W. R. and Wilson, R. P. (1984) Energy utilisation by free-ranging Jackass penguins, *Spheniscus demersus. Ecology*, **65:** 1648–55.

Nelson, R. A. (1980) Protein and fat metabolism in hibernating bears. *Fed. Proc.*, **123:** 892–4.

Newsome, A. E. (1975) An ecological comparison of the two arid-zone kangaroos of Australia, and their prosperity since the introduction of ruminant stock to their environment. *Quart. Rev. Biol.*, **50:** 389–424.

Nicholls, D. G., Murray, M. D., Butcher, E. and Moore, P. (1997) Weather systems determine the non-breeding distribution of Wandering albatrosses over Southern Oceans. *Emu*, **97:** 240–4.

Nicol, S. C. and Andersen, N. A. (2000) Patterns of hibernation of echidnas in Tasmania. In *Life in the Cold: Eleventh International Hibernation Symposium* (ed. G. Heldmaier and M. Klingenspor), pp. 21–8. Springer-Verlag, Berlin.

Nicol, S. C. and Anderson, N. A. (1996) Hibernation in the echidna: not an adaptation to cold? In *Adaptations to the Cold* (ed. F. Geiser, A. J. Hulbert and S. C. Nicol), pp. 7–12. University of New England Press, Armidale, NSW.

Obst, B. S., Nagy, K. A. and Ricklefs, R. E. (1987) Energy utilization by Wilson's Storm-petrel (*Oceanites oceanicus*). *Physiol. Zool.*, **60:** 200–10.

Oksche, A., Farner, D. S., Serventy, D. L., Wolff, F. and Nicholls, C. A. (1963) The hypothalamo-hypophysial neurosecretory system of the zebra finch, *Taeniopygia castanotis. Z. Zellforsch.*, **58:** 846–914.

O'Reilly, K. M. and Wingfield, J. C. (2001) Ecological factors underlying adrenocortical response to capture stress in Arctic-breeding shorebirds. *Gen. Comp. Endocrinol.*, **124:** 1–11.

Ortiz, R. M., Adams, S. H., Costa, D. P. and Ortiz, C. L. (1996) Plasma vasopressin levels and water conservation in fasting, postweaned Northern elephant seal pups (*Mirounga angustirostris*). *Mar. Mamm. Sci.*, **12:** 99–106.

Ortiz, R. M., Worthy, G. A. J. and Mackenzie, D. S. (1998) Osmoregulation in wild and captive West Indian manatees (*Trichechus manatus*). *Physiol. Zool.*, **71:** 449–57.

Packard, G. C. and Boardman, T. J. (1987) The misuse of ratios to scale physiological data that vary allometrically with body size. In *New Directions in Ecological Physiology* (ed. M. E. Feder, A. F. Bennett, W. W. Burggren and R. B. Huey), pp. 216–39. Cambridge University Press, Cambridge, New York.

Packard, G. C. and Boardman, T. J. (1988) The misuse of ratios, indices, and percentages in ecophysiological research. *Physiol. Zool.*, **61:** 1–9.

Packer, W. C. (1963) Dehydration, hydration, and burrowing behavior in *Heleioporus eyrei* (Gray). *Ecology*, **44:** 643–51.

Palmer, A. R. and Stobexk, C. (1986) Fluctuating asymmetry: measurement, analysis, patterns. *A. Rev. Ecol. Syst.*, **17:** 391–421.

Parmenter, C. A., Yates, T. L., Parmenter, R. R., Mills, J. N., Childs, J. E., Campbell, M. L., Dunnum, J. L. and Milner, J. (1998) Small mammal survival and trapability in mark-recapture monitoring programs for hantavirus. *J. Wildl. Dis.*, **34:** 1–12.

Parsons, P. A. (1990) Fluctuating asymmetry: an epigenetic measure of stress. *Biol. Rev.*, **65:** 131–45.

Peaker, M. (1971) Avian salt glands. *Phil. Trans. R. Soc. Lond.* B, **262:** 289–300.

Peaker, M. and Linzell, L. J. (1975) *Salt Glands in Birds and Reptiles.* Cambridge University Press, Cambridge. 307pp.

Pearce, A. F. (1991) Eastern boundary currents of the southern hemisphere. *J. R. Soc. W. A.*, **74:** 35–45.

Pearce, A. F. and Walker, D. I. (1991) The Leeuwin current: an influence on the coastal climate and marine life of Western Australia. *J. Roy. Soc. W. A.*, **74:** 1–140.

Pennycuik, C. J. (1982) The flight of petrels and albatrosses (Procellariiformes) observed in South Georgia and its vicinity. *Proc. R. Soc. Lond.* B, **300:** 75–106.

Peters, E. L. (1996) Estimating energy metabolism of goldfish (*Carassius auratus*) and southern toads (*Bufo terrestris*) from ^{86}Rb elimination rates. *Copeia*, **4:** 791–804.
Peters, E. L., Shawki, A. I., Tracy, C. R., Whicker, F. W. and Nagy, K. A. (1995) Estimation of the metabolic rate of the desert iguana (*Dipsosuarus dorsalis*) by a radionuclide technique. *Physiol. Zool.*, **68:** 316–41.
Peters, R. H. (1983) *The Ecological Implications of Body Size*. Cambridge University Press, Cambridge. 329pp.
Peterson, C. C. (1995) Anhomeostasis: water and solute relations in two populations of the desert tortoise (*Gopherus agassizii*) during chronic drought. *Physiol. Zool.*, **69:** 1324–58.
Peterson, M. E., Krieger, D. T., Drucker, W. D. and Halmi, N. S. (1982) Immunocytochemical study of the hypophysis in 25 dogs with pituitary-dependent hyperadrenocorticism. *Acta Endocrinol. (Copenh.)*, **101:** 15–24.
Petter, F., Lachiver, F. and Chekir, R. (1984) Les adaptations des rongeurs Gerbilidés à la vie dans les régions arides. *Bull. Soc. Bot. Fr.*, **131:** 365–73.
Pettit, T. N., Nagy, K. A., Ellis, H. I. and Whittow, G. C. (1988) Incubation energetics of the Laysan albatross. *Oecologia*, **74:** 546–50.
Pietruszka, R. D. (1987) Maxithermy and the thermal biology of an herbivorous sand dune lizard. *J. Arid Envts*, **14:** 175–85.
Pinshow, B., Fedak, M. A., Battles, D. R. and Schmidt-Nielsen, K. (1976) Energy expenditure for thermoregulation and locomotion in Emperor penguins. *Am. J. Physiol.*, **231:** 903–12.
Ponganis, E. P. and Kooyman, G. L. (2000) The diving physiology of birds: a history of studies on polar species. *Comp. Biochem. Physiol.*, **126A:** 143–51.
Ponganis, E. P., Kooyman, G. L. and Castellini, M. A. (1993) Determinants of the aerobic dive limit of Weddell seals: analysis of diving metabolic rates, post-dive end tidal PO_2s, and blood and muscle oxygen stores. *Physiol. Zool.*, **66:** 732–49.
Ponganis, E. P., Kooyman, G. L., Starke, L. N., Kooyman, C. A. and Kooyman, T. G. (1997) Post-dive blood lactate concentrations in Emperor penguins, *Aptenodytes forsteri*. *J. Exp. Biol.*, **200:** 1623–6.
Ponganis, P. J., Kooyman, G. L., van Dam, R. and LeMaho, Y. (1999) Physiological responses of king penguins during simulated diving to 136 m depth. *J. Exp. Biol.*, **202:** 2819–22.
Popovic, V. (1960) Endocrines in hibernation. *Bull. Mus. Comp. Zool.*, **124:** 105–30.
Poppitt, S. D., Speakman, J. R. and Racey, P. A. (1993) The energetics of reproduction in the common shrew, *Sorex araneus*. *Physiol. Zool.*, **66:** 964–82.
Pough, F. H. (1980) The advantages of ectothermy for tetrapopds. *Am. Nat.*, **115:** 92–112.
Pough, F. H. (1983) Amphibians and reptiles as low-energy systems. In *Behavioral Energetics: the Cost of Survival in Vertebrates* (ed. W. P. Aspey and S. I. Lustick), pp. 141–88. Academic Press, New York.

Pravosudov, V. V., Kitaysky, A. S., Wingfield, J. and Clayton, N. S. (2001) Long-term unpredictable foraging conditions and physiological stress response in Mountain Chickadees (*Poecile gambeli*).*Gen. Comp. Endocrinol.*, **123:** 324–31.

Prévost, J. (1961) *Écologie du Manchot Empereur*. Hermann, Paris. 294pp.

Prévost, J. and Vilter, V. (1963) Histologie de la sécretion oesophagienne du Manchot empereur. *Proc. 13th Intern. Ornithol. Congr., Ithaca.* pp. 1085–94.

Purohit, K. G. (1974) Observations on size and relative medullary thickness in kidneys of some Australian mammals and their ecophysiological appraisal. *Z. Angew. Zool.*, **4:** 495–505.

Rankin, C. J. and Davenport, J. A. (1981) *Animal Osmoregulation*. Blackie, Glasgow, London. 202pp.

Rautenstrauch, K. R., Rager, A. L. H. and Rekestraw, D. L. (1998) Winter behavior of desert tortoises in southcentral Nevada. *J. Wildl. Man.*, **62:** 98–104.

Reed, J. Z., Butler, P. J. and Fedak, M. A. (1994) The metabolic characteristics of the locomotory muscles of grey seals (*Halichoerus grypus*), harbour seals (*Phoca vitulina*) and Antarctic fur seals (*Arctocephalus gazella*). *J. Exp. Biol.*, **194:** 33–46.

Reed, J. Z., Chambers, C., Fedak, M. A. and Butler, P. J. (1994) Gas exchange of captive freely diving grey seals (*Halichoerus grypus*). *J. Exp. Biol.*, **191:** 1–18.

Reed, J. Z., Chambers, C., Hunter, C. J., Lockyer, C., Kastelein, R., Fedak, M. A. and Boutilier, R. G. (2000) Gas exchange and heart rate in the harbour porpoise, *Phocoena phocoena*. *J. Comp. Physiol.* B, **170:** 1–10.

Reese, J. B. and Haines, H. (1978) Effects of dehydration on metabolic rate and fluid distribution in the Jackrabbit, *Lepus californicus*. *Physiol. Zool.*, **51:** 155–65.

Reid, I. R. and McDonald, I. R. (1967) Renal function in the marsupial *Trichosurus vulpecula*. *Comp. Biochem. Physiol.*, **25:** 1071–9.

Reilly, J. J. and Fedak, M. (1991) Rates of water turnover and energy expenditure of free-living male common seals (*Phoca vitulina*). *J. Zool., Lond.*, **223:** 461–8.

Reidesel, M. L. (1960) The internal environment during hibernation. *Bull. Mus. Comp. Zool.*, **124:** 421–35.

Reist, J. D. (1985) An empirical evaluation of several univariate methods that adjust for size variation in morphometric data. *Can. J. Zool.*, **230:** 513–28.

Reiter, R. J. (1978) *The Pineal and Reproduction*. S. Karger, Basel. 223pp.

Reiter, R. J. and Follett, B. K. (1980) *Seasonal Reproduction in Higher Vertebrates.* S. Karger, Basel.

Rice, G. E. and Bradshaw, S. D. (1980) Changes in dermal reflectance and vascularity and their effects on thermoregulation in *Amphibolurus nuchalis* (Reptilia: Agamidae). *J. Comp. Physiol.* B, **135:** 139–46.

Richmond, C. R., Trujillo, T. T. and Martin, D. W. (1960) Volume and turnover of body water in *Dipodomys deserti* with tritiated water. *Proc. Soc. Exp. Biol. Med.*, **104:** 9–11.

Ridgway, S. H., Bowers, C. A., Miller, D., Schultz, M. L., Jacobs, C. A. and Dooley, C. A. (1984) Diving and blood oxygen in the White whale. *Can. J. Zool.*, **62:** 2349–51.

Risimiller, P. D. and Heldmaier, G. (1988) How photoperiod influences body temperature in *Lacerta viridis*. *Oecologia*, **75:** 125–31.

Rismiller, P. D. and Heldmaier, G. (1991) Seasonal changes in daily metabolic patterns of *Lacerta viridis*. *J. Comp. Physiol.*, **161:** 482–8.

Rismiller, P. D. and McKelvey, M. W. (2000) Spontaneous arousal in reptiles? Body temperature ecology of Rosenberg's goanna, *Varanus rosenbergi*. In *Life in the Cold: Eleventh International Hibernation Symposium* (ed. G. Heldmaier and M. Klingenspor), pp. 57–64. Springer-Verlag, Berlin.

Robertson, C. R. and Gales, R. P. (eds) (1998) *Ecology and Conservation of Albatrosses*. Surrey Beatty and Sons, Chipping Norton.

Robertson, G. and Newgrain, K. (1992) Efficacy of the tritiated water and ^{22}Na turnover methods in estimating food and energy intake by Emperor penguins *Aptenodytes forsteri*. *Physiol. Zool.*, **65:** 933–51.

Robertson, G. L., Mahr, E. A., Athar, S. and Sinha, T. (1973) Development and clinical application of a new method for the radioimmunoassay of arginine vasopressin in human plasma. *J. Clin. Invest.*, **52:** 2340–52.

Robin, J.-P. (1984) Relation entre les modifications de l'uricacidémie et du catabolisme protéique au cours du jeune prolongé chez le Manchot Empereur et l'Oie domestique. *Bull. Soc. Ecophysiol. (Paris)*, **9:** 201–8.

Rochford, D. J. (1980) Nutrient status of the oceans around Australia. *CSIRO Division of Fisheries and Oceanography Report*, **1977–79:** 9–20.

Rodbard, D. (1978) Data processing for radioimmunoassays: an overview. In *Clinical Immunochemistry: Chemical and Cellular Bases and Applications in Disease. Current Topics in Clinical Chemistry* (ed. S. J. Natelson, A. J. Pesce and A. A. Dietz), pp. 477–94. American Association for Clinical Chemistry, Washington, DC.

Rodbard, D., Munson, P. and De Lean, A. (1978) *Radioimmunoassay and related procedures in medicine: improved curve-fitting, parallelism testing, characterization of sensitivity and specificity, validation, and optimization of radioligand assays*, vol. 1, pp. 469–504. I.A.E.A., Vienna.

Roe, L. J., Thewissen, J. G. M., Quade, J., O'Neil, J. R., Bajpai, S., Sahni, A. and Hussain, S. T. (1998) Isotopic approaches to understanding the terrestrial to marine transition of the earliest cetaceans. In *The Emergence of Whales, Evolutionary Patterns in the Origin of Cetacea* (ed. J. G. M. Thewissen), pp. 399–421. Plenum Press, New York.

Rolfe, D. F. S. and Brand, M. D. (1997) The physiological significance of mitochondrial proton leak in animal cells and tissues. *Biosci. Rep.*, **17:** 9–16.

Rooke, I. J. (1984) *Research into the biology of the Silvereye leading to methods for minimizing grape damage in vineyards of south-west Australia*. Agricultural Protection Board of Western Australia, Perth, WA 6058.

Rooke, I. J., Bradshaw, S. D. and Langworthy, R. A. (1983) Aspects of water, electrolyte and carbohydrate physiology of the silvereye (*Zosterops lateralis*). *Aust. J. Zool.*, **31:** 695–704.

Rooke, I. J., Bradshaw, S. D., Langworthy, R. A. and Tom, J. A. (1986) Annual cycle of physiological stress and condition of the silvereye, *Zosterops lateralis* (Aves). *Aust. J. Zool.*, **34:** 493–501.

Ruben, J. A. (1995) The evolution of endothermy in mammals and birds: from physiology to fossils. *A. Rev. Physiol.*, **57:** 69–95.

Rundel, P. W. (1970) Ecological impact of fires on mineral and sediment pools and fluxes. In *Fire and Fuel Management in Mediterranean-Climate Ecosystems: Research Priorities and Programmes*, ed. J. K. Agee. MAB Technical Note no. 11, pp. 17–21. UNESCO, Paris.

Saint Girons, H. and Bradshaw, S. D. (1981) Preliminary observations of behavioural thermoregulation in an elapid snake, the dugite *Pseudonaja affinis* Günther. *J. Roy. Soc. W.A.*, **64:** 13–16.

Saltz, D. (1994) Reporting error measures in radio location by triangulation: a review. *J. Wildl. Man.*, **58:** 181–3.

Sapirstein, L. A., Vitt, D. G., Mandel, M. J. and Hanusk, G. (1955) Volumes of distribution and clearance of intramuscularly-injected creatinine in the dog. *Am. J. Physiol.*, **181:** 330–6.

Sapolsky, R. M., Romero, L. M. and Munck, A. U. (2000) How do corticoids influence stress responses? Integrating permissive, suppressive, stimulatory and preparative actions. *Endocr. Rev.*, **21:** 55–89.

Sarre, S. and Dearn, J. M. (1991) Morphological and fluctuating asymmetry among insular populations of the sleepy lizard, *Trachydosuarus rugosus* Gray (Squamata: Scincidae). *Aust. J. Zool.*, **39:** 91–104.

Sarre, S., Dearn, J. M. and Georges, A. (1994) The application of fluctuating asymmetry in the monitoring of animal populations. *Pacif. Cons. Biol.*, **1:** 118–22.

Schamberger, M. L. and Turner, F. B. (1986) The application of habitat modeling to the desert tortoise (*Gopherus agassizii*). *Herpetologica*, **42:** 134–8.

Schleucher, E. (1993) Life in extreme dryness and heat: a telemetric study of the behaviour of the Diamond dove *Geopelia cuneata* in its natural habitat. *Emu*, **93:** 251–8.

Schleucher, E., Prinzinger, R. and Withers, P. C. (1991) Life in extreme environments: investigations on the ecophysiology of a desert bird, the Australian Diamond dove (*Geopelia cuneata* Latham). *Oecologia*, **88:** 72–6.

Schmid, J. R. and Speakman, J. R. (2000) Daily energy expenditure of the grey mouse lemur (*Microcebus murinus*): a small primate that uses torpor. *J. Comp. Physiol.* B, **170:** 633–41.

Schmidt-Nielsen, B., Schmidt-Nielsen, K., Houpt, T. R. and Jarnum, S. A. (1956) Water balance of the camel. *Am. J. Physiol.*, **185:** 185–94.

Schmidt-Nielsen, K. (1960) The salt gland of marine birds. *Circulation*, **21:** 955–67.

Schmidt-Nielsen, K. (1964) *Desert Animals: Physiological Problems of Heat and Water.* Oxford University Press, Oxford.

Schmidt-Nielsen, K., Dawson, T. J., Hammel, H. T., Hinds, D. and Jackson, D. C. (1965) The Jack rabbit-a study in desert survival. *Hvalr. Skrif.*, **48:** 125–42.

Schmidt-Nielsen, K., O'Dell, R. and Osaki, H. (1961) Interdependence of urea and electrolytes in production of a concentrated urine. *Am. J. Physiol.*, **200:** 1125–32.

Schmidt-Nielsen, K. and Robinson, R. R. (1970) Contribution of urea to urinary concentrating ability in the dog. *Am. J. Physiol.*, **218:** 1363–9.

Schmidt-Nielsen, K. and Schmidt-Nielsen, B. (1953) The desert rat. *Scient. Am.*, **189:** 73–8.

Schmidt-Nielsen, K., Schmidt-Nielsen, B., Jarnum, S. A. and Houpt, T. R. (1957) Body temperature of the camel and its relation to water economy. *Am. J. Physiol.*, **186:** 103–12.

Schnabel, Z. E. (1938) The estimation of the total fish population of a lake. *Am. Math. Month.*, **45:** 348–52.

Schoeller, D. A. (1993) Development of the Doubly Labeled Water Method for Measuring Energy Expenditure. *Proceedings A. O. Nier Symposium on Inorganic Mass Spectrometry, Durango, Colorado.*

Schoeller, D. A., Leitch, C. A. and Brown, C. (1986) Doubly labelled water method: *In vivo* oxygen and hydrogen isotope fractionation. *Am. J. Physiol.*, **251:** R1137–43.

Schoeller, D. A., Van Santen, E. and Peterson, D. W. (1980) Total body water measurement in humans with ^{18}O and ^{3}H labeled water. *Am. J. Clin. Nutr.*, **33:** 2686–93.

Scholander, P. F. (1940) Experimental investigations on the respiratory function in diving mammals and birds. *Hvalr. Skrif.*, **22:** 1–131.

Scholander, P. F., Irving, L. and Grinnell, S. W. (1942) Aerobic and anaerobic changes in seal muscle during diving. *J. Biol. Chem.*, **142:** 431–40.

Schumaker, F. X. and Eschmeyer, R. W. (1943) The estimate of fish population in lakes and ponds. *J. Tenn. Acad. Sci.*, **18:** 228–49.

Schwarzenberger, F., Mostl, E., Palme, R. and Bamberg, E. (1996a) Faecal steroid analysis for non-invasive monitoring of reproductive status in farm, wild and zoo animals. *Anim. Reprod. Sci.*, **42:** 515–26.

Schwarzenberger, F., Tomasova, K., Holeckova, D., Matern, B. and Mostl, E. (1996b) Measurement of fecal steroids in the Black rhinocerous (*Diceros bicornis*) using group-specific enzyme immunoassays for 20-oxo-pregnanes. *Zoo Biol.*, **15:** 159–71.

Seely, M. K., Roberts, C. S. and McClain, E. (1988) Microclimate and activity of the lizard *Angolosaurus skoogi* on a dune slipface. *S. Afr. J. Zool.*, **23:** 92–102.

Sellami, A., Marcilhac, A., Koza, E. and Slaud, P. (2003) Hypothalamo-posthypophysial axis activity of a water-deprived desert rodent, *Meriones shawi shawi*: comparison with the rat. *J. Comp. Physiol.* B (in press).

Selye, H. (1936) A syndrome produced by diverse nocuous agents. *Nature (Lond.)*, **138:** 32.

Selye, H. (1946) The general adaptation syndrome and the diseases of adaptation. *J. Clin. Endocr. Metab.*, **6:** 117–230.

Selye, H. (1976) *Stress in Health and Disease*. Butterworths, Montreal.

Serventy, D. L. (1971) Biology of desert birds. In *Avian Biology* (ed. D. S. Farner and J. R. King), vol. 1, pp. 287–339. Academic Press, New York.

Serventy, D. L. and Whittell, H. M. (1967) *Birds of Western Australia*, 3rd edn. Lamb Publications, Perth, Western Australia. 431pp.

Seymor, A. M., Montgomery, M. E., Costello, B. H., Ihle, S., Johnsson, G., St. John, B., Taggart, D. A. and Houlden, B. A. (2001) High effective inbreeding coefficients correlate with morphological abnormalities in populations of South Australian koalas (*Phascolarctos cinerus*). *Anim. Cons.*, **4:** 211–19.

Shaffer, S. A., Costa, D. P. and Weimerskirch, H. (2001) Comparison of methods for evaluating energy expenditure of incubating Wandering albatrosses. *Physiol. Biochem. Zool.*, **74:** 823–31.

Shaffer, S. A., Costa, D. P., Williams, T. M. and Ridgway, S. H. (1997) Diving and swimming performance of White whales, *Delphinapterus leucas*: an assessment of plasma lactate and blood gas levels and respiratory rates. *J. Exp. Biol.*, **200:** 3091–9.

Shoemaker, V. H. (1988) Physiological ecology of amphibians in arid environments. *J. Arid Envts*, **14:** 145–53.

Shoemaker, V. H., Hillman, S. S., Hillyard, S. D., Jackson, D. C., McClanahan, L. L., Withers, P. C. and Wygoda, M. L. (1992) Exchange of water, ions, and respiratory gases in terrestrial amphibians. In *Environmental Physiology of the Amphibia* (ed. M. E. Feder and W. W. Burggren), pp. 125–50. University of Chicago Press, Chicago.

Shoemaker, V. H. and Nagy, K. A. (1977) Osmoregulation in amphibians and reptiles. *A. Rev. Physiol.*, **39:** 449–71.

Shoemaker, V. H., Nagy, K. A. and Costa, W. R. (1976) Energy utilisation and temperature regulation by jackrabbits (*Lepus californicus*) in the Mojave Desert. *Physiol. Zool.*, **49:** 364–75.

Short, J., Bradshaw, S. D., Giles, J., Prince, R. I. T. and Wilson, G. R. (1992) Reintroduction of macropods (Marsupialia: Macropodoidea) in Australia - A review. *Biol. Cons.*, **62:** 189–204.

Sibley, R. M. and Calow, P. (1989) A life-cycle theory of responses to stress. *Biol. J. Linn. Soc.*, **37:** 101–16.

Sicard, B. and Fuminier, F. (1996) Water redistribution and the life cycle in sahelo-sudanese rodents. *Mammalia*, **60:** 231–8.

Siebert, B. D. and Macfarlane, W. V. (1971) Water turnover and renal function of dromedaries in the desert. *Physiol. Zool.*, **44:** 225–40.

Silverin, B. (1985) Cortical activity and breeding success in the pied flycatcher, *Ficedula hypoleuca*. In *Current Trends in Comparative Endocrinology* (ed. B. Lofts and W. N. Holmes), pp. 429–31. University of Hong Kong Press, Hong Kong.

Silverin, B. (1986) Corticosterone-binding proteins and behavioral effects of high plasma levels of corticosterone in the Pied flycatcher. *Gen. Comp. Endocrinol.*, **64:** 67–74.

Simmons, L. W., Tomkins, J. L., Kotiaho, J. S. and Hunt, J. (1999) Fluctuating paradigm. *Proc. R. Soc. Lond. B*, **266:** 593–5.

Skadhauge, E. (1981) *Osmoregulation in Birds*. Springer-Verlag, Berlin. 203pp.

Skadhauge, E. and Bradshaw, S. D. (1974) Drinking of saline, and cloacal excretion of salt and water in the Australian Zebra Finch. *Am. J. Physiol.*, **52A:** 1236–67.
Slip, D. J., Hindell, M. A. and Burton, H. R. (1994) Diving behaviour of southern Elephant seals from Macquarie Island: an overview. In *Elephant Seals: Population Ecology, Behavior and Physiology* (ed. B. J. Le Boef and R. M. Laws), pp. 253–70. University of California Press, Berkeley, California.
Smith, A. P. and Lee, A. K. (1984) The evolution of strategies for survival and reproduction in possums and gliders. In *Possums and Gliders* (ed. A. P. Smith and A. K. Lee), pp. 17–33. Surrey Beatty and Sons, Sydney.
Smith, F. A. and Charnov, E. L. (2001) Fitness trade-offs select for semelparous reproduction in an extreme environment. *Evol. Ecol. Res.*, **3:** 595–602.
Smith, H. (1951) *The Kidney, Structure and Function in Health and Disease.* Oxford University Press, Oxford, UK.
Smith, H. W. (1932) Water regulation and its evolution in the fishes. *Q. Rev. Biol.*, **7:** 1–26.
Song, X., Körtner, G. and Geiser, F. (1995) Reduction of metabolic rate and thermoregulation during daily torpor. *J. Comp. Physiol.* B, **165:** 291–7.
Spanner, A., Stone, G. M. and Schultz, D. (1997) Excretion profiles of reproductive steroids in the faeces of captive Nepalese Red panda (*Ailurus fulgens fulgens*). *Reprod. Fert. Dev.*, **9:** 565–70.
Speakman, J. R. (1997) *Doubly Labelled Water: Theory and Practice.* Chapman & Hall, London. 399pp.
Speakman, J. R. and Racey, P. A. (1988) Consequences of non steady-state CO_2 production for accuracy of the doubly labelled water technique: the importance of recapture interval. *Comp. Biochem. Physiol.*, **90A:** 337–40.
Spencer, B. (1896) Amphibia. In *Reports of the Horn Expedition to Central Australia*, pp. 152–75. Melville, Mullen and Slade, Melbourne.
Spencer, B. and Gillen, F. (1912) *Across Australia.* Macmillan, London.
Sperber, I. (1944) Studies on the mammalian kidney. *Zool. Bidr., Upps.*, **22:** 249–432.
Stafford Smith, D. M. and Morton, S. R. (1990) A framework for the ecology of arid Australia. *J. Arid Envts*, **18:** 255–78.
Stallone, J. N. and Braun, E. J. (1988) Regulation of plasma antidiuretic hormone in the dehydrated kangaroo rat (*Dipodomys spectabilis* M.). *Gen. Comp. Endocrinol.*, **69:** 119–27.
Stead-Richardson, E. J., Bradshaw, S. D., Bradshaw, F. J. and Gaikhorst, G. (2001) Monitoring the oestrous cycle of the chuditch (*Dasyurus geoffroii*: Marsupialia, Dasyuridae): non-invasive analysis of faecal oestradiol-17β. *Aust. J. Zool.*, **49:** 183–93.
Stonehouse, B. (1975) *The Biology of Penguins.* Macmillan Press, London.
Storey, K. B. and Storey, J. M. (1990) Metabolic rate depression and biochemical adaptation in anaerobiosis, hibernation and estivation. *Q. Rev. Biol.*, **65:** 145–74.
Storey, K. B. and Storey, J. M. (1992) Natural freeze tolerance in ectothermic vertebrates. *A. Rev. Physiol.*, **54:** 619–37.

Storr, G. M. (1961) Microscopic analyses of faeces: a technique for ascertaining the diet of herbivorous mammals. *Aust. J. Biol. Sci.*, **14:** 157–64.

Storr, G. M. (1963) Estimation of dry matter intake in wild herbivores. *Nature (Lond.)*, **197:** 307–8.

Storr, G. M. (1964) Studies on marsupial nutrition. IV. Diet of the quokka, *Setonix brachyurus* (Quoy et Gaimard), on Rottnest Island, Western Australia. *Aust. J. Biol. Sci.*, **17:** 469–81.

Storr, G. M. (1967) Geographic races of the agamid lizard *Amphibolurus caudicinctus. J. R. Soc. W. A.*, **50:** 49–56.

Suber, R. L. and Kodell, R. L. (1985) The effect of three phlebotomy techniques on hematological and clinical chemical evaluation in Sprague-Dawley rats. *Veter. Clin. Pathol.*, **14:** 23–30.

Suomalainen, P. (1960) Stress and neurosecretion in the hibernating hedgehog. *Bull. Mus. Comp. Zool.*, **124:** 271–83.

Swihart, R. K. and Slade, N. A. (1985a) Influence of sampling interval on estimates of home- range size. *J. Wildl. Man.*, **49:** 1019–25.

Swihart, R. K. and Slade, N. A. (1985b) Testing for independence of observations in animal movements. *Ecology*, **66:** 1176–84.

Tang, P.L., Pang, S.F. and Reiter, R.J. (1996) *Melatonin: a universal photoperiodic signal with diverse actions: International Symposium on Melatonin – A Photoperiodic Signal, Hong Kong, September 18–20, 1995.* Karger, Basel, New York. vii + 208pp.

Tarasoff, F. and Toews, D. (1972) The osmotic and ionic regulatory capacities of the kidney of the Harbor seal, *Phoca vitulina. J. Comp. Physiol.*, B **81:** 121–32.

Tatner, P. (1990) Deuterium and oxygen-18 abundance in birds: implications for DLW energetic studies. *Am. J. Physiol.*, **258:** R804–12.

Taylor, P. (1978) Radioisotopes as metabolic labels for *Glossina* (Diptera: Glossinidae). II. The excretion of ^{137}Cs under field conditions as a means of estimating energy utilisation, activity, and temperature regulation. *Bull. Entom. Res.*, **68:** 331–40.

Tedman, R. A. (1991) The female reproductive tract of the Australian sea lion, *Neophoca cinerea* (Peron, 1816). *Aust. J. Zool.*, **39:** 351–72.

Telfer, N., Cornell, L. H. and Prescott, J. H. (1970) Do dolphins drink water? *J. Am. Vet. Med. Ass.*, **157:** 555–8.

Thewissen, J. G. M., Roe, L. J., O'Neil, J. R., Hussain, S. T., Sahni, S. and Bajpai, S. (1996) Evolution of cetacean osmoregulation. *Nature (Lond.)*, **381:** 379–80.

Thomas, D. H., Pinshow, B. and Degen, A. A. (1987) Renal and lower intestinal contributions to the water economy of desert-dwelling phasianid birds: comparison of free-living and captive Chukars and Sand partridges. *Physiol. Zool.*, **57:** 128–36.

Thompson, D. and Fedak, M. A. (1993) Cardiac responses of grey seals during diving at sea. *J. Exp. Biol.*, **174:** 139–54.

Tiebout, H. M. I. and Nagy, K. A. (1991) Validation of the doubly-labeled water method (^{3}HH^{18}O) for measuring water flux and CO_2 production in the tropical hummingbird *Amazilia saucerottei. Physiol. Zool.*, **64:** 362–74.

Tieleman, B. I. and Williams, J. B. (1999a) The evolution of rates of metabolism and water flux in desert birds. *Acta Ornithol.*, **34:** 173–4.
Tieleman, B. I. and Williams, J. B. (1999b) The role of hyperthermia in the water economy of desert birds. *Physiol. Biochem. Zool.*, **72:** 87–100.
Tieleman, B. I. and Williams, J. B. (2000) The adjustment of avian metabolic rates and water flux to desert environments. *Physiol. Biochem. Zool.*, **73:** 461–79.
Tomkins, J. L. and Simmons, L. W. (2003) Fluctuating asymmetry and sexual selection: paradigm shifts, publication bias and observer expectation. In *Developmental Stability: Causes and Consequences* (ed. M. Polak). Oxford University Press, New York (in press).
Tracy, C. R. and Sugar, J. (1989) Potential misuse of ANCOVAR: comment on Packard and Boardman. *Physiol. Zool.*, **62:** 993–7.
Tracy, R. L. and Walsberb, G. E. (2001) Developmental and acclimatory contributions to water loss in a desert rodent: investigating the time course of adaptive change. *J. Comp. Physiol.* B, **171:** 669–79.
Tucker, V. A. (1973) Metabolism during flight: evaluation of a theory. *J. Exp. Biol.*, **58:** 689–709.
Turlejska, E. and Baker, M. A. (1986) Elevated CSF osmolality inhibits thermoregulatory heat loss responses. *Am. J. Physiol.*, **251:** R749–54.
Turner, F. B., Hayden, P., Burge, B. L. and Roberson, J. B. (1986) Egg production by the desert tortoise (*Gopherus agassizii*) in California. *Herpetologica*, **42:** 93–104.
van Berkum, F. H., Huey, R. B. and Adams, B. A. (1986) Physiological consequences of thermoregulation in a tropical lizard (*Ameiva festiva*). *Physiol. Zool.*, **59:** 464–72.
van Beurden, E. (1982) Desert adaptations of *Cyclorana platycephala:* a holistic approach to desert-adaptation in frogs. In *Evolution of the Flora and Fauna of Arid Australia* (ed. W. R. Barker and P. J. M. Greenslade), pp. 235–40. Peacock Publications, South Australia, Adelaide.
van Devender, T. R. and Moodie, K. B. (1977) The desert tortoise in the late Pleistocene with comments about its earlier history. *Proc. Symp. Desert Tortoise Council, 1977, Las Vegas, Nevada*, pp. 41–45. The Council, San Diego, California.
van Devender, T. R., Moodie, K. B. and Harris, A. H. (1976) The desert tortoise (*Gopherus agassizi*) in the Pleistocene of the northern Chihuahuan desert. *Herpetologica*, **32:** 298–304.
Vanherck, H., Baumans, V., Brandt, C., Hesp, A. P. M., Sturkenboom, J. H., Vanlith, H. A., Vantintelen, G. and Beynen, A. C. (1998) Orbital sinus blood sampling in rats as performed by different animal technicians – the influence of technique and expertise. *Lab. Anim.*, **32:** 377–86.
Vardy, P. H. and Bryden, M. M. (1981) The kidney of *Leptonychotes weddelli* (Pinnipedia: Phocidae) with some observations on the kidneys of two other southern phocid seals. *J. Morphol.*, **167:** 13–34.
Vitt, L. J. and Pianka, E. R. (1994) *Lizard Ecology. Historical and Experimental Perspectives.* Princeton University Press, Princeton, New Jersey.

Waite, E. (1929) *The Reptiles and Amphibians of South Australia.* British Science Guild, South Australia, Adelaide.

Wallis, I. R., Henen, B. T. and Nagy, K. A. (1999) Egg size and annual egg production by female desert tortoiscs (*Gopherus agassizii*): the importance of food abundance, body size, and date of egg shelling. *J. Herp.*, **33:** 394–408.

Walter, A. and Hughes, M. R. (1978) Total body water volume and turnover rate in fresh-water and sea-water adapted glaucous-winged gulls, *Larus glaucescens. Comp. Biochem. Physiol.*, **61A:** 233–7.

Wang, L. C. H. (1989) Ecological, physiological and biochemical aspects of torpor in mammals and birds. In *Advances in Comparative and Environmental Physiology* (ed. L. C. H. Wang), pp. 361–401. Springer-Verlag, Berlin.

Warburg, M. R. (1997) *Ecophysiology of Amphibians Inhabiting Xeric Environments.* Springer, Berlin, New York. xv + 182pp.

Wasser, S. K., Hunt, K. E., Brown, J. L., Cooper, K., Crockett, C. M., Bechert, U., Millspaugh, J. J., Larson, S. and Monfort, S. L. (2000) A generalised fecal glucocorticoid assay for use in a diverse array of nondomestic mammalian and avian species. *Gen. Comp. Endocrinol.*, **120:** 260–75.

Wasser, S. K., Monfort, S. L. and Wildt, D. E. (1991) Rapid extraction of faecal steroids for measuring reproductive cyclicity and early pregnancy in free ranging yellow baboons (*Papio cynocephalus cynocephalus*). *J. Reprod. Fert.*, **92:** 415–23.

Weathers, W. W. and Nagy, K. A. (1980) Simultaneous doubly labelled water ($^3HH^{18}O$) and time-budget estimates of daily energy expenditure in *Phainopepla nitens. Auk*, **97:** 861–7.

Weaver, D., Walker, L., Alcorn, D. and Skinner, S. (1994) The contributions of renin and vasopressin to the adaptation of the Australian spinifex hopping mouse (*Notomys alexis*) to free water deprivation. *Comp. Biochem. Physiol.*, **108A:** 107–116.

Wegener, A. (1966) *The Origin of Continents and Oceans.* (Translated from the 1929 German edition by J. Biram.) Methuen, London.

Weimerskirch, H. (1987) Population dynamics of the Wandering albatross (*Diomedea exultans*) of the Crozet Islands; causes and consequences of the population decline. *Oikos*, **49:** 315–22.

Weimerskirch, H. (1998) Foraging strategies of Indian Ocean Albatrosses and their relationship with fisheries. In *Ecology and Conservation of Albatrosses* (ed. G. C. Robertson and R. P. Gales), pp. 168–79. Surrey Beatty and Sons, Chipping Norton.

Weimerskirch, H., Brothers, N. and Jouventin, P. (1997) Population dynamics of Wandering albatross, *Diomedea exulans*, and Amsterdam albatross, *D. amsterdamensis*, in the Indian Ocean and their relationships with long-line fisheries: conservation implications. *Biol. Cons.*, **79:** 257–70.

Weimerskirch, H., Catard, A., Prince, P. A., Cherel, Y. and Croxall, J. (1999) Foraging White-chinned petrels *Procellaria aequinoctialis* at risk: from the tropics to Antarctica. *Biol. Cons.*, **87:** 273–5.

Weimerskirch, H., Doncaster, C. P. and Cuenot-Chaillet, F. (1994) Pelagic seabirds and the marine environment: foraging patterns of Wandering albatrosses in relation to prey availability and distribution. *Proc. R. Soc. Lond.* B, **225:** 91–7.
Weimerskirch, H., Guionnet, T., Martin, J., Shaffer, S. A. and Costa, D. P. (2000) Fast and fuel efficient? Optimal use of wind by flying albatrosses. *Proc. R. Soc. Lond.* B, **267:** 1869–74.
Weimerskirch, H. and Wilson, R. P. (2000) Oceanic respite for Wandering albatrosses. *Nature (Lond.)*, **406:** 955–6.
Weimerskirch, H., Wilson, R. P. and Lys, P. (1997) Activity pattern of foraging in the Wandering albatross: a marine predator with two modes of searching prey. *Mar. Ecol. Prog. Ser.*, **151:** 245–54.
Weins, J. A. (1991) The ecology of desert birds. In *The Ecology of Desert Communities* (ed. G. A. Polis), pp. 278–310. University of Arizona Press, Tucson, Arizona.
Whitfield, C. L. and Livezey, R. L. (1973) Thermoregulatory patterns in lizards. *Physiol. Zool.*, **46:** 285–96.
Whitelaw, T. G., Brockway, J. M. and Reid, R. S. (1972) Measurement of carbon dioxide production in sheep by isotope dilution. *Q. J. Exp. Physiol.*, **57:** 37–55.
Wilkes, G. E. and Jannsens, P. A. (1986) Development of urine concentrating ability in pouch of a young marsupial, the tammar wallaby (*Macropus eugeneii*). *J. Comp. Physiol.* B, **156:** 573–82.
Williams, C. K. and Green, B. (1982) Ingestion rates and aspects of water, sodium and energy metabolism in caged swamp buffalo *Bubalus bubalis*, from isotopic dilution and material balance. *Aust. J. Zool.*, **30:** 779–90.
Williams, C. K. and Ridpath, M. G. (1983) Rates of herbage ingestion and turnover of water and sodium in feral swamp buffalo, *Bubalus bubalis*, in relation to primary production in a cyperaceous swamp in monsoonal northern Australia. *Aust. Wildl. Res.*, **9:** 397–408.
Williams, E. E. (1958) Rediscovery of the Australian chelid genus *Pseudemydura* Siebenrock (Chelidae, Testudines). *Breviora*, **84:** 1–11.
Williams, G. C. (1966) *Adaptation and Natural Selection: A Critique of some Current Evolutionary Thought.* Princeton University Press, Princeton, New Jersey. 307pp.
Williams, J. B. (1996) A phylogenetic perspective of evaporative water loss in birds. *Auk*, **113:** 457–72.
Williams, J. B. (2001) Energy expenditure and water flux of free-living Dune larks in the Namib Desert: A test of the re-allocation hypothesis. *Funct. Ecol.*, **15:** 175–85.
Williams, J. B., Bradshaw, S. D. and Schmidt, L. (1995) Field metabolism and water requirements of Spinifex Pigeons (*Geophaps plumifera*) in Western Australia. *Aust. J. Zool.*, **43:** 1–15.
Williams, J. B. and Nagy, K. A. (1984) Daily energy expenditure of Savannah sparrows: comparisons of time-energy budget and doubly-labeled water estimates. *Auk*, **101:** 221–9.

Williams, J. B., Ostrowski, S., Bedin, E. and Ismail, K. (2001) Seasonal variation in energy expenditure, water flux and food consumption of Arabian Oryx, *Oryx leucoryx*. *J. Exp. Biol.*, **204:** 2301–11.

Williams, J. B., Siegfried, W. R., Milton, S. J., Adams, N. J., Dean, W. R. J., du Plessis, M. A., Jackson, S. and Nagy, K. A. (1993) Field metabolism, water requirements, foraging behavior, and diet of wild ostriches in the Namib Desert. *Ecology*, **74:** 390–404.

Williams, J. B. and Tieleman, B. I. (2000) Flexibility in basal metabolic rate and evaporative water loss among Hoopoe larks exposed to different environmental temperatures. *J. Exp. Biol.*, **203:** 3153–9.

Williams, J. B. and Tieleman, B. I. (2001) Physiological ecology and behavior of desert birds. In *Current Ornithology* (ed. V. J. Nolan and C. F. Thompson), pp. 299–353. Kluwer Academic/Plenum Publishers, New York.

Williams, J. B. and Tieleman, B. I. (2002) Ecological and evolutionary physiology of desert birds: a progress report. *Am. Zool.*, **42:** 68–75.

Williams, J. B., Withers, P. C., Bradshaw, S. D. and Nagy, K. A. (1991) Metabolism and water flux of captive and free-ranging Australian parrots. *Aust. J. Zool.*, **39:** 131–42.

Wilson, D. S., Morafka, D. J., Tracy, C. R. and Nagy, K. A. (1999a) Winter activity of juvenile desert tortoises (*Gopherus agassizii*) in the Mojave desert. *J. Herp.*, **33:** 496–501.

Wilson, D. S., Tracy, C. R., Nagy, K. A. and Morafka, D. J. (1999b) Physical and microhabitat chracteristics of burrows used by juvenile desert tortoises (*Gopherus agassizii*). *Chelon. Cons. Biol.*, **3:** 448–53.

Wilson, E. A. (1907) Aves. British National Antarctic Expedition 1901–1904. *Bull. Br. Mus. Nat. Hist. Zool.*, **2:** 1–121.

Wilson, R. T. (1989) *Ecophysiology of the Camelidine and Desert Ruminants.* Springer-Verlag, Berlin, New York. x + 120pp.

Wilz, M. and Heldmaier, G. (2000) Comparison of hibernation, aestivation and daily torpor in the edible dormouse, *Glis glis*. *J. Comp. Physiol.* B, **170:** 511–21.

Windle, R. J., Forsling, M. L., Smith, C. P., and Balment, R. J. (1993) Patterns of neurohypophysial release during dehydration in the rat. *J. Endocrinol.*, **137:** 311–19.

Wingfield, J. (2001) Coping with unpredictable environmental events: Mechanisms to avoid and resist stress. In *Perspectives in Comparative Endocrinology: Unity and Diversity* (ed. H. J. T. Goos, R. K. Rastogi, H. Vaudry and R. Pierantoni), pp. 501–8. Monduzzi Editore, Sorrento, Italy.

Wingfield, J. and Silverin, B. (1986) Effects of corticosterone on territorial behavior of free-living male Song sparrows *Melospiza melodia*. *Horm. Behav.*, **20:** 405–17.

Wingfield, J. C. (1994) Modulation of the adrenocortical response to stress in birds. In *Perspectives in Comparative Endocrinology* (ed. K. G. Davey, R. E. Peter and S. S. Tobe), pp. 520–8. National Research Council of Canada, Ottawa.

Wingfield, J. C., Breuner, C., Jacobs, J., Lynn, S., Maney, D., Ramenofsky, M. and Richardson, R. (1998) Ecological bases of hormone-behavior interactions: the "Emergency Life History Stage". *Am. Zool.*, **38:** 191–206.

Wingfield, J. C., Hegner, R. E., Dufty, A. M. J. and Ball, G. F. (1990) The "Challenge Hypothesis": theoretical implications for patterns of testosterone secretion, mating systems and breeding strategies. *Am. Nat.*, **136:** 829–46.

Wingfield, J. C. and Romero, L. M. (2000) Adrenocortical responses to stress and their modulation in free-living vertebrates. In *Handbook of Physiology, Section 7: The Endocrine System, Volume 4: Coping with the Environment* (ed. B. S. McEwen), pp. 211–36. Oxford University Press, Oxford.

Wingfield, J. C., Smith, J. P. and Farmer, D. S. (1982) Endocrine responses of white-crowned sparrows to environmental stress. *Condor*, **84:** 399–409.

Withers, P. C. (1995) Cocoon formation and structure in the aestivating Australian desert frogs, *Neobatrachus* and *Cyclorana*. *Aust. J. Zool.*, **43:** 429–41.

Withers, P. C. (1998a) Evaporative water loss and the role of cocoon formation in Australian frogs. *Aust. J. Zool.*, **46:** 405–18.

Withers, P. C. (1998b) Urea: diverse functions of a 'waste' product. *Clin. Exp. Pharm. Physiol.*, **25:** 722–7.

Withers, P. C. and Guppy, M. (1996) Do Australian frogs co-accumulate counteracting solutes with urea during aestivation? *J. Exp. Biol.*, **199:** 1809–16.

Withers, P. C., Louw, G. N. and Henschel, J. (1980) Energetics and water relations of Namib desert rodents. *S. Afr. J. Zool.*, **15:** 131–45.

Withers, P. C., Richardson, K. C. and Wooller, R. D. (1990) Metabolic physiology of euthermic and torpid honey possums, *Tarsipes rostratus*. *Aust. J. Zool.*, **37:** 685–93.

Withers, P. C. and Thompson, G. G. (2000) Cocoon formation and metabolic depression by aestivating hylid frogs *Cyclorana australia and Cyclorana cultripes* (Amphibia: Hylidae). *J. R. Soc. W. A.*, **83:** 39–40.

Withers, P. C. and Williams, J. B. (1990) Metabolic and respiratory physiology of an arid-adapted Australian bird, the Spinifex pigeon. *Condor*, **92:** 961–9.

Wolfe, R. R. (1992) Isolation of urinary NH_3. In *Radioactive and Stable Isotope Tracers in Biomedicine* (ed. R. B. Wolfe), pp. 430–1. Wiley-Liss, Chichester.

Wong, W. W. and Klein, P. D. (1987) A review of the techniques for the preparation of biological samples for mass spectrometric measurements of hydrogen-2/hydrogen-1 and oxygen-18/oxygen-16 isotope ratios. *Mass. Spectrom. Rev.*, **5:** 313–42.

Wood, R. A., Nagy, K. A., MacDonald, N. S., Wakakuwa, S. T., Beckman, R. J. and Kaaz, H. (1975) Determination of oxygen-18 in water contained in biological samples by charged particle activation. *Anal. Chem.*, **47:** 646–50.

Wooller, R. D., Dunlop, J. N., Klomp, N. I., Meathrel, C. E. and Wienecke, B. C. (1991) Seabird abundance, distribution and breeding in relation to the Leeuwin Current. *J. R. Soc. W. A.*, **74:** 129–32.

Wooller, R. D., Renfree, M. B., Russell, E. M., Dunning, A., Green, S. W. and Duncan, P. (1981) Seasonal changes in a population of the nectar feeding marsupial *Tarsipes spenserae* (Marsupialia: Tarsipedidae). *J. Zool., Lond.*, **195:** 267–79.

Wooller, R. D., Russell, E. M. and Renfree, M. B. (1983) A technique for sampling pollen carried by vertebrates. *Aust. Wildl. Res.*, **10:** 433–4.
Wright, J. W. and Harding, J. W. (1980) Body dehydration in xeric adapted rodents: does the renin-angiotensin system play a role? *Comp. Biochem. Physiol.*, **66A:** 181–8.
Yagil, R. and Etzion, Z. (1979) The role of antidiuretic hormone and aldosterone in the dehydrated and rehydrated camel. *Comp. Biochem. Physiol.*, **63:** 275–8.
Yallow, R. S. and Berson, S. A. (1960) Immunoassay of endogenous plasma insulin in man. *J. Clin. Invest.*, **39:** 1157–75.
Zimmerman, L. C. and Tracy, C. R. (1989) Interactions between the environment and ectothermy and herbivory in reptiles. *Physiol. Zool.*, **62:** 374–409.

Index